The Pygmy Chimpanzee

Evolutionary Biology and Behavior

Hal Coolidge on the Asiatic Primate Expedition, 1937.

The Pygmy Chimpanzee

Evolutionary Biology and Behavior

Edited by

Randall L. Susman

School of Medicine
State University of New York at Stony Brook
Stony Brook, New York

PLENUM PRESS • NEW YORK AND LONDON

Library of Congress Cataloging in Publication Data

Main entry under title:

The Pygmy chimpanzee.

Includes bibliographical references and index.
1. Pygmy chimpanzee. I. Susman, Randall L., 1948–
QL737.P96P94   1984   599.88′44   84-13236
ISBN 0-306-41595-X

©1984 Plenum Press, New York
A Division of Plenum Publishing Corporation
233 Spring Street, New York, N.Y. 10013

Printed in the United States of America

To Hal Coolidge

Contributors

Alison Badrian Department of Anthropology, State University of New York at Stony Brook, Stony Brook, New York 11794

Noel Badrian Department of Anthropology, State University of New York at Stony Brook, Stony Brook, New York 11794

Harold Jefferson Coolidge Museum of Comparative Zoology, Harvard University, Cambridge, Massachusetts 02138

Raymond C. Heimbuch Department of Medical Biostatistics, Ortho Pharmaceutical Company, Raritan, New Jersey 08869

William L. Jungers Department of Anatomical Sciences, School of Medicine, State University of New York at Stony Brook, Stony Brook, New York 11794

Kabongo Ka Mubalamata Institut de Recherche Scientifique, Centre de Recherche de Lwiro, Bukavu, Zaire

Takayoshi Kano Department of Human Ecology, College of Medicine, University of the Ryukyus, 207 Uehara, Nishihara, Okinawa, Japan

Warren G. Kinzey Department of Anthropology, City College of New York, City University of New York, New York, New York 10031

Suehisa Kuroda Laboratory of Physical Anthropology, Faculty of Science, Kyoto University, Sakyo-ku, Kyoto, 606 Japan

Jeffrey T. Laitman   Department of Anatomy, Mount Sinai School of Medicine of the City University of New York, New York, New York 10029

Richard K. Malenky   Department of Ecology and Evolution, State University of New York at Stony Brook, Stony Brook, New York 11794

Henry M. McHenry   Department of Anthropology, University of California at Davis, Davis, California 95616

Mbangi Mulavwa   Institut de Recherche Scientifique, Centre de Mabali, B.P. 36 Bikoro, Equateur, Zaire

Vincent M. Sarich   Departments of Anthropology and Biochemistry, University of California, Berkeley, California 94720

E. S. Savage-Rumbaugh   Language Research Center, Yerkes Regional Primate Research Center, Emory University, and Georgia State University, Atlanta, Georgia 30322

Brian T. Shea   Department of Anthropology, American Museum of Natural History, New York, New York 10024. *Present address:* Departments of Anthropology and Cell Biology and Anatomy, Northwestern University, Evanston, Illinois 60201

Wladyslaw W. Socha   Primate Blood Group Reference Laboratory and WHO Collaborating Centre for Haematology of Primate Animals, Laboratory for Experimental Medicine and Surgery in Primates (LEMSIP), New York University School of Medicine, New York, New York 10016

Randall L. Susman   Department of Anatomical Sciences, School of Medicine, State University of New York at Stony Brook, Stony Brook, New York 11794

Nancy Thompson-Handler   Department of Anthropology, Yale University, New Haven, Connecticut 06520

Dirk F. E. Thys Van den Audenaerde   Vertebrate Section, Musée Royale de l'Afrique Centrale, B-1980 Tervuren, Belgium

Adrienne L. Zihlman   Department of Anthropology, University of California at Santa Cruz, Santa Cruz, California 95064

Historical Remarks Bearing on the Discovery of *Pan paniscus*

Whether by accident or by design, it was most fortunate that Robert M. Yerkes, the dean of American primatologists, should have been the first scientist to describe the characteristics of a pygmy chimpanzee, which he acquired in August 1923, when he purchased him and a young female companion from a dealer in New York. The chimpanzees came from somewhere in the eastern region of the Belgian Congo and Yerkes estimated the male's age at about 4 years.

He called this young male Prince Chim (and named his female, common chimpanzee counterpart Panzee) (Fig. 1). In his popular book, *Almost Human,* Yerkes (1925) states that in all his experiences as a student of animal behavior, "I have never met an animal the equal of this young chimp . . . in approach to physical perfection, alertness, adaptability, and agreeableness of disposition" (Yerkes, 1925, p. 244). Moreover,

> It would not be easy to find two infants more markedly different in bodily traits, temperament, intelligence, vocalization and their varied expressions in action, than Chim and Panzee. Here are just a few points of contrast. His eyes were black and in his dark face lacked contrast and seemed beady, cold, expressionless. Hers were brown, soft, and full of emotional value, chiefly because of their color and the contrast with her light complexion. Chim's ears were small, set close to his head and fringed with black hair, whereas Panzee's stood out conspicuously, and were light in color and hairless. Their foreheads, noses, lips and head conformations also differed noticeably. (Yerkes, 1925, p. 245.)

When it comes to animal temperament, we are at a loss for descriptive terms. Little Chim was notable for his bold, aggressive manner, his constant alertness

Figure 1. A portrait of Prince Chim.

and eagerness for new experiences. Seldom daunted, he treated the mysteries of life as philosophically as any man. Panzee was timid, nervous, hesitant before anything novel or new. When there was anything to learn by "trial and error," he took the lead and she followed at an eminently safe distance. Chim also was even-tempered and good-natured, always ready for a romp; he seldom resented by word or deed unintentional rough handling or mishap. Never was he known to exhibit jealousy. . . . [By contrast,] Panzee could not be trusted in critical situations. Her resentment and anger were readily aroused and she was quick to give them expression with hands and teeth. (Yerkes, 1925, p. 246.)

Everything seems to indicate that Chim was extremely intelligent. His surprising alertness and interest in things about him bore fruit in action, for he was constantly imitating the acts of his human companions and testing all objects. He rapidly profited by his experiences. . . . Never have I seen a man or beast take greater satisfaction in showing off than did little Chim. The contrast in intellectual qualities between him and his female companion may briefly, if not entirely adequately, be described by the term "opposites." (Yerkes, 1925, p. 248.)

In group behavior studies his actions "contrasted almost as markedly with that of his chimpanzee associates as did his appearance, for he was a little black-face with rather conspicuous nose, small ears, and a heavy coat of fine black hair, whereas most of them were white-face specimens with the typical chimpanzee nose, large, conspicuous ears, and much coarser and generally less thick coat of hair" (Yerkes, 1925, p. 251).

Sadly, Prince Chim died of pneumonia while visiting Madam Abreu's colony in Havana in July 1924. Professor Yerkes states that he was actually a prince of his kind and his "behavior even in death goes far to justify the title of this story, *Almost Human*. Prince Chim seems to have been an intellectual genius. His remarkable alterness and quickness to learn were associated with a cheerful and happy disposition which made him the favorite of all, and gave him a place of distinction not only in their regard but in their memories" (Yerkes, 1925, p. 255).

I recall with great pleasure meeting Prince Chim at the Yerkes' farm in Franklin, New Hampshire, and remember his greeting me with a firm handshake. Little did I realize that he would open the way to the important studies to be described in this volume 60 years later.

It was my good fortune to have my own interest in the anthropoid apes and particularly gorillas, in a large measure sparked by my distinguished friend, Robert M. Yerkes, for many years professor of psychology at Yale University. I was fortunate to have served as assistant zoologist on the Harvard Medical School African Expedition in 1926–1927, where I collected and documented a good-sized mountain gorilla in the mountains of the eastern Belgian Congo for the Museum of Comparative Zoology at Harvard. This expedition was followed by a survey of gorilla material in European and U. S. museums and publication of a revision of gorilla classification (Coolidge, 1927–1929). While visiting European museums to gather data for my gorilla study in 1928, I noticed a specimen that was described to be from south of the Congo River and was somewhat smaller than other chimpanzee specimens.

I shall never forget, late one afternoon in Tervuren, casually picking up from a storage tray what clearly looked like a juvenile chimp's skull from south of the Congo and finding, to my amazement, that the epiphyses were totally fused. It was clearly adult. I picked up four similar skulls in adjoining trays and found the same condition, which I measured for a future paper on the revision of chimpanzees.

Just then, Dr. Schouteden, the museum director, came to tell me that it was time to close and I showed him what I had discovered. He seemed most interested.

Two weeks later Dr. Ernst Schwarz, who was aware that chimps on either side of the Congo differed, visited the Congo Museum and Dr. Schouteden told him about my discovery. In a flash Schwarz grabbed a pencil and paper, measured one small skull, wrote up a brief description, and named a new pygmy chimpanzee race: *Pan satyrus paniscus*. He asked Schouteden to have his brief account printed without delay in the *Revue Zoologique* (April 1929) of the Congo Museum (Schwarz, 1929). I had been taxonomically scooped.

When I learned about this article and read a further paper by Schoutenden on the chimps from the left bank of the Congo, I intensified my study of the pygmy chimpanzee and its distribution. In 1931 I undertook a comprehensive survey of chimpanzee museum collections with a view to revising their classifications as I had done for gorillas. My survey of ape skeletal material included collections in the American Museum of Natural History, the Riksmuseum in Stockholm, the British Museum in London, the Berlin Museum, the Paris Museum of Natural History, the Royal Museum in Brussels, and the Congo Museum in Tervuren, which made me feel very much at home with the rich chimp and gorilla collections from the Belgian Congo.

In the American Museum of Natural History I found a skin and skull that had been collected by Jim Chapin near Lukolela on the southeast bank of the Congo in 1930. It clearly belonged to this small race. I also found the skin and skeleton of Dr. Yerkes' Prince Chim, No. 54336, American Museum of Natural History. Although I had no juveniles to compare it with, its dark skin and small ears clearly indicated that it also was a "bonobo." The Chapin skeleton was compared with specimens of *Pan,* carefully measured by Dr. Adolph Schultz, and characteristics of possible taxonomic significance listed. Most consistent are the small bones throughout. There seems to be more evidence from the external appearance of the skeleton collected by Chapin that it is closer to *Pan schweinfurthii* than to *Pan satyrus*, as Schwarz had stated.

I believed that the pygmy chimpanzee had a sufficient difference in degree as well as in variation from other chimpanzees that it should be classified as a full species, making it *Pan paniscus* instead of *Pan satyrus paniscus*. This classification was documented in my report in the *American Journal of Physical Anthropology* (Coolidge, 1933).

It is highly gratifying that the greatly needed further studies of this important ape are being carried out by primatologists in many countries in cooperation with the Zaire government, and also that one or more national parks, including the pygmy chimps' present principal habitat south of the Congo River, will hopefully be established and protected with adequate guards to save the species from extinction, for the benefit of future generations, as well as the welfare of the animals themselves.

Some believe that this true paedomorphic species is the most important of the chimpanzees for the study of the phylogeny and relationships of the apes and man. This possibility makes this volume of particular significance. The research reported here and future studies will no doubt help us answer the question of whether or not the pygmy chimpanzee approaches most closely the common ancestor of chimpanzee and man.

Harold J. Coolidge

Museum of Comparative Zoology
Harvard University
Cambridge, Massachusetts

References

Coolidge, H. J., 1927–1929, *Memoirs of the Museum of Comparative Zoology at Harvard College*, Volume L, Museum Publication, Cambridge, Massachusetts.

Coolidge, H. J., 1933, *Pan paniscus:* Pygmy chimpanzee from south of the Congo River, *Am. J. Phys. Anthropol.* **XVIII**(1):1–57.

Schwarz, E., 1929, Das Vorkommen des Schimpansen auf den linken Kongo-Ufer, *Rev. Zool. Bot. Afr.* **XVI**(4):425–426.

Yerkes, Robert M., 1925, *Almost Human,* The Century Co., New York.

Preface

On August 9th, 1982, a symposium entitled The Evolutionary Morphology and Behavior of the Pygmy Chimpanzee was held in Atlanta in conjunction with the IXth Congress of the International Primatological Society. The symposium honored Harold Jefferson Coolidge, a man who has had a lifelong devotion to the study of primates and primate conservation. Hal Coolidge played a key role in the discovery and description of the pygmy chimpanzee in the late 1920s, and in 1933 it was he who elevated *Pan paniscus* to species status consequent on his detailed morphological study. All of us who participated in this symposium were honored to present the results of our respective work in the presence of one of the major figures in primatology in this century, the man who, over 50 years ago, brought *Pan paniscus* to the attention of the scientific world.

Hal Coolidge began his lifelong association with the Harvard Museum of Comparative Zoology (MCZ) at the beginning of his sophomore year. In 1926 Richard Strong and George Shattuck invited Hal to accompany them (in the role of assistant mammalogist) on the Harvard Medical Expedition across Africa (1926–1927). As a result of this experience Hal decided on a career in mammalogy. Under the tutelege of Prof. Robert Yerkes of Yale (Harvard did not have a primatologist at this time), Hal was advised to pursue an interest in the biology of gorillas, which had been sparked by his prior travels in the eastern Congo in 1927. The result was a comprehensive work, *A Revision of the Genus Gorilla,* published by Coolidge in 1929. From 1929 through 1940 Hal was an assistant curator of mammals at the MCZ. During this time he worked principally with primates, and conducted research on gorillas, pygmy chimpanzees, and gibbons. Beginning in 1931 Hal taught a course on the evolution of animal sociology with William Morton Wheeler, and he lectured on primatology in courses offered by Ernest Hooton.

In 1937 Hal organized, funded, and led the legendary Asiatic Primate Expedition (APE), which was sponsored by the MCZ, Columbia University, and Johns Hopkins. The members of the APE included, in addition to Hal, Clarence Ray Carpenter (who by this time had pioneered studies of howling monkeys and spider monkeys in Panama), Sherwood Washburn (a student of Hooton), Adolph Schultz (a preeminent physical anthropologist and anatomist), J. Augustus Griswold (zoologist and preparator interested in small mammals and birds), and John Coolidge (photographer and illustrator) (Fig. 1). This distinguished group spent 9 months in northern Siam (now Thailand), North Borneo, and Sumatra. Their work focused on primates, and their scientific legacy was profound. From the APE, Carpenter published his classic work, "A Field Study in Siam of the Behavior and Social Relations of the Gibbon"; Washburn obtained the specimens for his thesis, "A Preliminary Metrical Study of the Skeleton of Langurs and Macaques"; and Schultz acquired wet specimens and skeletal material that served him in his manifold publications over

Figure 1. Members of the Asiatic Primate Expedition (APE). Front row (left to right): Augustus Griswold, Sherwood Washburn. Back row: Harold Coolidge, Adolph Schultz, Clarence Carpenter.

the next 35 years. Other legacies of the expedition included extensive collections of monkeys and gibbons, most of which are housed in the MCZ, field notes, films, and photographs. All constitute a priceless resource for students of primatology. Following the expedition, Hal toured the United States giving popular lectures, which helped pay the remaining cost of the expedition. The success of the APE also resulted, in part, in the establishment of the laboratory of primatology at Harvard (1938–1939). Hal, working in cooperation with Carpenter and Washburn, and with advice from Robert Yerkes, was instrumental in planning the primate laboratory.

Hal's work at the MCZ represented only a small part of his professional activity. In 1930 Hal established the American Committee for International Wildlife Protection (now the Committee for International Conservation). This was the earliest international conservation organization in the United States and it helped draft the London Convention on African Wildlife Protection; the Committee also sponsored many other publications. The Committee took part in the establishment of the International Union for the Conservation of Nature and Natural Resources (IUCN), which was established in 1948 with the assistance of the French Government and then Director-General of UNESCO, Julian Huxley. (Today the IUCN has over 50 member governments and 250 member institutions in over 90 countries.) Hal founded IUCN's Survival Service and Parks Commission. He was IUCN's first Vice President until 1954, was President from 1966 to 1972, and is now Honorary President. In 1961 Hal helped to found the IUCN's sister organization, the World Wildlife Fund, which finances the IUCN and other conservation projects. Hal was a director of WWF-International until 1976 and a founding director of WWF-US Hal's specific contributions to wildlife conservation are manifested in the establishment of national parks in Japan, Korea, Taiwan, the Phillipines, New Caledonia, Vietnam, Malaysia, and Indonesia. He organized the First World Conference on National Parks (held in Seattle in 1962). His contributions to conservation in third world countries has earned him over 25 major honors, including the J. Paul Getty Wildlife Conservation Prize for 1979. Hal was the first American so honored.

Hal began the pygmy chimpanzee symposium with personal remarks and recollections of his first encounter with Prince Chim, a captive pygmy chimpanzee owned by Prof. Yerkes (but not known at the time to be a pygmy chimpanzee) (see Coolidge, Foreword to this volume). Hal also related his meeting with Dr. Ernst Schwarz in Berlin and the sequence of events leading to Schwarz' announcement of *Pan paniscus*. Following the symposium a reception sponsored by the L. S. B. Leakey Foundation and the World Wildlife Fund-US was held in honor of Hal Coolidge.

A great many people and organizations are responsible, both directly and indirectly, for the present volume. The Musée Royale de l'Afrique Centrale at Tervuren (now the Tervuren Museum) has provided all of those interested in the comparative anatomy of *Pan paniscus* with the only sizable collection of skeletal material available. Over the years the scientific and technical staff of the Museum has provided many a visiting scientist with a warm and friendly atmosphere, and one of the most hospitable working environments to be found anywhere. We are deeply indebted to the past and present staff of the Tervuren Museum. Those of us who have had the privilege of studying *Pan paniscus* in the field are deeply indebted to the Government of the Republic of Zaire. In particular, we are indebted to the Institut de Recherche Scientifique (IRS) in Kinshasa for its unwavering support. We are grateful to Dr. Iteke Bochoa, Dr. Kankwenda M'Baya, the present Director, Dr. Nkanza Ndolumingu, and the staff of the IRS for their advice, help, and colleagueship.

The editor would like to thank the authors who have contributed to this volume. Divergent perspectives and methods characterize the contributions that follow. Despite this and the controversy that has attended discussions of the phylogenetic affinities and relevance of *Pan paniscus* for modeling early human evolution, the papers presented in Atlanta represent a relatively uniform and consistent view. In some ways pygmy chimpanzees closely resemble the earliest hominids, but in others the common chimpanzee and other apes provide us with important comparative data for phylogeny reconstruction and functional inference. As editor I have chosen to retain the common name of "pygmy chimpanzee" for *Pan paniscus*. I have done so largely for historical reasons and for reasons of familiarity. Also, I personally do not feel that the present alternatives, bonobo or bonobo chimpanzee, are any more justified. Recent work, including the published body weights for common and pygmy chimpanzees, shows that in fact the eastern long-haired chimpanzee *Pan troglodytes schweinfurthii* has a mean body weight that is similar to that of *Pan paniscus* (Jungers and Susman, this volume). It is also true that, as judged by criteria for mammalian drawfs, *Pan paniscus* does not follow the trend for accelerated negative scaling of the skeleton relative to the postcanine dentition. However, it must also be noted that Schwarz, Coolidge, and later workers have demonstrated that *Pan paniscus* has a relatively small skull for its body size or trunk length. Remembering that the skull was the basis on which the taxon was diagnosed and that Coolidge in 1933 had only a single small (female) specimen with which to describe the postcranial skeleton, it is understandable that this species was considered to be a dwarf or pygmy form. It was not until the 1960s that additional skeletons of *Pan paniscus* became available, through the efforts of

Vandebroek (see van den Audernaerde, this volume). It was not until very recently that enough body weights of free-ranging common and pygmy chimpanzees have been available to allow for intergroup comparisons among the varieties of chimpanzees.

Thus it seems that in historical perspective the term "pygmy chimpanzee" is not an inappropriate one. From our present perspective it is preferable to the cumbersome term "bonobo chimpanzee" or "bonobo." The word bonobo itself is problematic. It most likely is the mispronounciation of the word Bolobo, which is a town on the Zaire River, between Kinshasa and Lukolela, from which specimens of *Pan paniscus* were collected (see van den Audernaerde, this volume). It is interesting to note that the Lingala word for the pygmy chimpanzee is "Mokumbusu"; the Limongo name is "Eja" or "Engombe." Last of all for those concerned, it seems that, in any event, the name "pygmy chimpanzee" is at least as appropriate as the common name Celebes "black ape" given to *Macaca nigra* and the other Celebes macaques, or the name "barbary ape" given to *Macaca sylvanus*. I expect that the systematics of the genus *Pan* will be a subject of discussion for some time to come (see chapters by Socha and Sarich, this volume); different preferences for common names will, no doubt, also remain.

Finally, it should be noted that at the very time that we are beginning to learn about the behavior and ecology of free-ranging pygmy chimpanzees, the species and its environment have become threatened with disruption and elimination. Whether we will continue to learn about pygmy chimpanzees in the wild depends on the political and economic situation in central Africa. At present commercial interests threaten the rain forest and the survival of its unique fauna. The only hope for the long term is an international conservation effort under the leadership of the Zairian government, with the help of scientists and conservationists. In the Lomako Forest and at Wamba the threat is primarily from foreign commercial logging. In other areas of Equateur the pygmy chimpanzee is threated by hunting and by human encroachment in the form of shifting, slash-and-burn agriculture. As late as the 17th century pygmy chimpanzees may have ranged into northwestern Angola and as far south as the Quanza River (V. Reynolds, 1967, *The Apes*, Dutton, New York). With the colonization of the Congo, the Belgians installed roads and river transport systems, which led to the formation of population centers. Since 1900 the large mammals that once were found throughout much of Zaire have been decimated. Today vast areas along the major waterways and roads and near the population centers of Lukolela, Mbandaka, Boende, Befale, Basankusu, and Lisala are devoid of pygmy chimpanzees and significant numbers of other large mammals. Salonga National Park, in lower Equateur, is also devoid of *Pan paniscus*. If we are to continue our field studies, the first fruits of which are seen in the present volume, then a major conservation effort must be made to establish a reserve free from the inevitable perturbations of human encroachment. Already the government of the Republic of Zaire has expressed an interest in preserving its rich natural history. With help from the international scientific and conservation communities we are hopeful that the establishment of a protected area for pygmy chimpanzees will take place before it is too late.

Randall L. Susman

Contents

Chapter 6

An Allometric Perspective on the Morphological and Evolutionary Relationships between Pygmy (*Pan paniscus*) and Common (*Pan troglodytes*) Chimpanzees   89

Brian T. Shea

Chapter 7

Body Size and Skeletal Allometry in African Apes 131

William L. Jungers and Randall L. Susman

Chapter 8

Body Build and Tissue Composition in *Pan paniscus* and *Pan troglodytes*, with Comparisons to Other Hominoids 179

Adrienne L. Zihlman

Chapter 9

**The Common Ancestor: A Study of the Postcranium of *Pan paniscus*,
Australopithecus, and Other Hominoids**

Henry M. McHenry

Part II. Behavior of *Pan paniscus*

Chapter 10

**Feeding Ecology of the Pygmy Chimpanzees (*Pan paniscus*) of
Wamba**

Takayoshi Kano and Mbangi Mulavwa

Chapter 11

Feeding Ecology of *Pan paniscus* in the Lomako Forest, Zaire   275

Noel Badrian and Richard K. Malenky

Chapter 12

Interaction over Food among Pygmy Chimpanzees   301

Suehisa Kuroda

Chapter 13

Social Organization of *Pan paniscus* in the Lomako Forest, Zaire 325

Alison Badrian and Noel Badrian

Chapter 14

Sexual Behavior of *Pan paniscus* under Natural Conditions in the Lomako Forest, Equateur, Zaire 347

Nancy Thompson-Handler, Richard K. Malenky, and Noel Badrian

Chapter 15

The Locomotor Behavior of *Pan paniscus* in the Lomako Forest ...

Randall L. Susman

Chapter 16

***Pan paniscus* and *Pan troglodytes*: Contrasts in Preverbal Communicative Competence**

E. S. Savage-Rumbaugh

PART I

Molecular Biology, Systematics, and Morphology

CHAPTER 1

The Tervuren Museum and the Pygmy Chimpanzee

DIRK F. E. THYS VAN DEN AUDENAERDE

The discovery of the existence of a peculiar form of chimpanzee in the rain forest on the left bank of the Zaire River certainly is one of the major faunistic events of the 20th century. The history of this discovery and our knowledge of the pygmy chimp was, until the 1960s, strongly influenced in Europe by the temper of two men: Henri Schouteden and G. Vandebroek. These two great Belgian zoologists, both of whom I came to know well, were to make important contributions to the knowledge of the pygmy chimpanzee, although their names are not usually linked to *Pan paniscus* by the international forum of scientists who have worked on this species.

I first met Dr. Schouteden in 1954, after he had retired from the Museum. We remained good friends until his death in 1972. Through the years I came to know him very well, and by his numerous tales and stories I also learned the history of the Tervuren Museum and of the zoological discoveries in Central Africa from 1900 through the 1960s.

While still a zoology student, Schouteden became associate scientist of the Congo Museum* upon its foundation in 1897. He became director of the Museum in 1927 and he retired in 1946. His work and association continued with the Museum until his death in 1972. Schouteden was small in stature, somewhat timid, but a very friendly man. At the time he started

*This was the original name of the Tervuren Museum.

DIRK F. E. THYS VAN DEN AUDENAERDE ● Vertebrate Section, Musée Royale de l'Afrique Centrale, B-1980, Tervuren, Belgium.

3

working in the Museum, he was aware that the Congo represented a blank on the zoological map. He was stimulated to take an active role in collecting by a remark made to him about the poor insect collections at the Congo Museum. He became determined to win honor and fame for the Museum by cataloguing thoroughly the whole Central African fauna and by establishing the largest African zoological collection both in number of specimens and in geographic diversity of the localities sampled. Often writing up to 20 letters a day in his minute handwriting, he continuously encouraged a wide array of Belgian nationals working in the Congo to make natural history collections. This correspondence and the specimens it attracted became the basis of the rich collections of the Tervuren Museum. In 1920 Schouteden created the Cercle Zoologique Congolais as a link between the collectors in the Belgian Congo and the Museum. Until his retirement Schouteden remained the Cercle's inspiring and active Secretary-President and editor of its bulletin.

Schouteden's timid nature also had a strong influence on his zoological research. Having started as an entomologist, he felt most at ease with the order Hemiptera, where he described numerous new genera and species. Early in his career with the Museum, around 1910, he also became involved by professional obligation with birds. Although he was to describe several new species, Schouteden never felt as comfortable in his bird studies as he did in his entomological enterprises. This was the reason, no doubt, that he failed to recognize the Congo peacock (*Afropavo congensis*) as a new species, even though a stuffed pair had been present in the Museum since 1898. It was his American friend, James A. Chapin, who later was to recognize this taxon (in 1936) as a new genus and species. A very similar situation had occurred a decade earlier with the discovery of the pygmy chimp. The recognition of this important taxon also was to elude Schouteden due to his tentativeness in the field of mammalogy.

Officially in charge of mammals in Tervuren since 1910, Schouteden was as uneasy with this group as with birds. He was given to basing taxonomic diagnosis almost exclusively on pelage and superficial characters, leaving the craniology and other anatomy for specialists. For primate systematics he deferred to his German friend, Dr. Ernst Schwarz, who regularly visited Tervuren. In the 1920s Schouteden was devoting himself mainly to the enlargement of the mammal collections. To this end he traveled and collected animals, plants, and geological samples as well as ethnographic materials in the Congo in 1920–1922 and 1924–1926. In 1920 it was generally believed that no chimpanzee existed on the left bank (south) of the Congo River, and as such Schouteden did not concern himself with the question of apes in this area. He spent his time and efforts gathering new material and locality data for *Pan troglodytes schweinfurthii* on the right bank of the river. One wonders now how a prominent zoologist and explorer could travel all along the Congo River from Kinshasa (formerly Leopoldville), to Kisangani (formerly Stanleyville), penetrating on the left bank at Kunungu, Bolobo, along Lake Mai-Ndombe (formerly Leopold II), and at Lake Tumba, without discovering the existence of pygmy chimpanzees there. Even more curious is the fact that as early as 1910 a chimpanzee skull from northern Kasai was received at Tervuren (did Schouteden consider it as a labeling error?). Lastly, Schouteden in 1920 had received a picture of a chimpanzee taken near Bongo at Lake Leopold II by a Mr. Molin (see Schouteden, 1944), and in 1921 a wet specimen that died in the Antwerp Zoo was received at Tervuren. It was said to be from Ibembo. Nobody seemed to notice the obvious differences in these few specimens. Perhaps people were misled by the range of skin colors that normally occurred in common chimpanzees. Schouteden was not alone, however; a pygmy chimpanzee collected at Bongandanga (south of the Congo River) had been in the British Museum of Natural History since 1881. As an entomologist with a shy and retiring nature and one who was conservative in scientific matters, Schouteden's failure to recognize this new variety of great ape can be understood. Others, however, in Europe and the United States were on the trail of the left-bank chimpanzee.

On December 6, 1927, the Congo Museum received the skull and skin of an adult female chimpanzee collected near Befale. This animal was obtained by Dr. Ghesquière, an entomologist, together with specimens of *Allenopithecus nigroviridis, Cercopithecus neglectus,* and *Cercopithecus mona wolfi.* The female chimpanzee skull was destined to become the type of *Pan satyrus paniscus,* but Dr. Schouteden registered it simply as "*Pan*" (#9338). I never could find out why he did not add a specific name. On 14 January 1928, at a meeting of the Cercle Zoologique Congolais, he nevertheless mentioned this skull as ". . . curieusement petit pour une bête de semblables dimensions" (Schouteden, 1928). On 11 December 1927, another skull was received at the Museum, this time from Kunungu, and sent by Mr. Ngwe, the technical assistant who accompanied Schouteden on his two Congo expeditions. Still later in December 1927 (no exact day of arrival is known) another male skull, that became the "co-type, "was received from Djombo, where the animal was named "Buhumbusu," according to the field notes of the collector, Ghesquière. In the autumn of 1928 Schwarz spent several weeks in Tervuren where he saw skulls mentioned heretofore. It was at this time that he confirmed the novelty of this left bank pygmy chimp and proposed the name *Pan satyrus paniscus.* The announcement of this new taxon was made at the October 13th meeting of the Cercle Zoologique Congolais (Schwarz, 1928). The scientific diagnosis was published by Schwarz on

April 1, 1929. It is evident from that paper that Schwarz had, by that time, seen the skull from Bongandanga in the British Museum.

Dr. Schouteden never told me about this episode in the Museum's history. Only once did he briefly allude to it, dismissing it by saying that he never had felt qualified for conducting craniological work on primates. I am sure he must have been saddened that such an important discovery had escaped him. Nonetheless, his reaction was characteristic of his personality. He immediately began writing letters to his correspondents in the Congo and to members of the Cercle Zoologique Congolais urging them to collect all possible information on this animal from south of the Congo River. As a result by February of 1930, less than one year after the diagnosis of the new chimpanzee, Schouteden was able to add 13 new collecting sites to the distribution map and to publish photographs of live animals. He also added data on their behavior and external morphology (Schouteden, 1930). Early in 1931 he reported again on *Pan satyrus paniscus* including most of the data from his 1930 paper and mentioning more skull material received at Tervuren. He added two new localities and typically noted ". . . moi-meme, je savais des longtemps qu'en region de Lukolela le Chimpansee existait" (Schouteden, 1931), as if he could not resist the temptation to mention at least once, and somewhat cryptically, that he had been the first zoologist to suspect the existence of this animal. Knowing him very well, I can imagine (based on his silence on the subject) that he never forgave himself for missing this discovery. Outwardly unperturbed, however, he continued his own collecting and encouraging others. By 1940, at the beginning of World War II, contact between Belgium and the Belgian Congo lessened. At the time the Congo Museum was already in possession of 21 pygmy chimpanzee skulls from 17 localities, all from the left bank of the Congo River. The distribution map published by Schouteden (1944) was based on approximately 25 known localities and once more clearly demonstrated that *Pan paniscus* and *Pan troglodytes* were allopatric. Schouteden's work at this time confirmed the first major study of the pygmy chimpanzee by Coolidge, who had elevated the taxon to species rank, calling it *Pan paniscus* (Coolidge, 1933). At the time Dr. Schouteden retired from the Museum directorship in 1946, Dr. M. Poll, an ichthyologist, became curator of vertebrates in the Museum. Poll was to have only a tangential influence on our knowledge of *Pan paniscus*. After 1946, the man who was to play the next leading role was Prof. G. Vandebroek (1900–1977), Professor at the University of Leuven.

Vandebroek was originally an embryologist. Between 1925 and 1940 he worked on embryos of sea urchins, where he did some important marking experiments. He then concentrated his research on the origin, formation, and evolution of the excretory system in craniate chordates. A brilliant zoologist, a perfectionist, and an esteemed teacher, he produced meticulous, well-illustrated papers on these topics. Interest in a genetic disorder that was present in his family changed the orientation of his research from the late 1930s on. His research interests then shifted from mammalian and human genetics to the human racial characters and to modes of human evolution and the morphology of the great apes. Vandebroek conducted an exhaustive study of the molar dentition of apes wherein he sought to explain dental evolution from the Prosimii through the apes and man. He formulated a new theory of the formation and evolution of the mammalian dentition and a new nomenclature for the dentition of mammals (Vandebroek, 1961). This dental theory was both controversial and, due to its complexity, a nightmare for his students.

In his ontogenetic research on apes, Vandebroek soon felt the need for additional cranial material of the pygmy chimpanzee, especially for juvenile and fetal specimens. From the Belgian Science Foundation (NFWO) he received funds for an expedition to the Congo. The Ministry of Colonies granted him the scientific permission to collect pgymy chimpanzee material, but on condition that all specimens collected would be housed at Tervuren. Thus Vandebroek visited the Congo in 1955. He was not a hunter himself, but he traveled extensively on the left bank of the river in search of pygmy chimpanzees. He also encouraged local Africans and Europeans to preserve skulls and skeletons for him. In one place Vandebroek left a petrol drum with a small quantity of formaldehyde. When he returned 3 months later, he found the petrol drum filled with more than 20 decomposing chimpanzee skulls. Many of the teeth had fallen out of their sockets and Vandebroek related to me how it took him several days to replace all the loose teeth into their proper sockets (this was a task he performed apparently with great accuracy, for of all the subsequent primatologists who have visited Tervuren, none has questioned the dental associations in Vandebroek's material). Another important aspect of Vandebroek's field research was to reaffirm the geographic distribution of the pygmy chimpanzee. He explored the Congo River throughout Equateur and concluded that the river was everywhere broad enough to serve as a geographic barrier to nonswimming animals. Vandebroek reckoned that at only in one place, in the vicinity of Wanie Rukula, a little upstream from Kisangani (formerly Stanleyville), could chimpanzees possibly cross the River. In the dry seasons, low water levels might allow river crossing. Vandebroek therefore organized chimpanzee collecting at Wanie Rukula on both sides of the river (see area on map, Fig. 1). He stated afterward that apparently no crossing of the river by

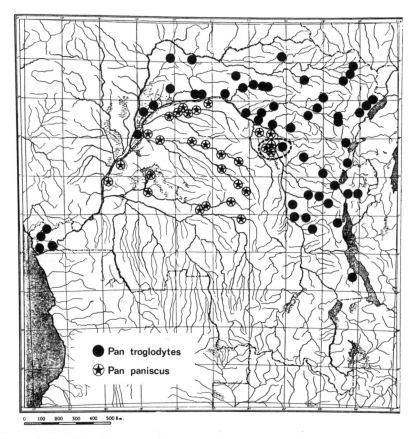

Figure 1. The distribution in Zaire of the chimpanzee (*Pan troglodytes schweinfurthii* and *P. t. troglodytes*) and of the pygmy chimpanzee (*Pan paniscus*) according to the locality records of the Tervuren Museum. The area encircled by an interrupted line was especially explored by G. Vandebroek.

any chimp had occurred, as on both sides not any trace of genetic mixing with the chimp race of the opposite bank could be found (Vandebroek, 1959).

In Belgium almost all material collected was registered in the Museum. Through Vandebroek's work the skull collection at Tervuren nearly doubled and the postcranial material of *Pan paniscus* almost tripled. Most of the skulls were then loaned to Vandebroek's lab at the University of Leuven. Always a perfectionist, he was slow to publish his new data before checking and rechecking his observations. As time passed, Vandebroek developed health problems, which drained his energy, and alas the wonderful pygmy chimpanzee collection he had made remained largely unknown to the scientific community. Some data on the dentition and on the frontal-squamosal suture were published in his textbook on the evolution of vertebrates (Vandebroek, 1969), but this was done without mentioning the source of these data. It was not generally known that these data were the initial observations on the more recent *Pan paniscus* skeletons that he himself had collected. As the early 1960s passed, it was increasingly clear that the long-awaited monograph on the pygmy chimpanzee skull and skeleton might never be finished. Dr. Poll, Curator of Vertebrates, became concerned that this very important collection might remain unpublished and unknown. He then summoned Vandebroek to finish a monograph or to return all skulls and skeletons to the Museum. This was a first refused by Vandebroek, but finally Vandebroek relinquished most of the skulls to the Museum. Meanwhile, Dr. Poll arranged for Prof. R. Fenart (of the University of Lille, France), who had been working on the craniology of the gorilla, to conduct a comprehensive, comparative craniologic study of common and pygmy chimpanzees. The Lille University team (which also included F. Deblock) started their work. Deblock, who carried out the major part of the study, became a regular visitor to Tervuren. All skulls were measured and X-rayed in Lille. Over a 3-year period all skulls were measured, except for the juvenile skulls in the possession of Vandebroek at Leuven. Dr. Poll then again futilely urged Vandebroek to return all skulls to Tervuren, and again feelings between the two men hardened. I still recall both men taking me aside and explaining their views of the problem as if they both needed some moral support. As a young researcher I hesitated to interpose myself between the two venerable scientists. I did suggest, nevertheless, a possible remedy, whereby R. Deblock would go to see Vandebroek personally and without Poll's knowledge. Vandebroek, actually warm and friendly, agreed to a personal loan of the skulls to Deblock separate from the Museum and Dr. Poll. In this way Deblock was able to examine all skull material available and to finish his work (Fenart and Deblock, 1973). Afterward, Vandebroek's health continued to deteriorate and his assistant, who had been working on the postcranial material, died accidentally. Vandebroek's own condition became acute and, suffering from Parkinson's disease, he was forced to retire from his professorship in 1975. He died in 1977. After Vandebroek's retirement nearly all of the skulls and skeletons at Leuven were returned to the Museum.

Since 1960 relations between Congo (now Zaire) and Belgium have been strained, and the special scientific relationship between the Tervuren Museum and Zaire virtually ceased to exist. As a result, little chimpanzee material has been received in the past 20 years and the collections have

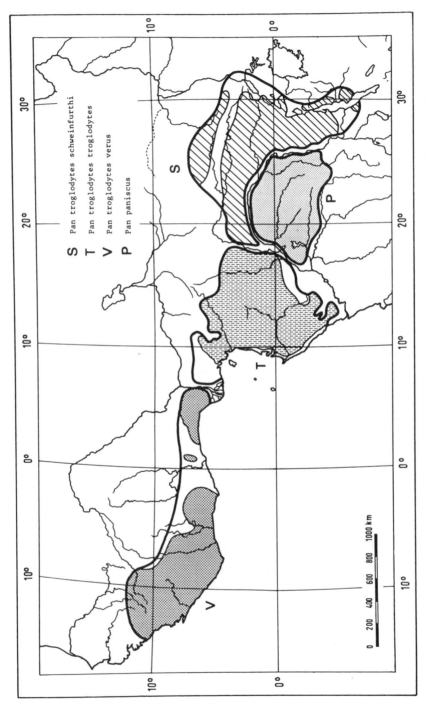

Figure 2. The probable distribution of chimpanzees around 1900, based on old locality records and museum data.

remained what they were following Vandebroek's expedition. The Museum today has a total of 155 skulls and postcranial remains of 17 specimens. Since the early 1970s this material has been extensively studied by scientists from many countries (as the following chapters will attest). My own contribution to the knowledge of *Pan paniscus* is very modest, and remains restricted to the curatorial charge of this collection since Dr. Poll's retirement. Aware of the value of this rare collection, I have tried to provide strong encouragement to those who have worked on these collections. We have attempted to provide maximum facilities, as well as housing, for both the cranial and postcranial material. I continue my keen interest in the zoogeography of Central Africa, and I have spent time checking the locality data and plotting the coordinates of the pygmy chimpanzee specimens. As a result I was able to produce both of the distribution maps published by Fenart and Deblock (see Fig. 2).

I hope that the information related here will contribute a bit to the understanding of the Tervuren pygmy chimpanzee collection and that it will heighten appreciation of the work of Henri Schouteden and G. Vandebroek, two scientists who have contributed greatly to our present knowledge of *Pan paniscus* through their committment to the pygmy chimpanzee collection of the Tervuren Museum.

References

Coolidge, H. J., 1933, *Pan paniscus:* Pigmy chimpanzee from south of the Congo River, *Am. J. Phys. Anthropol.* **XVIII**(1):1–57.

Fenart, R., and Deblock, R., 1973, *Pan paniscus* et *Pan troglodytes* Craniométrie: Étude comparative et ontogénique selons les méthodes classiques et vestibulaire. Tome I, *Mus. R. Afr. Cent. Tervuren Belg. Ann. Ser. Octavo Sci. Zool.* **204**:1–473.

Schouteden, H., 1928, Deux singes interessants, *Bull. Cercle Zool. Cong.* **V**(1):9.

Schouteden, H., 1930, Le chimpanzé de la rive gauche du Congo, *Bull. Cercle Zool. Cong.* **VI**(4):114–119.

Schouteden, H., 1931, Quelques notes sur le chimpanzé de la rive gauche du Congo, *Pan satyrus paniscus, Rev. Zool. Bot. Afr.* **XX**(3):310–314.

Schouteden, H., 1944, De Zoogdieren van Belgisch Congo en Rwanda-Urundi. I. Primates, Chiroptera, Insectivora, Pholidota, *Ann. K. Mus. Belg. Congo, C., Zool. Ser. II,* **III**(1):168.

Schwarz, E., 1928, Le chimpanzé de la rive gauche du Congo, *Bull. Cerc. Zool. Cong.* **V**(3):70–71.

Schwarz, E., 1929, Das Vorkommen des Schimpansen auf den linken Kongo-Ufer, *Rev. Zool. Bot. Afr.* **XVI**(4):425–426.

Vandebroek, G., 1959, Notes ecologiques sur les anthropoides africains, *Ann. Soc. R. Zool. Belg.* **LXXXIX**(1):203–211.

Vandebroek, G., 1961, The Comparative Anatomy of the Teeth of lower and non specialized mammals, in: *International Colloqium on the Evolution of Mammals,* Publ. Kon. Vl. Acad. Wet. Lett. Sch. Kun. Belg., Brussels, Part I, pp. 215–320, Part II, 181 pp.

Vandebroek, G., 1969, *Evolution des Vertebres de leur origine á l'homme,* Masson, Paris.

CHAPTER 2

Blood Groups of Pygmy and Common Chimpanzees

A Comparative Study

WLADYSLAW W. SOCHA

1. Introduction

The discovery of the blood groups of man (Landsteiner, 1901) and, particularly, confirmation of their hereditary nature (Bernstein, 1925) added a new dimension to taxonomic studies. The importance of blood groups for systematics was demonstrated quite early by Hirszfeld and Hirszfeld (1919), who carried out analyses of blood group frequencies among human races. The first comparative study of hominoid (ape and human) red cell antigens was undertaken by Landsteiner and Miller (1925). Thirty-five years later the first attempts were made to compare blood groups among various chimpanzee species, including the pygmy chimpanzee (André *et al.*, 1961). Although at present impressive numbers of specificities have been identified on the red cells of common chimpanzees (*Pan troglodytes*), and hundreds of animals of this species have been tested for their blood groups, the testing of blood groups in pygmy chimpanzees has been infrequent, due to the relatively few specimens in captivity. Testing done thus far on the blood groups of *Pan paniscus* has revealed interesting differences between pygmy and common chimpanzees and shed light on discussions of the phylogenetic position of *Pan paniscus*. The purpose of

WLADYSLAW W. SOCHA ● Primate Blood Group Reference Laboratory and WHO Collaborating Centre for Haematology of Primate Animals, Laboratory for Experimental Medicine and Surgery in Primates (LEMSIP), New York University School of Medicine, New York, New York 10016.

this chapter is to highlight the results of our ongoing research on blood group serology of the pygmy and common chimpanzees.

2. Methodology of Blood Grouping and General Classification of Blood Groups of Primates

When the red cells of one individual are mixed *in vitro* with the serum of another individual of the same or different species, very often, though not always, clumping (agglutination) of the red cells occurs as evidence of the presence in the serum of specific antibodies. In that way, by comparing the reactions of red cells and sera of various persons, the first blood groups of man, the A-B-O groups, were defined. Shortly thereafter, the discoveries of other blood group systems quickly followed, and, at present, hundreds of various specificities can be detected on human erythrocytes if proper reagents (antisera) are available for carrying out the tests.

While the discovery of the A-B-O groups was originally made with antisera normally occurring in the sera of most humans (except those belonging to group AB), the usefulness of such *"natural" antibodies* was limited by their low titers and avidity. This is why routine blood grouping work is carried out by means of potent antisera obtained either from the blood of persons or animals purposely immunized with red cell antigen or from persons accidentally sensitized as the result of transfusion of incompatible blood or in the course of incompatible pregnancy.

The antigen–antibody reaction is a complex phenomenon that depends on a variety of factors, among which, in the case of red cell antigens, the temperature of reaction, the size and character of antibody molecules, as well as the condition of the red cell membrane are the most important. In order to optimize the activity of antibodies and enhance their reactions with red blood cells, blood grouping techniques are adapted to the requirements of the given antigen or antibody. For example, the tests for the A-B-O groups are, by and large, carried out by the so-called saline agglutination technique, in which the red cells to be blood-grouped are suspended in normal saline (or a low-ionic strength medium) and mixed with anti-A or anti-B reagents to produce direct agglutination. The same technique has proved adequate for tests with anti-M and anti-N sera and any other antibodies of the so-called IgM class. When, however, antibodies of smaller size (by and large of IgG class) are used, agglutination does not occur until a third component, the antiglobulin serum, is added. The latter brings about the clumping of the red cells by specifically reacting with immunoglobulin antibody molecules with which the red cells were previously coated. This *antiglobulin technique* is employed for tests with some of the human anti-Rh sera and with the majority of reagents produced by immunization of primates. In some instances, the *enzyme-treated red cell method* is preferable for tests with small-molecule antibodies of IgG class. Proteolytic enzymes (e.g., ficin, bromelin, papain) modify the erythrocyte membrane in such a way as to produce clumping of the red cells by antibodies that normally only attach themselves to the red cells without causing their visible agglutination.

Some group specificities are detected not only on the red blood cells, but also in body fluids and secretions of those individuals known as *secretors*. In human populations, the frequency of secretors varies from 61% to 98% (Socha, 1966). All apes and monkeys are secretors and their saliva, urine, seminal fluid, gastric juice, and amniotic fluid can be tested for the A, B, and H specificities by the *hemagglutination inhibition technique*. With this technique the substance investigated for the presence of soluble A, B, and H group substances is incubated with anti-A, anti-B and anti-H reagents, and then the indicator cells of group A, B, and O are added to that mixture and their agglutination or absence of agglutination observed. When the indicator cells of group A, for example, are not agglutinated by anti-A antiserum, it is concluded that the latter is inhibited by the A substance presumably present in the substance tested.

The A-B-O testing of the human and nonhuman primate red cells and secretions is usually complemented and confirmed by the so-called *serum tests*. The serum to be tested is mixed with control red cells of known types A, B, and O. When, for example, the clumping of group B cells takes place, but not those of group A or O, one concludes that the investigated serum contained anti-B antibodies. According to *Landsteiner's rule,* the normal serum contains antibodies of specificities for which corresponding antigens are absent from red cells and from secretions. For example, the red cells and/or saliva of a group A individual contain A but not B agglutinogen, and therefore the serum has anti-B but not anti-A agglutinins. This reverse reciprocal relationship between antigens and antibodies, shown in Table I, is at the basis of the A-B-O blood grouping principle, which is valid not only for human but also for ape and monkey testing.

For details of blood grouping techniques in human and nonhuman primates, the reader is referred to specialized articles and textbooks (Socha *et al.,* 1972; Erskine and Socha, 1978; Socha and Ruffié, 1983).

The study of blood groups of nonhuman primates was a logical extension of human serohematology, and, understandably, the first attempts to define specificities of primate red cells were carried out using the reagents originally prepared for typing human erythrocytes. The blood groups,

Table I.   The Four Blood Groups of the Human A-B-O System: Reciprocal
Relationship between Antigens and Antibodies

Blood group (phenotype)	Red cell antigens (agglutinogens)	Serum isoagglutinins	Remarks
O	None	Anti-A and anti-B	—
A	A	Anti-B	—
B	B	Anti-A	Not observed in chimpanzees
AB	A and B	None	Not observed in chimpanzees

such as A-B-O, M-N, Rh-Hr, Lewis, and I-i, defined in that way in non-human primates were called *human-type blood groups* and were considered homologues of the human red cell antigens. The second category of red cell specificities were those detected by antisera specifically produced for typing primates and obtained either by immunizing laboratory animals with the red cells of apes and monkeys or, preferably, by iso- or cross-immunizations of primates. These so-called *simian-type blood groups* were believed to be primate's own specificities, some of which could be analogues of the human red cell antigens. The term "human-type specificity" may not be justified, since some of those specificities probably existed well before the emergence of man (Ruffié and Socha, 1980). Sharp separation of the second class of blood groups also appears unfounded; the so-called simian-type specificities, originally defined on the red cells of apes, were later detected, in polymorphic form, on human red cells as well (Socha and Moor-Jankowski, 1979). I propose to abandon the earlier classification and instead discuss the genetically separated blood group systems of chimpanzees in chronological order of their discovery.

3. The A-B-O Blood Group System

Similar to the situation in humans, when the serum of a chimpanzee is mixed with red cells of other chimpanzees, agglutination of the red cells occurs in some but not all combinations. Based on these isoagglutination reactions, two blood groups can be distinguished in chimpanzees, designated group A and group O, respectively, because of their correspondence with the human blood groups A and O. In fact, chimpanzee red cells can be typed by using reagents prepared for human blood.

When anti-A and anti-B agglutinins of human origin are used for testing chimpanzee red cells, the sera must first be absorbed with chimpanzee group O red blood cells to remove nonspecific heteroagglutinins that react with erythocytes of *all* chimpanzees (Wiener *et al.*, 1963). High-titered human anti-A and anti-B reagents, which are now readily available commercially for tests in humans, can be used for work on chimpanzee blood without prior absorptions if they are first diluted 1:10 in saline. At this dilution, the generally low-titered, nonspecific heteroagglutinins are not reactive. Some seed extracts (lectins) known to have type-specific agglutinating activity are also useful for typing chimpanzee red cells. They have the advantage that no prior absorption to remove anti-chimpanzee heteroagglutinins is required. The anti-A reagent made of saline extracts of lima beans (*Phaseolus vulgaris*) was successfully used by Wiener *et al.* (1969c) and Wiener and Moor-Jankowski (1971) for A-B-O blood grouping of chimpanzee blood. Another reagent of anti-A specificity, free of nonspecific heteroagglutinins, and thus suitable for blood grouping chimpanzees, was obtained from the snail, *Helix pomatia* (Prokop *et al.*, 1965), and *Achatina granulata* (Wiener *et al.*, 1969a).

Table II summarizes the results of tests for the A-B-O blood groups carried out over the last 20 years on common and pygmy chimpanzees. It must be stressed that the early findings reported by André and differing significantly from the later observations must be qualified, since the blood samples he tested were often hemolyzed and were grouped not by direct hemagglutination, but by the absorption method, which has been shown to be less exact. Results reported by Schmitt *et al.* (1962) and Schmitt (1968) are based on the elution technique, first applied in tests on non-human primate red cells by Landsteiner and Miller (1925) to overcome the interference of heteroagglutinins in the reagents of human origin. Low titer, low avidity, as well as lack of stability of such eluates often rendered the tests difficult to carry out, and yielded results of questionable reliability. In view of these shortcomings of the testing techniques employed in earlier studies, the question of the occurrence of group O in pygmy chimpanzees, which we cannot confirm in our studies, remains open. The results in Table II do not reflect separation of the three subspecies of the common chimpanzee, *Pan troglodytes*. It appears, however, that there are significant differences in frequencies of the O and A blood groups among various common chimpanzee subgroups (*P. t. versus, P. t. troglodytes*, and *P. t. schweinfurthii*). According to Moor-Jankowski and Wiener (1967), the frequency of O can vary from as low as 9.5% in *P. t. versus* to as high as 39.5% in *P. t. schweinfurthii*.

3.1. Subgroups of A

The anti-A agglutinins used for hemagglutination tests are not homogeneous but are composed of many, qualitatively different fractions.

Table II. A-B-O Blood Groups of Chimpanzees

Species	Number tested	Group A		Group O		Reference
		Number	Frequency	Number	Frequency	
Pan paniscus	17	15	—	2	—	André *et al.* (1961)
	6	5	—	1	—	Schmitt (1968)
	14	14	—	0	—	Present series
Pan troglodytes	58	101	0.639	57	0.361	André *et al.* (1961)
	143	123	0.860	20	0.140	Schmitt *et al.* (1962)
	158	135	0.854	23	0.145	Schmitt, 1968.
	228	195	0.855	33	0.145	Moor-Jankowski and Wiener (1972)[a]
	60	56	0.929	4	0.066	Wiener *et al.* (1972)
	497	462	0.929	35	0.071	Present series[b]

[a]Combined from three earlier studies.
[b]Includes also blood samples of animals reported previously (Moor-Jankowski *et al.*, 1972, 1975).

The two principal varieties of anti-A are: (1) *anti-A proper*, which reacts with all group A red cells with about equal intensity, and (2) *anti-A_1*, which reacts with some, but not all, red blood cells of group A. By testing group A red cells (of both humans and apes) side by side with the two kinds of anti-A reagents, one can divide these cells into two subgroups of A, namely A_1 (which reacts with both anti-A proper and anti-A_1), and A_2 (which react only with anti-A, but not with anti-A_1). For blood grouping, the anti-A_1 reagent is obtained either from a human group B serum absorbed with A_2 red cells or from a seed extract of *Dolichos biflorus*. In human populations, about one third of group A Caucasoids and Negroids are group A_2, but no subgroup A_2 is found among Mongoloids. Among 190 common chimpanzees tested by us as group A, 22 (11.5%) were found to be subgroup A_2, while all pygmy chimpanzees investigated so far are of the A_1 type.

The absence of A_2 type among pygmy chimpanzees (which may be an artifact due to the small sample size) is not the only peculiarity that distinguishes pygmy chimpanzees from group A, common chimpanzees. When the red cells of pygmy and common chimpanzees and humans are titrated with a variety of anti-A and anti-A_1 reagents, significant differences are observed (Table III). Although the red cells from group A, *P. troglodytes* are regularly agglutinated by human anti-A as well as by the lectin *Dolichos biflorus*, the reactions are distinctly weaker than with

Table III.   Comparison of the Reactions of the Red Cells of *Pan paniscus*, *Pan troglodytes*, and Humans with Various Anti-A Reagents[a]

| | Titers with human anti-A | | Titers with anti-A_1 reagents | |
	Unabsorbed	Absorbed group A red cells of *P. paniscus*	Human anti-A absorbed with human A_2 red blood cells	*Dolichos biflorus* lectin
Pan paniscus[b]				
Annemie	64	0	32	48
Kitty	96	1	36	32
Lanie	64	1	36	24
Camille	64	0	32	32
Pan troglodytes				
Ch-211, Oscar, group A_1^{ch}	48	0	8	6
Ch-192, Lindsay, group A_1^{ch}	48	0	8	6
Ch-639, Dina, group A_1^{ch}	48	0	4	3
Ch-9, Amos, group $A_{1,2}$	32	0	2	2
Ch-332, Tang, group A_2	24	0	0	0
Ch-116, Shirley, group O	0	0	0	0
Ch-168, Walter, group O	0	0	0	0
Human				
Group A_1 (SL)	64	2	32	24
Group A_1 (FS)	64	1	24	20
Group A_2 (DP)	16	0	0	0
Group O (HB)	0	0	0	0

[a]The titers are the reciprocal of the highest serum dilution giving a distinct (one plus) reaction.
[b]All *Pan paniscus* subjects are type A.

lectin *Dolichos biflorus*, the reactions are distinctly weaker than with human red cells of subgroup A_1. Moreover, red cells from about one tenth of all group A, common chimpanzees gave especially weak reactions with anti-A_1 reagents though distinct clumping could still be observed. The animals of the latter type were considered to be of intermediate type $A_{1,2}$ (for example, Ch-9 Amos in Table III). Finally, the red cells of some common chimpanzees react, as do the human A_2 cells, only with anti-A but not with anti-A_1 reagents, and their reactions are much weaker than those of human A_2 blood. Thus, the two main subgroups of group A appear to exist in *Pan troglodytes* as in humans, but the distinction between these subgroups in chimpanzees is not as sharp as in humans. To emphasize the difference between the blood of *P. troglodytes* and humans, the subgroups of A in chimpanzees have been designated as A_1^{ch} and A_2^{ch}, respectively, where "ch" stands for chimpanzee.

   In contrast, the red cells of all pygmy chimpanzees thus far tested yield reactions indistinguishable (both in avidity and titer, as shown in Table III) from those of human subgroup A_1. Similar observations were reported by Schmitt *et al.* (1962). The relative agglutinability of human and chimpanzee group A red cells by anti-A reagents is therefore as follows:

human A_1, *P. paniscus* A > *P. troglodytes* A_1 > *P. troglodytes* $A_{1,2}$

> *P. troglodytes* A_2 > human A_2

In the case of *P. paniscus,* group A red cells, though reacting like human subgroup A_1, are designated simply as A because subgroup A_2 has so far not been found.

It may be of interest that the red cells of newborn chimpanzees of group A usually give only very weak reactions with the most potent anti-A reagents, and only part of their red cells are agglutinated, while many remain unagglutinated. This indicates that in chimpanzees, as in humans, the A-B red cell agglutinogens are incompletely developed at birth (Wiener *et al.*, 1974*b*).

3.2. The H Specificity

Group O red blood cells of humans and chimpanzees are characterized not merely by the absence of agglutinogens A and B but also by the regular presence of specificity H, at one time mistaken for the expected specificity O. As it turned out, however, all A-B-O agglutinogens on human red cells were found to contain greater or lesser amounts of the H substance detected by anti-H lectin (*Ulex europaeus*) or by an eel serum. In fact, there is a reciprocal relationship between H and A, B specificities, as shown by the fact that the red cells of heterozygotes *AO* react more strongly with anti-H than do those of homozygotes *AA*, while the strongest reactions are observed with the red cells of homozygotes *OO*. The human A_2 red cells give, by and large, stronger reactions with anti-H than do subgroup A_1 erythrocytes. These findings led to the hypothesis that the H substance is a precursor of A and B (Watkins, 1966).

In adult and juvenile *P. troglodytes,* group O red cells generally are agglutinated by anti-H lectin as in humans, but the reactions are weaker than for human group O adults, and red cells of occasional group O chimpanzees failed to react with this reagent. Thus, the reactions of group O red cells of *P. troglodytes* with anti-H resemble those of human newborn group O red cells. Group A red cells of these apes give only very weak or negative reactions with all anti-H reagents.

All pygmy chimpanzee red cells tested by us gave negative reactions with anti-H lectin (*Ulex europaeus*), as expected, since they all belong to subgroup A_1.

3.3. Secretion of A-B-H Substances

Inhibition tests carried out on saliva of over 100 *Pan troglodytes* yielded positive results, thus indicating that all animals were secretors of the group substances. As a rule, the animals whose blood red cells were typed as group A (A_1 or A_2) proved to be secretors of A and H, while all group O chimpanzees secreted only H substance. By the same token, all pygmy chimpanzees whose saliva became available for testing were found to be secretors of A and H. Table IV shows some representative inhibition titers obtained with saliva of *P. troglodytes*, *P. paniscus*, and human

Table IV. Hemagglutination Inhibition of Saliva of Chimpanzees and Human Donors

	Titers[a] of saliva that inhibited specific reactions of given reagents with proper indicator red blood cells (human)		
	Anti-A and group A_2 red blood cells	Anti-B and group B red blood cells	Anti-H and group O red blood cells
Pan paniscus[b]			
Annemie	256	0	256
Kitty	128	0	256
Lanie	128	0	128
Basondjo	256	0	256
Linda	256	0	128
Pan troglodytes			
Ch-48, group A	128	0	128
Ch-1, Buddha, group A	256	0	256
Oliver, group A	256	0	256
Ch-191, Kurt, group O	0	0	256
Ch-168, Walter, group O	0	0	256
Human controls			
AJ, group A, secretor	128	0	64
MS, group B, secretor	0	256	128
WK, group O, secretor	0	0	128
BF, nonsecretor	0	0	0

[a]Reciprocal of the highest dilution of saliva neutralizing (inhibiting) the reaction of the reagent prepared as to yield 8–16 agglutinating doses of antibody.
[b]All *Pan paniscus* subjects are type A.

donors. It is evident that there are no differences as to the specificity and strength of reactions among common and pygmy chimpanzees and human group A or O secretor salivas. Inhibition titers roughly express relative concentrations of the group substance in the fluid tested.

3.4. Serum Isoagglutinins

In chimpanzees, as in humans, anti-B isoagglutinins are regularly present in the sera of group A individuals, while anti-A and anti-B isoagglutinins are detected, as a rule, in the sera of group O animals. In selected cases, the level of isoagglutinins has been established by titrations using human A_1, A_2, B, and O cells as targets. Table V gives representative titers of these antibodies in the sera of pygmy and common chimpanzees, as well as in human sera. As can be seen, neither sera of pygmy chimpanzees nor those of *P. troglodytes* agglutinated the human group O red cells, which indicates that normal chimpanzee sera does not contain nonspecific agglutinins against human red cells. As mentioned earlier, the reverse is not true: most normal human sera react with chimpanzee cells,

Table V. Representative Titrations of Anti-A and Anti-B Isoagglutinins in Chimpanzee Sera

	Titers[a] of the serum[b] with the red blood cells					
	Human group				Gibbon group B	Orangutan group A_1B
	O	A_1	A_2	B		
Pan paniscus[c]						
Bosondjo	0	0	0	32	32	8
Laura	0	0	0	16	8	4
Linda	0	0	0	16	20	3
Lisa	0	0	0	2	3	½
Pan troglodytes						
Ch-192, Lindsay, group A	0	0	0	4	5	4
Ch-363, Tommy, group A	0	0	0	8	8	8
Ch-467, Tabletop, group A	0	0	0	2	2	1
Ch-168, Walter, group O	0	6	1½	16	20	8
Human controls						
Group A	0	0	0	16	NT	NT
Group B	0	6	4	0	NT	NT
Group O	0	12	4	20	NT	NT

[a]See footnote to Table III.
[b]Preabsorbed with human group O red blood cells.
[c]All *Pan paniscus* subjects are type A. NT, not tested. All titrations by saline method.

irrespective of their blood group. On the other hand, the normal sera of pygmy and common chimpanzees contain anti-A and/or anti-B isoagglutinins of titers comparable to those found in normal human sera. The similarity goes even further, in that in the sera of group O chimpanzees, as in the sera of group O humans, the titer of anti-B agglutinins is, in general, higher than that of anti-B in the sera of group A individuals (Weiner et al., 1974b). The possible reason is that the group O sera in chimpanzees, as in humans, contain not only the anti-A and anti-B isoagglutinins, but also isoagglutinins designated anti-C for a specificity shared by group A and group B red cells, but lacking in group O (Socha and Wiener, 1973). Wiener and Ward (1966) provided convincing evidence for the role of the factor C in pathogenesis of the A-B-O hemolytic disease of the human newborn, and possibly also in chimpanzees.

A summary of the findings on the A-B-O blood groups of chimpanzees is as follows: Of the four groups of the A-B-O blood group system, only two, A and O, were found in *P. troglodytes;* the latter group was quite infrequent (less than 10%). All 14 pygmy chimpanzees tested were found to be group A. This casts doubt on earlier reports (based on unreliable blood grouping techniques) of group O in *P. paniscus.* Both common and pygmy chimpanzees are secretors of the A and/or H group substances and concentrations of those substances detected in saliva were comparable to those usually found in saliva of human secretors. A reciprocal relationship between red cell antigens and serum anti-A and anti-B isoagglutinins holds in both types of chimpanzee, as it does in humans, and the titers of these antibodies are similar to those detected in human sera. The main difference between A-B-O groups of *P. troglodytes* and *P. paniscus* concerns the nature of the A red cell agglutinogen. In the common chimpanzee this antigen is in general weaker than in humans, and its division into subgroups A_1 and A_2 is not as sharp as in human red cells. The A agglutinogen of the pygmy chimpanzee is, however, serologically indistinguishable from human A_1 antigen.

4. The M-N Blood Group System

The two major cell agglutinogens of this system, namely M and N, discovered in humans by Landsteiner and Levine (1927), are defined by the reagents obtained from the sera of rabbits immunized with M- or N-bearing red cells. By and large the human M or N red blood cells are used for immunization, but injections of rabbits with chimpanzee, baboon, or rhesus monkey blood can also yield specific and potent anti-M reagents (Weiner et al., 1964b). An anti-N reagent, useful also for typing red cells of nonhuman primates, can be obtained from extracts of plants such as *Vicia graminea* (Ottensooser and Silberschmidt, 1953) or *Vicia unijuga* (Moon and Wiener, 1974).

The anti-M and anti-N reagents define three M-N types, two of which are encountered in chimpanzees (Table VI). It is noteworthy, however, that while rabbit anti-M sera give consistently very strong reactions with both common and pygmy chimpanzee red blood cells, the rabbit anti-N reagents either fail to react with any of the chimpanzee blood or only weakly clump the red cells of some of *P. troglodytes*. The anti-N reagents do not react with the red cells of pygmy chimpanzees. Much better and consistent reactions are observed when anti-N lectins are used instead: clear agglutinations are obtained with blood of some, but not all common chimpanzees. The same lectins, however, failed to agglutinate the red cells of pygmy chimpanzees investigated thus far. One can therefore distinguish two M-N types among common chimpanzees, namely M and MN, but all pygmy chimpanzees so far tested are exclusively of type M. Among 497 *P. troglodytes* included in the present series, there were 204 (41%) of type M and 293 (59%) of type MN. Although the red blood cells of *P. troglodytes* and *P. paniscus* are agglutinated by the same anti-M antisera, the M agglutinogens on the erythrocytes of both species are not necessarily identical. This was shown in absorption–fractionation experiments carried out by Moor-Jankowski *et al.* (1972). Results of a similar study are reproduced in Table VII. Direct tests reveal that red cells of all three pygmy chimpanzees were agglutinated by a rabbit anti-M serum to a titer only one-third to one-sixth as high as human M or MN cells, and to generally lower titers than the red cells of three *P. troglodytes* used in the same experiment. Absorption of the anti-M serum with human M cells removed all reactivity not only for the human blood, but also for all the chimpanzee red cells. The experiment demonstrates that agglutination of the red cells of all chimpanzees was due to the presence of an

Table VI. Serological Definition of Blood Groups of the M-N System

Blood group designation	Reaction with the reagent		Remarks
	Anti-M	Anti-N	
M	Positive	Negative	—
N	Negative	Positive	Not observed in chimpanzees
MN	Positive	Positive	—

Table VII. Comparisons of the Reactions of Red Cells of *Pan paniscus, Pan troglodytes,* and Humans with a Particular Rabbit Anti-M Serum

| | Titer of anti-M rabbit serum | | | |
| | | After absorption with red cells of | | |
	Unabsorbed	Human type M	Pooled *Pan paniscus*	Pooled *Pan troglodytes*
Pan paniscus				
Annemie	7	0	0	0
Kitty	7	0	0	0
Lanie	7	0	0	0
Laura	7	0	0	0
Pan troglodytes				
Ch-140, Brenda	14	0	3	0
Ch-85, Billy	12	0	½	0
Ch-169, Possum	5	0	0	0
Human				
Type M	20	0	2	0
Type N	0	0	0	0
Type MN	40	0	3	0

M-like agglutinogen, rather than to nonspecific heteroagglutinins. Absorption of the anti-M serum with pooled red cells of *P. paniscus* removed the activity for absorbing cells, but left behind a fraction of antibodies still reactive with human M red cells as well as with the red cells of *P. troglodytes*. Absorption with red cells of *P. troglodytes,* however, removed all reactivity from this anti-M reagent, as did absorption with human M red cells. Thus, these absorption and titration experiments showed that while the red cells of pygmy chimpanzees had an M-like agglutinogen, this agglutinogen differed qualitatively not only from the human M agglutinogen, but also from the M-like agglutinogen on the red cells of *P. troglodytes*.

In summary, while all chimpanzees, *P. troglodytes* as well as *P. paniscus,* have an M-like agglutinogen on their red cells, the M of pygmy chimpanzees is qualitatively different from that found on the red cells of *P. troglodytes* and on human erythrocytes. Among chimpanzees, only *P. troglodytes* is polymorphic with regard to N specificity; it is detectable on ape red cells by means of specific anti-N lectins, but not with rabbit antisera.

5. The V-A-B-D Blood Group System

This system, closely related to the human M-N groups, was the first to be defined in *P. troglodytes* by means of antisera produced by iso- or cross-immunization of chimpanzees (Wiener *et al.*, 1974*a*). The V-A-B-D system is built around a central antigen, the V^c antigen,* which has a unique position among specificities of this system. This is indicated by the fact that the V^c is demonstrable on the red cells of chimpanzees not only by isoimmune anti-V^c serum, but also by an anti-V^c reagent obtained from the serum of chimpanzees immunized with human red cells. Reagents of the two kinds give reactions paralleling those obtained with the anti-N^V lectin from the seeds of *Vicia graminea*, which, as explained earlier, reacts with human-type antigen N. All remaining specificities of this system are defined exclusively by chimpanzee isoimmune sera. A hypothesis postulating the existence of five multiple alleles (v, v^A, v^B, v^D, and V) to account for the main phenotypes of the system (see Table VIII) gave a satisfactory fit in earlier population studies (Socha and Moor-Jankowski, 1979). It has also been confirmed by as yet unpublished genealogical data. The heritable nature of the rare types shown in Table VIII is supported by observations of a few families of *P. troglodytes* in which the variants V^{pq} and V^q were transmitted to offspring from at least one of the parents. The very high polymorphism of the V-A-B-D blood groups in *P. troglodytes* (16 various V-A-B-D types have thus far been observed in these apes, as shown in Table VIII) sharply contrasts with the complete lack of such variability within *P. paniscus*. All pygmy chimpanzees tested thus far are invariably v.D, the rarest (1.4%) among regular V-A-B-D types of *P. troglodytes*. That the reactions of anti-D^c reagents with the red cells of pygmy chimpanzees were serologically identical with those obtained with the red cells of *P. troglodytes* was ascertained by cross-absorption experiments: absorption with red blood cells of *P. paniscus* inactivated the anti-D^c serum to the same extent as absorptions with the red cells of *P. troglodytes*. Allowing for the relatively small number of pygmy chimpanzees tested, it would seem that in the course of the speciation of the pygmy chimpanzee, the V-A-B-D locus developed as a single-gene locus, thus becoming a species characteristic of *P. paniscus*. If this was so, the rare appearance of the gene v^D among common chimpanzees would have to be considered the result of more or less recent hybridization with pygmy chimpanzees. Obviously, an opposite process cannot be excluded, namely, that the multiallelic V-A-B-D locus of *P. troglodytes* is an effect of multiple

---

*The superscript c included in the designation of a specificity stands for "chimpanzee."

Table VIII. Chimpanzee V-A-B-D Blood Group System: Serology and Distribution

					Distribution			
					Pan troglodytes ($N = 498$)		Pan paniscus ($N = 14$)	
	Reaction with serum of specificity							
Designation	Anti-V^c	Anti-A^c	Anti-B^c	Anti-D^c	Number	Frequency	Number	Frequency
v.A	−	+	−	−	93	0.1867	0	0.0
v.B	−	−	+	−	46	0.0924	0	0.0
v.D	−	−	−	+	7	0.0140	14	1.0
v.AB	−	+	+	−	103	0.2068	0	0.0
v.AD	−	+	−	+	26	0.0522	0	0.0
v.BD	−	−	+	+	20	0.0402	0	0.0
V.O	+	−	−	−	32	0.0642	0	0.0
V.A	+	+	−	−	85	0.1707	0	0.0
V.B	+	−	+	−	50	0.1004	0	0.0
V.D	+	−	−	+	14	0.0281	0	0.0
Rare types:								
V^q.A	$+^a$	+	−	−	3	0.0060	0	0.0
V^q.B	$+^a$	−	+	−	2	0.0040	0	0.0
V^{pq}.A	$+^b$	+	−	−	3	0.0060	0	0.0
V^{pq}.B	$+^b$	−	+	−	9	0.0181	0	0.0
V^{pq}.D	$+^b$	−	−	+	4	0.0080	0	0.0
V^{pq}.O	$+^b$	−	−	−	1	0.0020	0	0.0

[a]Positive reaction with two of three anti-V^c reagents.
[b]Positive reaction with one of three anti-V^c reagents.

mutations occurring at the v^D locus carried by the common ancestor of the two species of chimpanzees.

6. The Rh-Hr Blood Group System

The Rh system is of particular interest to students of blood groups of primates, since the discovery of the Rh blood factor on human red cells resulted from investigation of the properties of an anti-rhesus monkey serum (Landsteiner and Wiener, 1937; Wiener, 1938; Landsteiner and Wiener, 1940). In fact, the first description of the rhesus factor of human blood was made possible by the use of rabbit anti-rhesus monkey and guinea pig-anti-rhesus monkey sera. Later, sera of rabbits immunized with baboon blood also proved to be suitable for detecting Rh specificity on human red cells (Wiener *et al.*, 1969*b*). By now antisera of that kind have only historical and theoretical significance, having been replaced in practice by much more potent and reliable anti-Rh reagents obtained either from human volunteers immunized with human Rh-positive red blood cells, or from patients accidentally exposed to the Rh antigen. Reagents of the latter type were also used for blood grouping common chimpanzees (Wiener and Gordon, 1961; Wiener *et al.*, 1964*a*) and pygmy chimpanzees (Moor-Jankowski *et al.*, 1972).

The red cells of all *P. troglodytes* reacted with human anti-Rh_0(D) sera to titers as high as or one to two dilutions lower than human Rh-positive cells. Moreover, when chimpanzee red cells were coated with human anti-Rh_0 antibody and then titrations were done with rabbit anti-human globulin serum, they reacted to the same or almost the same titer as sensitized human Rh-positive cells. These results showed that the red cells of *P. troglodytes* have Rh-like agglutinogens and that the number of Rh combining sites on the surface of these red cells is of the same order of magnitude as in humans. In further experiments it was found, however, that when parallel tests were run using various human anti-Rh_0 reagents, fine differences between red cells of *P. paniscus* and *P. troglodytes* were observed, indicating that the Rh_0 agglutinogens of the two chimpanzee species were not identical (Wiener *et al.*, 1971). In fact, red cells of pygmy chimpanzees appeared to have the Rh_0 agglutinogen somewhat less developed than that on the red cells of *P. troglodytes*.

7. The R-C-E-F Blood Group System

Although some of the antisera that define blood groups of this system were obtained by isoimmunization of chimpanzees as early as 1965 (Wiener *et al.*, 1965) and the relationship of the then discovered chimpanzee blood specificities to the human Rh-Hr system was indicated shortly thereafter (Wiener *et al.*, 1966), the present concept of the R-C-E-F blood group system of chimpanzees resulted from more recent studies (Socha and Moor-Jankowski, 1978, 1980).

Reagents of six specificities, namely, anti-R^c, anti-C^c, anti-E^c, anti-F^c, anti-c^c, and anti-c_f^c, define 22 blood groups of the R-C-E-F system. Of these, 20 have been actually observed and are listed in Table IX. In addition, some rare, irregular forms are recognized by unusually weak reactions with standard antisera or by the absence of parallel reactions when tests are carried out side by side with several reagents of identical specificity. To account for 22 regular types theoretically expected to occur in *P. troglodytes*, a series of ten allelic genes is postulated: r^1, r^2, r^C, r^{CF}, R^1, R^2, R^C, R^{CF}, R^{CE}, and R^{CEF}. Heritability of some of the irregular forms of R^C was also confirmed in some of the family studies (Socha and Moor-Jankowski, 1978; Socha, 1981). Similar to the human Rh-Hr system, the chimpanzee R-C-E-F system is built around a central antigen, R^c, which is not only hierarchically analogous to human Rh_0, but also shares with the latter some of its specificities (Socha and Moor-Jankowski, 1978). The R^c or R^c-like structure is present not only on the red cells of R^c-positive chimpanzees and Rh-positive humans, but also is detectable on the red cells of lowland and mountain gorillas and gibbons (Socha and Moor-Jankowski, 1980). It has also been found recently on the red cells of some (but not all) orangutans (Socha and van Foreest, 1981). The R^c antigen is also present on the red cells of all pygmy chimpanzees tested so far. However, as Table IX indicates, the R of these animals is of irregular, incomplete type, since, unlike the blood of most common chimpanzees, the red cells of *P. paniscus* react with only one of the three anti-R antisera that are now routinely used for parallel testing. This is comparable to serological reactions obtained with rare human blood called by Unger and co-workers "cognates of Rh" (Unger *et al.*, 1959). Using the same terminology as the one proposed by Unger, we assigned the R^c antigen of *P. paniscus* the symbol R_{ab} to indicate that at least two specificities (R^A and R^B) are missing from this particular antigen. Furthermore, the R^c occurs in all pygmy chimpanzees in combination with specificities C^c and E^c, and thus as type RCE. We found this type in less than 1.5% of *P. troglodytes* (see Table IX).

Recently, while testing red cells of pygmy chimpanzees with the unique anti-R^c serum that agglutinated the red cells of *P. paniscus*, we noticed individual differences in the strength of reactions: blood of one of the animals gave reactions much weaker than the red cells of other animals of the same species. The difference was even more remarkable when comparative titrations were carried out; the titer with red cells of that particular individual was 8–10 times lower than with blood of any other pygmy chimpanzee. These observations prompted us to perform cross-absorption experiments, the results of which are summarized in Table X. As can be seen, absorption of anti-R^c serum with strongly reacting red cells of a pygmy chimpanzee (e.g., red cells of pygmy chimpanzee Laura) removed the activity of the serum against red cells of *all* pygmy chimpanzees while leaving behind a weak fraction of antibodies reactive with the red cells of all R^c-positive *P. troglodytes*. When, however, absorption was carried out with weakly reacting red cells of pygmy chimpanzee Bosondjo, this removed from the serum the antibodies reactive exclusively with the absorbing cells but left behind fractions of

Table IX. Chimpanzee R-C-E-F Blood Group System: Serology and Distribution of Types

Designation	Reaction with serum of specificity[a]						Pan troglodytes (N = 570)		Pan paniscus (N = 14)	
	Anti-R^c	Anti-C^c	Anti-E^c	Anti-F^c	Anti-c^c	Anti-c_f^c	Number	Frequency	Number	Frequency
rc_1	−	−	−	−	+	+	105	0.2234	0	0.00
rc_2	−	−	−	−	+	−	1	0.0021	0	0.00
(rc)	−	−	−	−	+	NT	16	0.0340	0	0.00
rCc_1	−	+	−	−	+	+	1	0.0021	0	0.00
rCF	−	+	−	+	−	−	5	0.0106	0	0.00
$rCFc_1$	−	+	−	+	+	+	35	0.0745	0	0.00
$rCFc_2$	−	+	−	+	+	−	2	0.0042	0	0.00
$(rCFc)$	−	+	−	+	+	NT	2	0.0042	0	0.00
Rc_1	+	−	−	−	+	+	80	0.1702	0	0.00
Rc_2	+	−	−	−	+	−	26	0.0553	0	0.00
(Rc)	+	−	−	−	+	NT	34	0.0723	0	0.00
RC	+	+	−	−	−	−	2	0.0042	0	0.00
RCc_1	+	+	−	−	+	+	4	0.0085	0	0.00
RCc_2	+	+	−	−	+	−	2	0.0042	0	0.00
RCF	+	+	−	+	−	−	21	0.0046	0	0.00

RCFc$_1$	+	+	–	–	+	+	+	34	0.0723	0	0.00
RCFc$_2$	+	+	–	–	+	+	–	26	0.0553	0	0.00
(RCFc)	+	+	–	–	+	+	NT	18	0.0382	0	0.00
RCE	+	+	+	–	–	–	–	7	0.0148	0	0.00
RCEc$_1$	+	+	+	–	–	+	+	15	0.0319	0	0.00
RCEc$_2$	+	+	+	–	–	+	–	4	0.0085	0	0.00
(RCEc)	+	+	+	–	–	–	NT	4	0.0085	0	0.00
RCEF	+	+	+	+	–	–	–	10	0.0213	0	0.00
RCEFc$_1$	+	+	+	+	+	+	+	1	0.0021	0	0.00
RCEFc$_2$	+	+	+	+	+	+	–	1	0.0021	0	0.00
(RCEFc)	+	+	+	+	+	–	NT	3	0.0064	0	0.00
Irregular forms:											
R$_{var}$Cc$_1$	+[b]	+	–	–	–	+	+	2	0.0042	0	0.00
R$_{var}$CF	+[b]	+	–	–	+	–	–	6	0.0128	0	0.00
R$_{var}$Cfc$_1$	+[b]	+	–	–	+	+	+	2	0.0042	0	0.00
R$_{ab}$CE	+[c]	+	+	+	–	–	–	0	0.00	14	1.00
R$_{var}$CEF	+[b]	+	+	+	+	–	–	1	0.0021	0	0.00

[a]NT, not tested.
[b]Weak reactions with all three anti-R^c reagents.
[c]Positive reaction with only one of three anti-R^c reagents.

Table X. Fractionation of the Antibodies in a Chimpanzee anti-R^c Isoimmune Serum (Ch-11, Tom) by Absorption with Red Cells of Chimpanzees (*P. troglodytes* and *P. paniscus*) and Humans

		Titers[a] of anti-R^c isoimmune chimpanzee (Ch-11, Tom) serum[b]						
	Unabsorbed	Absorbed with red cells of						
		Pan paniscus		*Pan troglodytes*			Human	
		Bosondjo	Laura	Andy	Walter	ShuShu	O,M,Rh$_2$rh	O,M,rh
Pan paniscus								
Bosondjo	4	0^c	0	2	2	0	0	2
Kitty	48	16	0	16	16	0	8	32
Laura	32	12	0^c	16	16	0	8	24
Pan troglodytes								
Ch-225, Andy (Rc$_1$)	32	20	4	0^c	4	0	4	16
Ch-168, Walter (RCF)	32	12	4	0	0^c	0	4	16
Ch-322 ShuShu (RCEFc$_2$)	64	32	4	24	16	0^c	16	32
Ch-67, Jigs (rc$_1$)	0	0	0	0	0	0	0	0
Human								
OC (O,M,Rh$_2$rh)	10	10	4	0	0	0	0^c	8
WK (O,M,rh)	6	2	0	0	0	0	0	0^c

[a]See footnote to Table III.
[b]Preabsorbed with red cells of chimpanzee Ch-15 to yield possibly pure anti-R^c reagent.
[c]Titer with absorbing red cells.

antibodies still capable of agglutinating the red cells of other pygmy chimpanzees as well as those of common chimpanzees of R^c-positive type. When, on the other hand, red cells of *P. troglodytes* were used for absorptions of the anti-R^c serum, various results were obtained, depending on the R-C-E-F type of the absorbing red cells. Thus, when red cells of types Rc_1 or RCF were used for absorptions, the resulting reagent remained active against red cells of all pygmy chimpanzees, but it lost its agglutinating activity against red cells of some but not all *P. troglodytes*. Absorption with the red cells of *P. troglodytes* of type RCF or type $RCEFc_2$ removed the activity of the anti-R^c serum against red blood cells of all *P. paniscus* as well as all *P. troglodytes*.

The results of these tests indicate that, although in direct tests all pygmy chimpanzees appear as R^c-positive, the R^c antigen on their red cells is different from that encountered on the red cells of the majority of the R^c-positive *P. troglodytes*. This difference consists in the absence from pygmy chimpanzee red cells of some of the associated specificities that make up the complete R^c antigen. In addition, taking into account the differences in titers resulting from absorption experiments, it appears that the "incomplete" R^c antigen of *P. paniscus* exists in two forms, one of which is more or less comparable to that found on the erythrocytes of *P. troglodytes* of types RCE and/or RCEF, and a second that could be assigned the symbol $R_{var}CE$ or rCE and has no counterpart among the R-C-E-F types of *P. troglodytes*.

To summarize: The study of blood groups of the R-C-E-F system in *P. paniscus* and *P. troglodytes* reveals sharp and significant differences between the two kinds of chimpanzees. While 20 R-C-E-F types have been observed among *P. troglodytes* and at least two more are assumed to exist in this species, pygmy chimpanzees are monomorphically type RCE, a form quite infrequently observed in common chimpanzees. Only by absorption-fractionation procedures was it possible to further subdivide the RCE blood group of *P. paniscus* by identifying a weak form $R_{var}CE$ or rCE that has no counterpart among R-C-E-F blood groups of *P. troglodytes*. The R^c antigen of pygmy chimpanzee red cells is different from that found on red cells of R^c-positive *P. troglodytes* in that it lacks some of the associated specificities that are the usual constituents of the R^c antigen of the common chimpanzee.

8. Other Blood Group Systems

The available iso- and cross-immune chimpanzee antisera define a number of other specificities on chimpanzee red cells that have not been genetically connected with any of the complex blood group systems just discussed. These specificities are assumed to be expressions of independently inherited systems, each composed of a dominant gene, which determines the presence of the given specificity on the red cell membrane, and of its silent, recessive allele. Table XI lists these unrelated specificities and gives their frequencies in *P. paniscus* and *P. troglodytes*.

Blood groups of this category prove to be of importance in the work on pygmy chimpanzees since these are the only "simian-type" red cell specificities clearly polymorphic in this species. As shown in Table XI, three out of six unrelated specificities routinely tested on the red cells of chimpanzees display individual differences among *P. paniscus*. One of these specificities, the so-called N^c, which is serologically similar to specificities of the V-A-B-D system, was found on the red blood cells of one-third of the pygmy chimpanzees, but only in less than 10% of *P. troglodytes*. The other two polymorphic traits, namely, O^c and T^c, appear to be serologically related to the R-C-E-F system, since the treatment of the red cell with proteolytic enzymes enhances reactions with anti-O^c and anti-T^c reagents. They occur in pygmy chimpanzees with frequencies of 64% and 27%, respectively. Distributions of the three polymorphic traits appear favorable for the purpose of paternity investigations (Wiener and Socha, 1976).

9. Genealogical Studies

Genealogical studies constitute the ultimate confirmation of the hypothesis of the mode of inheritance of blood groups. Numerous "family"

Table XI. Unrelated Specificities Detected on the Red Cells of Chimpanzees by Means of Isoimmune and Cross-immune Agglutinating Antisera

	Distribution					
	Pan troglodytes			*Pan paniscus*		
		Frequency			Frequency	
Designation of specificity	Number tested	Positive	Negative	Number tested	Positive	Negative
G^c	483	0.7785	0.2215	14	0.00	1.00
H^c	498	0.5683	0.4317	14	1.00	0.00
K^c	445	0.9900	0.0100	14	1.00	0.00
N^c	470	0.0936	0.9064	14	0.36	0.64
O^c	498	0.6822	0.3178	11	0.64	0.36
T^c	290	0.5689	0.4311	11	0.27	0.73

studies on blood groups of *P. troglodytes* have been carried out (Wiener *et al.*, 1965, 1975; Socha and Moor-Jankowski, 1979, 1980) supporting genetic hypotheses based on population analysis. Thus far, however, studies have included only one filial generation and therefore are of limited value for genetic considerations.

As for pygmy chimpanzees, blood grouping tests have been performed to date only on the members of one family. However, unlike families of *P. troglodytes*, this set also comprises the first representative of the second filial generation. Pedigree and blood groups of this "family" are shown in Fig. 1.

Of the three independent polymorphic specificities of *P. paniscus* red cells listed in the previous section, only one, namely N^c, showed individual differences among members of this family. It is obvious that all N^c-positive individuals in this family must be heterozygotes *Nn*.

It is of interest that the only member of this family differing from the others with respect to the R-C-E-F type was a recently (mid-1970s) wild-caught male (Bosondjo). The remaining animals in this figure were either colony-born or introduced into the colony many years ago (early 1950s).

10. Summary and Conclusions

Blood and/or saliva of 14 colony-born and wild-caught pygmy chimpanzees has been tested over the last 10 years. Individuals and data are summarized in Table XII. The origin of subject animals and the time span over which samples were collected preclude at least close consanguinity among the 14 pygmy chimpanzees tested. These animals show surprisingly few individual differences in their blood groups. This is in striking contrast to the situation observed in *P. troglodytes*, where serological polymorphism reaches levels similar to those in humans. In addition, red cell antigens and specificities of blood of *P. paniscus* are often serologically different from those of *P. troglodytes*. Specifically, while common chimpanzees are either group A or O, and the A antigen of their red cells differs from human red cells, all pygmy chimpanzees have an A antigen that is serologically indistinguishable from human A_1.

While *P. troglodytes* can be either M or MN, all pygmy chimpanzees proved to be type M. The M antigen on the red cells of the latter species differs qualitatively not only from the M on human erythrocytes but also from the M-like antigen detectable on the red cells of *P. troglodytes*.

In addition to M-N types shared by humans and chimpanzees, there is a complex serological system, the V-A-B-D system, which is the chimpanzee extension of the M-N groups but without a direct counterpart on

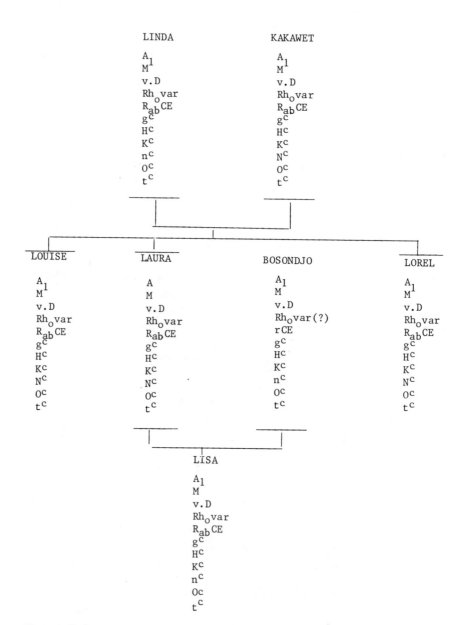

Figure 1. Pedigree and blood groups of a pygmy chimpanzee family.

Table XII. Summary of Blood Group Findings in 14 Pygmy Chimpanzees

| | Date blood obtained | A-B-O | | M-N | V-A-B-D | Rh-Hr | R-C-E-F | G^c | H^c | K^c | N^c | O^c | T^c |
Origin		Blood	Saliva										
Annemie — Antwerp Zoo	1972	A_1	Sec	M	v.D	$\bar{R}h_0$	$R_{ab}CE$	−	+	+	−	−	+
Kitty — Antwerp Zoo	1972	A_1	Sec	M	v.D	$\bar{R}h_0$	$R_{ab}CE$	−	+	+	−	−	+
Lanie — Antwerp Zoo	1972	A_1	Sec	M	v.D	$\bar{R}h_0$	$R_{ab}CE$	−	+	+	+	−	+
Hannimickie — Antwerp Zoo	1973	A_1	NT	M	v.D	$\bar{R}h_0$	$R_{ab}CE$	−	+	+	−	−	−
Linda — San Diego Zoo	1973	A_1	Sec	M	v.D	$\bar{R}h_0$	$R_{ab}CE$	−	+	+	+	+	+
Lorel — San Diego Zoo	1975	A_1	NT	M	v.D	$\bar{R}h_0$	$R_{ab}CE$	−	+	+	+	+	−
Laura — San Diego Zoo	1975	A_1	NT	M	v.D	$\bar{R}h_0$	$R_{ab}CE$	−	+	+	+	+	−
Kakawet — San Diego Zoo	1975	A_1	NT	M	v.D	$\bar{R}h_0$	$R_{ab}CE$	−	+	+	+	NT	NT
Louise — San Diego Zoo	1975	A_1	NT	M	v.D	$\bar{R}h_0$	$R_{ab}CE$	−	+	+	−	NT	NT
Camille — San Diego Zoo	1975	A_1	NT	M	v.D	$\bar{R}h_0$	$R_{ab}CE$	−	+	+	+	NT	NT
Kevin — San Diego Zoo	1977	A_1	NT	M	v.D	$\bar{R}h_0$	$R_{ab}CE$	−	+	+	−	+	−
Vernon — San Diego Zoo	1981	A_1	NT	M	v.D	$\bar{R}h_0$	$R_{ab}CE$	−	+	+	−	+	−
Bosondjo — San Diego Zoo[b]	1982	A_1	Sec	M	v.D	$\bar{R}h_0$	$R_{ab}^{var}CE$ or rCE	−	+	+	−	+	−
Lisa — Yerkes/San Diego	1982	A_1	NT	M	v.D	$\bar{R}h_0$	$R_{ab}CE$	−	+	+	−	+	−

[a]NT, not tested.
[b]Wild caught.

human red cells. Of 16 V-A-B-D types already identified on the red cells of *P. troglodytes* by means of chimpanzee immune sera, only one, namely v.D, was observed in *P. paniscus* and it occurs on the red cells of all 14 pygmy chimpanzees. The type v.D could be considered, therefore, a species-specific characteristic of *P. paniscus*.

Although in direct tests with human anti-Rh reagents all *P. troglodytes* and *P. paniscus* red cells behave similarly (namely, as the so-called $\overline{Rh}$ type), more refined comparative tests point to the existence of subtle differences in expressivity of the Rh (D) antigen in the two chimpanzees. By and large, the Rh_0 of pygmy chimpanzees gives somewhat weaker reactions than Rh_0 (D) antigen on *P. troglodytes* red blood cells; it also displays some qualitative distinctness as demonstrated by absorption-fractionation experiments.

Striking differences between the two species are discerned in the appearance of red cell antigens of the R-C-E-F system, which is the chimpanzee extension of the Rh-Hr blood group system. While in direct tests with various chimpanzee immune antisera at least 24 R-C-E-F types were identified on the red cells of *P. troglodytes*, all pygmy chimpanzees so far tested have been classified as $R_{ab}CE$ type, an irregular type not observed in common chimpanzees. By additional adsorption and comparative titrations, it became possible to detect one pygmy chimpanzee whose blood was slightly different from others and tentatively classified as $R_{ab}^{var}CE$ or rCE.

The only clearly polymorphic specificities thus far observed on the red cells of pygmy chimpanzees are those defined by chimpanzee isoimmune reagents yet not classified as part of any of the known chimpanzee blood group systems. Three of six specificities of this kind showed individual differences among pygmy chimpanzees. Inheritance as a simple dominant character of one of these antigen traits, N^c, has been traced through three generations of one captive pygmy chimpanzee group.

Before the first pygmy chimpanzees were investigated for their blood groups, a host of data were already available on the distribution of various red cell antigens in a few subspecies of *P. troglodytes*. Although significant differences were observed in frequencies of the A-B-O blood groups and "simian-type" groups of *P. troglodytes troglodytes, P. t. schweinfurthii,* and *P. t. verus* (Moor-Jankowski and Wiener, 1972), all or almost all blood types were represented in each of these chimpanzee subspecies. Blood group phenotypes of a single animal could therefore hardly be used as a criterion of the individual's taxonomic classification. It could, at most, support or strengthen, in terms of statistical probability, the classification based on morphological criteria.

Unlike the situation among subspecies of *P. troglodytes*, the red cell antigens of pygmy chimpanzees prove in many respects qualitatively different from those found on erythrocytes of common chimpanzees. Some of the red cell characteristics appear exclusively on the red cells of *P. paniscus,* to the extent that, for example, the presence of human-like A_1 or of $R_{ab}CE$ specificities in the blood of a chimpanzee of questionable origin could be regarded as conclusive for classifying this animal as *P. paniscus.*

Another striking difference between blood groups of the two species of chimpanzees is the relative monomorphism of the red cell antigens of *P. paniscus* compared to the remarkable serological diversity of *P. troglodytes.* There is no single explanation for this phenomenon, although, as improbable as it may appear, we may be dealing with a highly inbred isolate from which all the animals investigated by us originated. Tests for other genetic markers, such as the major histocompatibility complex, or serum and red cell isozymes, could be of great importance. One cannot exclude another, more remote possibility, that the pygmy chimpanzee developed its own specific red cell membrane antigenic structures, which do not fit precisely the combining sites of the immunoglobulin molecules of our agglutinating reagents obtained from the sera of common chimpanzees. Ill fitting of antigens and antibodies may cause the minute differences among antigens that are not recognized in our tests. If this were the case, isoimmunization of pygmy chimpanzees would be necessary to produce type-specific reagents capable of detecting other red cell polymorphisms existing in the blood of pygmy chimpanzees but as yet unrecognized.

Whatever the nature of the serological differences between the two kinds of chimpanzees, there is no doubt that from the point of view of blood group serology, pygmy chimpanzees constitute a species apart, and are even as distant from *P. troglodytes* as to be placed in a separate genus.

ACKNOWLEDGMENTS. The research on which this chapter is based was supported in part by a donation from Ortho Diagnostic Systems, Raritan, New Jersey.

References

André, A., Courtois, G., Lennes, G., Ninane, G., and Osterrieth, P. M., 1961, Mise en evidence d'antigènes de groupes sanguins, A, B, O et Rh chez les singes chimpanzés, *Ann. Inst. Pasteur* **101**:82–95.

Bernstein, G., 1925, Zusammenfassende Betrachtungen über die erblichen Blutstrukturen des Menschen, *Z. Indukt. Abstammungs. Vererbungsl.* **37**:233.

Erskine, A. G., and Socha, W. W., 1978, *The Principles and Practice of Blood Grouping,* 2nd ed., C. V. Mosby, Saint Louis.

Hirszfeld, L., and Hirszfeld, H., 1919, Serological differences between the blood of different races, *Lancet* **ii**:675–679.

Landsteiner, K., 1901, Ueber Agglutinationserscheinungen normalen menschlichen Blutes, *Klin. Wochenschr.* **14**:1132.

Landsteiner, K., and Levine, P., 1927, A new agglutinable factor differentiating individual human blood, *Proc. Soc. Exp. Biol. Med.* **24**:600.

Landsteiner, K., and Miller, C. P., 1925, Serological studies on the blood of the primates. I. The differentiation of human and anthropoid bloods *J. Exp. Med.* **42**:841–852.

Landsteiner, K., and Wiener, A. S., 1937, On the presence of M agglutinins in the blood of monkeys, *J. Immunol.* **33**:19–25.

Landsteiner, K., and Wiener, A. S., 1940, An agglutinable factor in human blood recognizable by immune sera for rhesus blood, *Proc. Soc. Exp. Biol. (NY)* **43**:223.

Moon, G. J., and Wiener, A. S., 1974, A new source of anti-N lectin: Leaves of the Korean *Vicia unijuga*, *Vox Sang* **26**:167–170.

Moor-Jankowski, J., and Wiener, A. S., 1967, Sero-primatology, a new discipline, in: *Progress in Primatology*, Fischer, Stuttgart, pp. 378–381.

Moor-Jankowski, J., and Wiener, A. S., 1972, Red cell antigens of primates, in: *Pathology of Simian Primates* (R. N. T.-W. Fiennes, ed.), S. Karger, Basel, Part 1, pp. 270–317.

Moor-Jankowski, J., Wiener, A. S., Socha, W. W., Gordon, E. B., and Mortelmans, J., 1972, Blood groups of the dwarf chimpanzee (*Pan paniscus*), *J. Med. Primatol.* **1**:90–101.

Moor-Jankowski, J., Wiener, A. S., Socha, W. W., Gordon, E. B., Mortelmans, J., and Sedgwick, C. J., 1975, Blood groups of pygmy chimpanzees (*Pan paniscus*), *J. Med. Primatol.* **4**:262–267.

Ottensooser, F., and Silberschmidt, K., 1953, Haemagglutinin anti-N in plant seeds, *Nature* **172**:914.

Prokop, O., Rachwitz, A., and Schlesinger, D., 1965, A "new" human blood group receptor A_{hel} tested with saline extracts from *Helix hortensis* (garden snail), *J. Forensic Med. (S. Afr.)* **12**:108–111.

Ruffié, J., and Socha, W. W., 1980, Les groupes sanguins érythrocytaires des primates non-hominiens, *Nouv. Rev. Fr. Hématol.* **22**:147–200.

Schmitt, J., Immunobiologische Untersuchungen bei Primaten. Ein Beitrag zur Evolution der Blut- und Serumgruppen, 1968, *Bibl. Primatol.* **8**:1–146.

Schmitt, J., Spielman, W., and Weber, M., 1962, Serologische Untersuchungen zur Frage der verwandtschaftlichen Beziehungen von *Pan paniscus* Schwarz 1929 zu anderen Hominoiden, *Z. Saugetierkd.* **27**:45–61.

Socha, W., 1966, *Problems of Serological Differentiation of Human Population*, Polish State Medical Publishers, Warsaw.

Socha, W. W., 1981, Blood groups as genetic markers in chimpanzees: Their importance for the National Chimpanzee Breeding Program, *Am. J. Primatol.* **1**:3–13.

Socha, W. W., and Moor-Jankowski, J., 1978, Rh antibodies produced by an isoimmune chimpanzee: Reciprocal relationship between chimpanzee isoimmune sera and human anti-Rh₀ reagents, *Int. Arch. Allergy Appl. Immunol.* **56**:30–38.

Socha, W. W., and Moor-Jankowski, J., 1979, Blood groups of anthropoid apes and their relationship to human blood groups, *J. Hum. Evol.* **8**:453–465.

Socha, W. W., and Moor-Jankowski, J., 1980, Chimpanzee R-C-E-F blood group systems: A counterpart of the human Rh-Hr groups, *Folia Primatol.* **33**:172–188.

Socha, W. W., and Ruffie, J., 1983, *Blood Groups of Primates: Theory, Practice, Evolutionary Meaning*, G. Masson, Paris.

Socha, W. W., and van Foreest, A. W., 1981, Erythroblastosis fetalis in a family of captive orangutans, *Am. J. Primatol.* **1**:326.

Socha, W. W., and Wiener, A. S., 1973, Problem of blood factor C of A-B-O system, *N.Y. State J. Med.* **73**:2144–2156.

Socha, W. W., Wiener, A. S., Gordon, E. B. and Moor-Jankowski, J., 1972, Methodology of primate blood grouping, *Transplant. Proc.* **4**:107–111.

Unger, L. J., Wiener, A. S., and Katz, L., 1959, Studies on blood factors Rh^A, Rh^B, and Rh^C, *J. Exp. Med.* **110**:495–510.

Watkins, M. W., 1966, Blood group substances, *Science* **162**:172–181.

Wiener, A. S., 1938, The agglutinogens M and N in anthropoid apes, *J. Immunol.* **34**:11–18.

Wiener, A. S., and Gordon, E. B., 1961, The blood groups of chimpanzees: The Rh-Hr (CDE/cde) blood types, *Am. J. Phys. Anthropol.* **19**:35–43.

Wiener, A. S., and Moor-Jankowski, J., 1971, Blood groups of chimpanzees, in: *Chimpanzee: Immunological Specificities of Blood* (C. Kratochvil, ed.), *Primates in Medicine*, Volume 6, Karger, Basel, pp. 115–144.

Wiener, A. S., and Socha, W. W., 1976, Methods available for solving medicolegal problems of disputed parentage, *J. Forensic Sci.* **21**:42–64.

Wiener, A. S., and Ward, F. A., 1966, The serological specificity (blood factor) C of the A-B-O blood groups. Theoretical implications and practical applications, *Am. J. Clin. Pathol.* **46**:27–35.

Wiener, A. S., Moor-Jankowski, J., and Gordon, E. B., 1963, Blood groups of apes and monkeys. II. The A-B-O blood groups, secretor and Lewis types of apes, *Am. J. Phys. Anthropol.* **21**:271–281.

Wiener, A. S., Moor-Jankowski, J., and Gordon, E. B., 1964a, Blood groups of apes and monkeys. IV. The Rh-Hr blood types of anthropoid apes, *Am. J. Hum. Genet.* **16**:246–251.

Wiener, A. S., Moor-Jankowski, J., and Gordon, E. B., 1964b, Blood group antigens and crossreacting antibodies in primates including man. I. Production of antisera for agglutinogen M by immunization with blood other than human type M blood, *J. Immunol.* **92**:391–396.

Wiener, A. S., Moor-Jankowski, J., Riopelle, A. J., and Shell, N. F., 1965, Simian blood groups. Another blood group system, C-E-F, in chimpanzees, *Transfusion* **5**:508–515.

Wiener, A. S., Moor-Jankowski, J., Gordon, E. B., and Kratochvil, C. H., 1966, Individual differences in chimpanzee blood, demonstrated with absorbed human anti-Rh_0 sera, *Proc. Natl. Acad. Sci. USA* **56**:458–472.

Wiener, A. S., Brian, P., and Gordon, E. B., 1969a, Further observations on the hemagglutinins of the snail *Achatina granulata*, *Haematologia (Budapest)* **3**:9–16.

Wiener, A. S., Moor-Jankowski, J., and Brancato, G. J., 1969b, LW Factor, *Haematologia (Budapest)* **3**:389–393.

Wiener, A. S., Moor-Jankowski, J., and Gordon, E. B., 1969c, The specificity of hemagglutinating bean and seed extracts (lectins). Implication for the nature of A-B-O agglutinins, *Intl. Arch Allergy Appl. Immunol.* **36**:582–591.

Wiener, A. S., Socha, W. W., and Gordon, E. B., 1971, Fractionation of human anti-Rh_0 sera by absorption with red cells of apes, *Haematologia (Budapest)* **5**:227–240.

Wiener, A. S., Gordon, E. B., Socha, W. W., and Moor-Jankowski, J., 1972, Population genetics of chimpanzee blood groups, *Am. J. Phys. Anthropol.* **37**:301–310.

Wiener, A. S., Moor-Jankowski, J., Socha, W. W., and Gordon, E. B., 1974a, The chimpanzee V-A-B blood group system, *Am. J. Hum. Genet.* **26**:35–44.

Wiener, A. S., Socha, W. W., and Moor-Jankowski, J., 1974b, Homologues of the human A-B-O blood groups in apes and monkeys, *Haematologia (Budapest)* **8**:195–216.

Wiener, A. S. Socha, W. W., Moor-Jankowski, J., and Gordon, E. B., 1975, Family studies on the simian-type blood groups of chimpanzees, *J. Med. Primatol.* **4**:45–50.

CHAPTER 3

Pygmy Chimpanzee Systematics

A Molecular Perspective

VINCENT M. SARICH

1. Introduction

The biological understanding of any group of organisms is greatly facilitated by having its phylogeny available. That phylogeny, in both its cladistic and temporal aspects, is most readily derived, at least for living and recently extinct forms, from comparative studies of proteins and nucleic acids. The pygmy chimpanzee (*Pan paniscus*) provides a particularly intriguing test case for these claims, there having been a great deal of difficulty in understanding the species in its own right, as well as in developing an appreciation of what, if anything, it has to tell us about our own evolution.

The branching order, or cladistic aspect, of pygmy chimp phylogeny was first explored biochemically by Goodman (1961). Goodman showed that in simple two-dimensional (paper/starch gel) serum protein electrophoretic comparisons of the various hominoids the strongest qualitative similarities were shown by the two species of *Pan*. Later work has served to demonstrate the representativeness of those early results and to add greater precision.

VINCENT M. SARICH • Departments of Anthropology and Biochemistry, University of California, Berkeley, California 94720.

2. Results

2.1. Further Electrophoretic Studies

Goodman provided no quantitative genetic distance estimates, but more recent single-dimension, side-by-side comparisons in poly-acrylamide gels by our lab give the same qualitative results. Band counting (Sarich, 1977; Cronin *et al.*, 1980) revealed that 13 of 20 comparable bands, on the average, have the same mobility in paired *paniscus–troglodytes* comparisons. By contrast, intergeneric comparisons between *Pan, Gorilla,* and *Homo* reveal that only four to six of the 20 bands have the same mobility. These very simple tests strongly indicate that the two chimpanzees share a much more recent common ancestry with one another than either gorillas or humans.

2.2. Immunology

Similarly, the immunological comparisons of pygmy chimpanzee albumin and transferrins show definite derived character associations with those of *Pan troglodytes*. We know that change has taken place along the *Pan* albumin lineage from analyses of the microcomplement fixation data involving the various hominoid albumins (Sarich, 1970), yet *paniscus* and *troglodytes* albumins are immunologically identical to one another. They also have identical electrophoretic mobilities, even though *Pan, Homo,* and *Gorilla* albumins each have a unique mobility. The transferrin picture, on the other hand, is less definitive due to the presence in *Pan* of alleles that can be as different immunologically from one another as they are from those in *Homo* or *Gorilla* (Cronin, 1975). Therefore one cannot make a definite *paniscus–troglodytes* immunological comparison for the transferrins, because the species boundaries are not congruent with transferrin variation. This sort of "problem" is to be expected since polymorphisms have finite life spans, thus obviating typological association between species and allele.

2.3. Restriction Endonuclease Comparisons of Mitochondrial DNA

The mitochondrial genome is a closed circular structure of approximately 16,000 base pairs. It is now known that it accumulates substitutions some 10-fold more rapidly than nuclear DNA, and rapid comparisons of mtDNAs from different individuals can be made using a large number of available restriction endonucleases (Brown *et al.*, 1979). Comparisons among the mtDNAs of the various hominoids, including *Pan paniscus,* have recently been carried out (Ferris *et al.*, 1981*a,b*), and cladistic analysis of those data suggests that an average chimpanzee mtDNA lineage (evolving clonally) has accumulated about 6% change since its separation from the lineages leading to *Gorilla* and *Homo* mtDNAs. Of this 6%, about 4%, or two-thirds, is shared by *paniscus* and *troglodytes,* and 2% has accumulated along the two species lineages since their divergence.

3. Phylogenetic Implications of the Molecular Data

It is clear from the molecular data that *paniscus* and *troglodytes* share a common *Pan* lineage subsequent to the separation of the *Gorilla* and *Australopithecus–Homo* lines. The times involved are of some interest, though the absence of a relevant fossil record limits their application. The obvious recency of the *paniscus–troglodytes* separation (if indeed it is yet complete) precludes other than the serum protein and mtDNA data from having utility in timing this event. We start with the best current molecular estimate of the beginning of the *Pan–Homo–Gorilla* radiation at 4.5–5.0 MYA (million years ago) and the continuing inability of any molecular data set to resolve the trichotomy. Even the latest mtDNA sequence data do not provide anything resembling a convincing association of any pair of lineages of this trio to the exclusion of the third (Brown *et al.*, 1982). The extant *Pan* line is then some 4.5–5.0 million years old (avoiding the taxonomic question of the appropriate generic designation for the immediate ancestor of this trio) and the separation among those individuals labeled *troglodytes* and *paniscus* in our collections occurred about two-thirds of the way along it; that is, about 1.5 MYA, a figure suggested by both the mtDNA and serum protein electrophoretic data. Though one should avoid any temptation to derive taxonomic status from genetic distance, we note that mammalian lineages seem to take, on the average, at least 1 million years or so to differentiate sufficiently in morphology and/or behavior to be recognized as separate species [as based on serum protein electrophoretic comparisons (Sarich, 1977)]. The ultimate answer as to the taxonomic status of the pygmy chimpanzee must derive from populations studied in their native habitats, and we need, in particular, information as to variation within *paniscus* and the immediately adjacent *troglodytes* populations to be able to evaluate the degrees of morphological, behavioral, and molecular discontinuities across the proposed species boundaries.

The molecular phylogenetic picture is thus unambiguous. The pygmy chimpanzees studied are drawn from a population whose lineage separated from that leading to the *troglodytes* studied sometime in the early part of the Pleistocene, about 1.5 MYA. Accordingly, there cannot be any special evolutionary relationship between pygmy chimps and hominids. This is not to say, however, that the study of pygmy chimps will not provide special insights into the African ape radiation and hominid origins. That this has become, in some minds, an emotionally contentious issue in a peculiarly anthropological fashion is not our doing (Johnson, 1981; Latimer *et al.*, 1981). It might be useful to attempt to put the issues into a productive framework.

4. Origin and Adaptive Radiation of the African Apes (Including Hominids)

4.1. Developing an Understanding of Organismal Evolution

This dispute is symptomatic of a much broader malaise in evolutionary biology, which centers on the extent to which phylogenetic reconstruction is possible and justified, and on how much one can tell about the past without a fossil record of the group in which one is interested. This brief chapter is not the place to resolve these matters, which loom so large in so many minds, but it may prove useful to reiterate some simple points.

1. Evolutionary understanding of diversity requires, and is indeed almost synonymous with, evolutionary reconstruction. This would appear to be a truism beyond dispute; after all, how can one talk of the causes and mechanisms until one has identified the event—what has happened?

2. This reconstruction is aided by having fossils, but can certainly legitimately proceed in their absence. If there are any lingering doubts about this, consider that the first step in any evolutionary reconstruction must be a determination of the phylogeny involved. Now consider where that phylogeny is to come from. In one of the papers already referred to (Latimer *et al.*, 1981), we find Simpson (1975) quoted approvingly: "Fossils remain . . . the most direct and most important data bearing on phylogeny." And in the preceding paragraph, we find: "In fact, molecular analysis of many mammals has merely confirmed paleontologically derived phylogeny reconstructions." The remarkable thing is that statements of this sort are actually believed by the vast majority of evolutionary biologists, when there is, as pointed out so well by Patterson (1981), hardly a single example in which paleontological data have revealed the phylogenetic relationships of living forms. There is certainly no such case among the primates. If one is to rely on the fossil record for phylogenetic knowl-

edge concerning extant organisms, then the game is lost before it can ever begin.

3. The temporal gap between the molecule-dated origin of the hominid line and the earliest australopithecines is small and ever shrinking. It is even now no more than 1 million years.

4. The *Pan, Gorilla,* and *Australopithecus–Homo* lineages separated from one another at very nearly the same time, and thus existing chimpanzee–gorilla similarities are homoplasic or, more likely, primitive retentions.

5. Early australopithecine females were quite small, even though it is difficult to imagine a scenario in which an adaptation to an open country, terrestrial existence would tend to select for small size. Sexual dimorphism at this time is, as one might expect, marked, with australopithecines tending to conform to the condition found in terrestrial catarrhines.

6. The fundamental hominid adaptation is bipedalism and this would have been easiest to accomplish mechanically in a lightly built animal. As there is a strong positive allometry of upper body size relative to overall size in modern apes, the probability is then high that the immediate pongid ancestor of the earliest hominids would have been a small, lightly built animal. Clearly the living form that best fits those specifications is the pygmy chimpanzee. It is for these reasons that 16 years ago I stated (Sarich, 1968), "we begin the reconstruction and understanding of our recent history with a form not unlike a small chimpanzee." There is no reason to amend that view now.

4.2. The Role of the Pygmy Chimpanzee in Telling Us about Hominid Origins

It has been clear for more than 30 years that the basic change involved in the pongid–hominid transition was the development of habitual bipedalism. I have just pointed out why this was most likely to have occurred in a small, lightly built form—something "not unlike a small chimpanzee." The existence of the pygmy chimpanzee significantly extends the African pongid range of variation and thus the comparative data base so necessary to realistic phylogenetic reconstruction. The important thing in this reconstruction as it pertains to our lineage is to get that prototype African pongid up on its hind limbs. Once that is done, then we can begin worrying about some less important details, such as its teeth or range of sexual dimorphism. Having a small African pongid around makes it easier to think about getting our ancestors up on those hind limbs; it also greatly broadens the range of available information that is necessary for formulating and testing hypotheses concerning hominid origins. How much more it may have to say is surely, at this point, limited far more by our inability and, in many cases, unwillingness to make optimum use of anatomical and behavioral information on this least known of our closest relatives.

References

Brown, W. M., George, M., Jr., and Wilson, A. C., 1979, Rapid evolution of animal mitochondrial DNA, *Proc. Natl. Acad. Sci. USA* **76:**1967–1971.

Brown, W. M., Prager, E. M., Wang, A., and Wilson, A. C., 1982, Mitochondrial DNA sequences of primates: Tempo and mode of evolution, *J. Mol. Evol.* **18:**225–239.

Cronin, J. E., 1975, Molecular Systematics of the Order Primates, Ph. D. Dissertation, University of California, Berkeley.

Cronin, J. E., Cann, R., and Sarich, V. M., 1980, Molecular evolution and systematics of the Genus *Macaca,* in: *The Macaques* (D. G. Lindburg, ed.), Van Nostrand Reinhold, New York, pp. 31–51.

Ferris, S. D., Wilson, A. C., and Brown, W. M., 1981a, Evolutionary tree for apes and humans based on cleavage maps of mitochondrial DNA, *Proc. Natl. Acad. Sci. USA* **78:**2432–2436.

Ferris, S. D., Brown, W. M., Davidson, W. S., and Wilson, A. C., 1981b, Extensive polymorphism in the mitochondrial DNA of apes, *Proc. Natl. Acad. Sci. USA* **78:**6319–6323.

Goodman, M., 1961, The role of immunochemical differences in the phyletic development of human behavior, *Hum. Biol.* **34:**104–150.

Johnson, S. C., 1981, Bonobos: Generalized hominid prototypes or specialized insular dwarfs, *Curr. Anthropol.* **22:**363–375.

Latimer, B. M., White, T. D., Kimbel, W. H., Johanson, D. C., and Lovejoy, C. O., 1981, The pygmy chimpanzee is not a living missing link in human evolution, *J. Hum. Evol.* **10:**475–488.

Patterson, C., 1981, Significance of fossils in determining evolutionary relationships, *Annu. Rev. Ecol. Syst.* **12:**195–223.

Sarich, V. M., 1968, Human origins: An immunological view, in: *Perspectives on Human Evolution, 1* (S. L. Washburn and P. C. Jay, eds.), Holt, Rinehart, and Winston, New York, pp. 94–121.

Sarich, V. M., 1970, Primate systematics with special reference to Old World monkeys: A protein perspective, in: *Old World Monkeys: Evolution, Systematics, and Behavior* (J. R. Napier and P. H. Napier, eds.), Academic Press, New York, pp. 175–226.

Sarich, V. M., 1977, Rates, sample sizes, and the neutrality hypothesis for electrophoresis in evolutionary studies, *Nature* **265:**24–28.

Simpson, G. G., 1975, Recent advances in methods of phylogenetic inference, in: *Phylogeny of the Primates* (W. P. Luckett and F. S. Szalay, eds.), Plenum, New York, pp. 3–19.

CHAPTER 4

A Measure of Basicranial Flexion in Pan paniscus, the Pygmy Chimpanzee

JEFFREY T. LAITMAN AND RAYMOND C. HEIMBUCH

1. Introduction

The cranial base is located at one of the most sensitive areas of the body. Developmentally, it can affect or be affected by the brain and its associated vasculature from above; the vertebral column caudally; the dentognathic apparatus anteriorly; and the upper respiratory tract from below. The contributing influences of these differing regions have made the cranial base a prime site of investigation for those examining the evolution or development of cranial morphology. In particular, the various angles of exocranial and endocranial flexion of the base have been used frequently both as a means of monitoring functional development in primates as well as a vehicle for assessing patterns in primate and hominid evolution [see reviews by Schulter (1976) and Sirianni and Swindler (1979)].

This study focuses upon the basicranium of the least known of the living pongids, *Pan paniscus*, the "bonobo" or pygmy chimpanzee. The relatively recent discovery of the species in 1929 by Schwarz (1929), and the limited availability of both skeletal and cadaver specimens, have had the effect of restricting many researchers' familiarity with *Pan paniscus*. As a result, the species has often been precluded from consideration in comparative analyses. Interest in *Pan paniscus* has, however, increased

JEFFREY T. LAITMAN ● Department of Anatomy, Mount Sinai School of Medicine of the City University of New York, New York, New York 10029. *RAYMOND C. HEIM-BUCH* ● Department of Medical Biostatistics, Ortho Pharmaceutical Company, Raritan, New Jersey 08869.

considerably within the last decade. This has been due in large part to curiosity over how this gracile, "scaled down" version of the common chimpanzee, *Pan troglodytes,* might relate to our hominoid or hominid ancestors [see McHenry and Corruccini (1981) for review]. As a result of this concern, features of the skull of *Pan paniscus,* including the region of the cranial base, have begun to come under closer examination. Although aspects of the base were noted in some earlier studies, such as the original morphological description of the species by Coolidge (1933), Giles' allometric study of the pongids (1956), or the review of the chimpanzee skeleton by Schultz (1969), concentrated interest in the base has been relatively recent. The most extensive investigation of the basicranium of *Pan paniscus* can be found in Cramer's (1974, 1977) comparison of craniofacial morphology between *Pan paniscus* and *Pan troglodytes;* and in the works of Fenart, Deblock, and Cousin using the vestibular method of cranial orientation to compare the two species of chimpanzees (Fenart and Deblock, 1973; Deblock and Fenart, 1973, 1977; Cousin *et al.,* 1981).

In this study we have examined the basicranium of *Pan paniscus* through the use of a measure of basicranial flexion. This measure, which we call the basicranial line, was originally designed to describe flexion in our studies examining the relationship between the cranial base and upper respiratory region in extant and fossil primates (Laitman, 1977; Laitman and Crelin, 1976; Laitman *et al.,* 1978, 1979; Laitman and Heimbuch, 1982). The basicranial line reflects those portions of the base that are closely associated with contiguous areas of the upper respiratory tract. Although designed to aid in understanding cranial–soft tissue relationships, we soon found that the basicranial line provided us with considerable information on how this aspect of the base varied among different primate species, and what growth patterns were shown by each group. The application of this method to *Pan paniscus* provides the opportunity to examine both how this aspect of the basicranium changes during development, and how it compares to the other primate groups we have studied.

2. Methods and Materials

2.1. Craniometric Measurements and Statistical Methods

The basicranial shape of *Pan paniscus* was described using the aforementioned basicranial line. We have discussed this line in detail previously in our study of extant primates (Laitman *et al.,* 1978) and our investigations of fossil hominids (Laitman *et al.*, 1979; Laitman and Heimbuch, 1982). The data for this study on *Pan paniscus* were collected and analyzed using the same techniques as in our previous studies. We shall summarize the description here.

Our previous statistical analyses of extant species were based on measurements made on an age-graded series of 228 human and nonhuman primate crania. Genera and number of specimens studied consisted of: *Macaca* (14), *Hylobates* (26), *Symphalangus* (18), *Pongo* (29), *Pan (troglodytes)* (29), *Gorilla* (24), *Homo* (88). Specimens were separated into five age groups defined by the following dental stages: (1) prior to eruption of the deciduous dentition; (2) from the eruption of the deciduous first central incisor to completion of the deciduous dentition; (3) eruption of the first permanent molar; (4) eruption of the second permanent molar; (5) eruption of the third permanent molar.

A series of nine linear measurements were taken between five craniometric points identified on the midline of the exocranial surface of the basicranium (Fig. 1A). The craniometric points used were (A) prosthion, (B) staphylion, (C) hormion, (D) sphenobasion, and (E) endobasion. Measurements were taken of the distance between points A–B, A–C, B–C, B–D, B–E, C–D, D–E, C–E, and A–E. These measurements determine a basicranial line that indicates the degree of exocranial base flexion (Fig. 1, bottom). Measurements were taken directly upon specimens with hand calipers. The use of a craniometer, midsagittally sectioned skulls, or radiographic implants was unnecessary. The procedure does not damage the skulls.

As noted, the basicranial line was designed as a means to assess flexion in that portion of skull most closely related to the upper respiratory tract. The basicranial line should not be confused with classic measures of basicranial flexion, such as Huxley's (1867) "basicranial axis," which utilized the angle formed by nasion–prosphenion–basion; Cameron's (1927) nasion–pituitary point–basion angle; or Scott's (1958) anterior cribiform point–prosphenion–basion angle. These traditional measures use endocranial points (and thus require radiographs or midsagittally sectioned skulls) and are concerned with assessing a single, central angle of craniofacial flexion. The basicranial line analyzes flexion on the external surface of the base (or exocranial base), and thus approaches the question of flexion from a different perspective.

The central concern of this study was to compare the basicranial lines of *Pan paniscus* to those we previously described for extant primates. The primary technique used for this comparison was a classification criterion designed to identify similar forms (Tatsuoka, 1971). It does this by compiling multivariate information into a statement of distance. The less

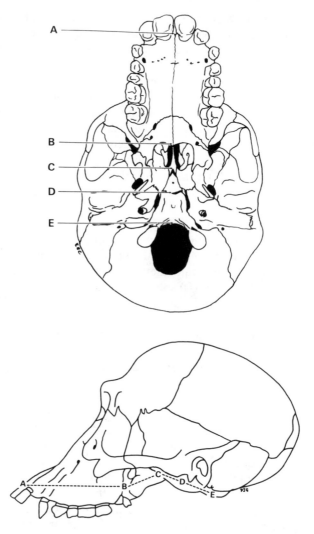

Figure 1. (Top) Craniometric points on the midline of the basicranium of an adult female common chimpanzee (*Pan troglodytes*). A, Prosthion; B, staphylion; C, hormion; D, sphenobasion; E, endobasion. (Bottom) Topographic projection of the basicranial line on the cranium of the same chimpanzee.

distance one specimen is from another, the more similar they are in form, and conversely, the greater in distance, the more disparate in form. One advantage of this classification criterion was that it allowed the investigation of the effect of alternate variance models upon the classification of the *Pan paniscus* specimens studied (Table I).

Table I. Variance Models Used in Assignments of Cranial Specimens

Label	Dimension	Variance model	Percentage of correct assignment over entire primate sample
A	7	Equal covariance for all genera	76.7
B	7	Equal covariance within family	71.2
C	4	Equal covariance for all genera	63.6
D	4	Equal covariance within family	63.2
E	3	Equal covariance for all genera	54.0

As in past studies, the multivariate data used in this analysis of *Pan paniscus* were coordinate values derived from the linear measurements taken on the specimens. Coordinates were chosen since they provide both easily interpretable results as well as a means of comparison with our previous studies on other extant forms. A major advantage of coordinate data is that they provide more easily interpretable statistical results than do linear data. While statistical analyses of linear data can demonstrate that there are significant differences between two populations, it is usually difficult to translate these differences into specific morphological terms. If coordinate values are used to describe morphology, then statistical differences are directly interpretable as the alteration in location of specific points. In addition, an accurate picture of the morphology being described is available by plotting the points on coordinate paper.

The purpose of this study was to explore the relationship between the shape of the basicranial lines in *Pan paniscus* and those in other extant groups. As shape was of primary concern in our investigation, it was necessary that statistical differences between groups did not reflect *both* size and shape differences. Therefore, basicranial coordinates were standardized for size using the method given by Sneath (1967). The points describing a specimen were translocated to an origin at the arithmetic mean of the coordinates. A size descriptive constant was derived, which was proportional to the area occupied by the points. Each point was then moved along a vector between the origin and the point by this proportional amount. In effect this reduced the specimen to a constant area, maintaining angular relationships and thus not altering shape. The new coordinates were scale free in a manner similar to Mosimann's measurement vector (1970), but were different in that they could assume negative values. The standardized coordinates, when multiplied by the size constant, give the coordinates of the points in centimeters. The specimens were then translocated back to prosthion as origin.

2.2. *Pan paniscus* Specimens

This study was based upon measurements made upon the crania of 45 specimens from the collection of the Musée de l'Afrique Centrale in Tervuren, Belgium. The major criteria for selecting a specimen were the completeness of its cranial base and the possibility of obtaining the necessary craniometric measurements. Accordingly, many specimens in the Tervuren collection exhibiting damage to the base were excluded. Of the 45 specimens chosen for study, no specimens were found in dental stage 1; seven were selected for inclusion in stage 2; 12 in stage 3; ten in stage 4, and 16 in stage 5. Collection numbers and raw data for the samples are listed in Table II.

Table II. Basicranial Measurements of *Pan paniscus*

	Specimen	AB	AC	BC	BD	BE	CD	DE	CE	AE	Sex	Collection number[a]
Stage 1	None											
Stage 2	1	3.26	4.25	1.08	1.96	3.35	1.13	1.58	2.70	6.61	M	11293
	2	3.42	4.55	1.20	2.06	3.36	1.15	1.64	2.78	6.83	M	12087
	3	3.20	4.35	1.17	2.05	3.23	1.10	1.27	2.35	6.40	M	18050
	4	3.82	5.20	1.52	2.34	3.58	1.17	1.56	2.72	7.39	?	29014
	5	2.79	3.62	0.90	1.65	2.69	0.89	1.25	2.14	5.47	F	29003
	6	3.13	4.08	1.05	2.03	3.08	1.08	1.33	2.39	6.17	F	29008
	7	3.15	4.11	1.07	1.86	2.94	1.05	1.33	2.37	6.03	M	29007
Stage 3	1	3.69	5.04	1.60	2.21	3.34	1.10	1.55	2.63	7.01	M?	26938
	2	3.21	4.50	1.38	2.20	3.35	1.11	1.42	2.52	6.49	?	23464
	3	4.32	5.84	1.79	2.63	4.02	1.26	1.73	2.96	8.31	M	27011
	4	3.90	5.02	1.26	2.20	3.50	1.28	1.81	3.06	7.33	F	27003
	5	4.13	5.47	1.40	2.55	3.86	1.45	1.54	2.97	7.91	?	28709
	6	3.89	4.83	1.21	2.04	3.32	1.18	1.59	2.74	7.16	?	29018
	7	3.81	5.01	1.36	2.16	3.41	1.27	1.68	2.87	7.22	M	29010
	8	3.85	4.90	1.17	2.23	3.40	1.25	1.43	2.67	7.22	?	29012
	9	4.21	5.39	1.48	2.49	3.74	1.41	1.60	2.99	7.95	?	29024
	10	4.71	5.98	1.44	2.42	3.95	1.36	1.83	3.19	8.65	M	29023
	11	4.37	5.42	1.22	2.40	3.85	1.46	1.67	3.13	8.18	?	29022
	12	4.26	5.51	1.40	2.53	4.08	1.36	1.72	3.07	8.29	?	29021
Stage 4	1	5.06	6.14	1.23	2.21	4.05	1.32	2.07	3.38	9.04	F	20529
	2	5.53	7.12	1.84	2.85	4.88	1.47	2.30	3.75	10.38	F	21742
	3	4.75	6.02	1.70	2.67	4.53	1.60	2.20	3.78	9.22	?	22908
	4	5.19	7.08	2.15	3.30	4.98	1.48	2.09	3.56	10.16	M	26971
	5	4.55	6.25	1.99	2.81	4.37	1.28	2.01	3.26	8.91	F	27010

Table II. Basicranial Measurements of *Pan paniscus*

Specimen	AB	AC	BC	BD	BE	CD	DE	CE	AE	Sex	Collection number[a]
6	4.90	6.26	1.62	2.63	4.40	1.52	2.07	3.58	9.30	F	29027
7	4.62	6.13	1.70	2.61	4.18	1.33	1.78	3.09	8.78	M	29028
8	4.71	6.02	1.46	2.54	4.13	1.21	1.98	3.15	8.84	?	29029
9	5.63	7.16	1.67	2.75	4.63	1.37	2.12	3.49	10.24	F	29033
10	4.94	6.37	1.59	2.55	4.58	1.43	2.24	3.66	9.46	F	29031
Stage 5 1	5.89	8.02	2.49	3.31	5.23	1.39	2.32	3.65	11.09	F	9338
2	6.03	7.82	2.01	3.11	5.15	1.38	2.28	3.65	11.17	F	11351
3	5.15	6.82	1.91	2.94	5.05	1.49	2.33	3.82	10.16	F	13201
4	6.18	7.71	1.76	2.89	5.15	1.36	2.50	3.85	11.31	?	15924
5	6.09	7.66	1.80	2.73	4.72	1.35	2.31	3.65	10.77	F	15296
6	6.39	8.25	1.91	2.91	5.03	1.30	2.28	3.57	11.38	M	15295
7	5.27	6.39	1.50	2.55	4.52	1.51	2.23	3.73	9.76	?	14738
8	6.00	7.55	1.70	2.75	4.77	1.50	2.24	3.76	10.69	F	20882
9	6.64	8.24	1.80	2.90	5.02	1.31	2.48	3.76	11.65	M	26939
10	5.62	7.70	2.29	3.41	5.56	1.44	2.72	4.15	11.17	M	26960
11	6.30	7.99	1.78	2.73	4.63	1.36	2.27	3.61	10.79	F	26989
12	6.03	8.06	2.26	3.07	5.00	1.35	2.23	3.55	10.95	F	26991
13	6.07	7.67	1.89	2.87	5.05	1.56	2.48	4.03	11.10	M	28712
14	5.80	7.66	2.03	3.02	4.95	1.29	2.26	3.52	10.74	M	27699
15	6.17	7.76	1.90	2.91	4.92	1.50	2.49	3.98	11.05	F	27698
16	5.65	7.51	2.02	3.07	4.96	1.48	2.15	3.63	10.58	M	29037

[a]All specimens from the collection at the Musée de l'Afrique Centrale, Tervuren, Belgium.

3. Results

This section examines the relationship between the basicranial lines of *Pan paniscus* to those previously described by us for other extant primates (Laitman *et al.*, 1978). The basicranial lines of *Pan paniscus* are compared to those of *Pan troglodytes*, *Gorilla*, *Pongo*, *Macaca*, and *Homo* in Fig. 2. Growth patterns for *Pan paniscus* and the above groups are presented in Fig. 3. The size-standardized coordinates and means for these groups are listed in Table III. The assignment of *Pan paniscus* under alternate variance models to the most similar groups (exclusive of other *Pan paniscus* stages) is shown in Table IV. Figure 4 illustrates the location of *Pan paniscus* relative to all other extant species studied in the first two dimensions of a seven-dimensional discriminant space. This figure represents the exact location of the specimens in only two-dimensional space.

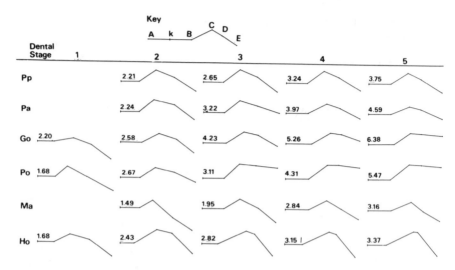

Figure 2. Group mean basicranial lines derived from coordinate values. (A) Prosthion; (B), staphylion; (C) hormion; (D) sphenobasion; (E) endobasion; k, size constant. Dental stages: (1) prior to eruption of the deciduous dentition; (2) from eruption of the first central incisor to completion of the deciduous dentition; (3) eruption of first permanent molar; (4) eruption of second permanent molar; (5) eruption of third permanent molar. Genera: Pp, *Pan paniscus* (pygmy chimpanzee); Pa, *Pan troglodytes* (common chimpanzee); Go, *Gorilla* (gorilla); Po, *Pongo* (orangutan); Ma, *Macaca* (macaque); Ho, *Homo* (human). As shown in the key, a standardized segment of the palate from A to the hash mark between A and B has been removed in the figure from all the basicranial lines.

The alternate variance models used information from as many as seven dimensions in identifying the specimens.

Two major observations can be made about the basicranial lines of *Pan paniscus*. First, the basicranial lines of *Pan paniscus* appear most similar to those of juvenile and subadult (stages 2–4) members of other pongid groups. As can be seen in Fig. 2 and by the assignments in Table III, stage 2 and stage 3 *Pan paniscus* are consistently assigned to stage 2 or stage 3 members of other groups. Stage 4 *Pan paniscus*, however, is most frequently assigned to stage 2 *Gorilla*, and only secondarily to stage 4 *Pan troglodytes*. The other major assignments of stage 4 *Pan paniscus* are to stage 3 *Pan troglodytes* and stage 3 *Gorilla*.

Stage 5 (fully adult) *Pan paniscus* is overwhelmingly assigned, on all variance models, to stage 4 *Pan troglodytes*, with assignment to stage 5 *Pan troglodytes* being a distant second. The other assignments of stage 5 *Pan paniscus*, with the exception of two assignments to stage 5 *Macaca*, are all to juvenile or subadult stages of the other groups. Subadult and

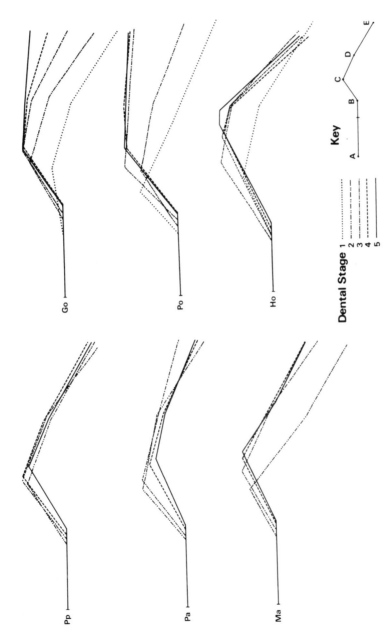

Figure 3. Development of the basicranial line in *Pan paniscus* and most closely related groups. (A) Prosthion; (B) staphylion; (C) hormion; (D) sphenobasion; (E) endobasion. Dental stages and genera are the same as in Fig. 2.

Table III. Group Means for Size Constant and Standardized Coordinates[a]

Group	Dental stage	Size constant	B[b] X	C X	C Y	D X	D Y	E X	E Y
Pan paniscus	2	2.21	1.47	1.81	0.28	2.36	0.11	2.90	−0.27
	3	2.65	1.59	1.94	0.32	2.39	0.13	2.87	−0.24
	4	3.24	1.54	1.96	0.31	2.35	0.15	2.91	−0.18
	5	3.75	1.59	2.03	0.27	2.37	0.11	2.90	−0.21
Pan troglodytes	2	2.24	1.49	1.86	0.30	2.38	0.18	2.89	−0.18
	3	3.22	1.52	1.92	0.30	2.38	0.17	2.92	0.00
	4	3.97	1.56	2.03	0.24	2.39	0.13	2.90	−0.13
	5	4.59	1.59	2.07	0.19	2.38	0.12	2.91	−0.11
Gorilla	1	2.20	1.44	1.92	0.08	2.38	−0.07	2.88	−0.42
	2	2.58	1.55	1.98	0.22	2.43	0.08	2.87	−0.25
	3	4.23	1.59	2.04	0.28	2.40	0.21	2.88	−0.06
	4	5.26	1.67	2.13	0.27	2.42	0.24	2.88	0.09
	5	6.38	1.66	2.08	0.28	2.37	0.26	2.91	0.21
Pongo	1	1.68	1.45	1.76	0.26	2.24	0.01	2.94	−0.33
	2	2.67	1.51	1.90	0.26	2.38	0.15	2.92	−0.10
	3	3.11	1.54	1.93	0.37	2.36	0.32	2.90	0.27
	4	4.31	1.60	2.06	0.35	2.36	0.36	2.89	0.30
	5	5.47	1.58	2.06	0.37	2.37	0.35	2.88	0.31
Macaca	2	1.49	1.49	1.83	0.19	2.34	−0.24	2.84	−0.55
	3	1.95	1.54	1.95	0.24	2.42	0.01	2.86	−0.33
	4	2.84	1.59	2.05	0.23	2.42	0.02	2.87	−0.23
	5	3.16	1.60	2.11	0.23	2.43	−0.01	2.86	−0.23
Homo	1	1.68	1.42	1.83	0.19	2.32	0.07	2.90	−0.35
	2	2.43	1.36	1.94	0.35	2.34	0.26	2.84	−0.24
	3	2.82	1.40	2.12	0.33	2.34	0.26	2.79	−0.28
	4	3.15	1.45	2.14	0.32	2.32	0.27	2.80	−0.31
	5	3.37	1.50	2.19	0.34	2.30	0.34	2.80	−0.26

[a]The coordinates for point A are equal to 0.
[b]The Y coordinate of B is equal to zero.

adult *Pan paniscus* thus appear most similar to the younger members of other groups.

The second major observation concerns developmental change in the basicranial line of *Pan paniscus*. As can be seen clearly in Fig. 3, there is little developmental variation in the basicranial lines within *Pan paniscus*. In the other groups studied, a recognizable growth pattern can be seen. In *Pan troglodytes*, there appears to be both a lowering and posterior migration of the vomer (represented by distance B–C) and a general "flat-

Table IV. Assignment of *Pan paniscus* under Alternate Variance Models

Group	Assignment[a]	A	B	C	D	E
				Variance mode used[b]		
Pan paniscus 2	Pa2	1	1	2	4	2
(*N* = 7)	Po2	3	2	1	2	
	Go1			3	1	3
	Ma3	2	1	1		1
	Pa3	1	2			
	Go2		1			1
Pan paniscus 3	Ma3	5	1	4	2	5
(*N* = 12)	Go2	2	5	2	4	1
	Pa3	3	3	2	2	2
	Go1			2	2	2
	Pa2	1	2		1	
	Po2			1	1	
	Ma4			1		1
	Ho2	1	1			
	Pa4					1
Pan paniscus 4	Go2	1	1	4	2	2
(*N* = 10)	Pa4	2	3		2	2
	Pa3	2	1	1	2	1
	Go3	1	3	1	1	1
	Ma4	1		1	1	2
	Po2	1		2		1
	Go1			1	1	1
	Ma3	2				
	Hy5		1		1	
	Pa2		1			
Pan paniscus 5	Pa4	5	6	6	6	9
(*N* = 16)	Pa5	3	3	5	6	
	Ma4	4	3	2	1	2
	Go2			2	3	1
	Po2	2	1			1
	Go3		3			1
	Ho2	1		1		1
	Ma5	1				1

[a]Pa, *Pan troglodytes*; Po, *Pongo*; Go, *Gorilla*; Ma, *Macaca*; Ho, *Homo*; Hy, *Hylobates*. Numerals refer to dental stage.
[b]See Table I for key to variance models.

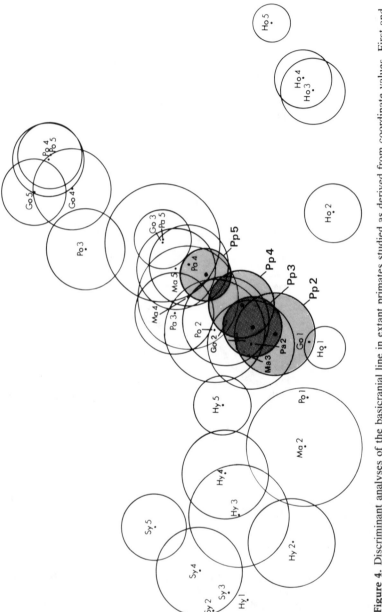

Figure 4. Discriminant analyses of the basicranial line in extant primates studied as derived from coordinate values. First and second axes. Circles represent 90% Bayesian credibility boundaries about the group mean. Boundaries for groups with a sample size of two or less are not represented. Genera: Pp, *Pan paniscus*; Pa, *Pan troglodytes*; Po, *Pongo*; Go, *Gorilla*; Ma, *Macaca*; Hy, *Hylobates*; Sy, *Symphalangus*; Ho, *Homo*. Numbers represent respective dental stages. No attempt has been made to ascribe biological meaning to these axes.

tening'' of the entire basicranial line from stage 2 through stage 5. *Macaca* shows considerable enlargement of the vomer height after stage 2, accompanied by a marked elevation of the entire base subsequent to this stage. Little change can be seen in *Macaca* following stage 3.

The most noticeable developmental changes can be seen in the growth patterns of *Pongo, Gorilla,* and *Homo.* Both *Pongo* and *Gorilla* show a number of similarities in the growth patterns of their basicranial lines. Both of these groups exhibit a pronounced developmental increase in vomer height accompanied by a marked elevation of the basisphenoid (represented by the distance C–D) and basiocciput (distance D–E). The growth pattern in *Homo* also exhibits considerable developmental change, although completely different from that seen in any of the other groups studied. *Homo* is characterized by an enlargement of the vomer height between stage 1 and stage 2, and than a unique posterior movement of the vomer base (point C) which brings the vomer and spheno-occipital synchondrosis (D) almost in proximity to each other by stage 5. Unlike all other groups, the basiocciput (D–E) of *Homo* flexes sharply in an anterior–interior direction from its point of union with the basisphenoid at the spheno-occipital synchondrosis (D), providing the characteristic exocranial flexion of the human basicranium [see Laitman *et al.,* (1978) for discussion].

The only noticeable developmental change in *Pan paniscus* can be seen in the slight posterior-inferior movement of the vomer (B–C) after stage 4. Aside from this one shift, the basicranial lines of all stages of *Pan paniscus* appear very similar. The species shows less developmental variation in this measure of the base than any of the other groups studied.

4. Discussion

This study provides our preliminary analysis of one aspect of the basicranium of *Pan paniscus* through the use of the basicranial line. Comparative examination of this measure shows that *Pan paniscus* exhibits relatively little exocranial flexion between the end of the hard palate (B) and the anterior-most point on the foramen magnum (E). In this regard, the skull base of the species closely approximates the general ''non-flexed'' pattern shown by the other nonhuman primates studied. Although considerable interspecific variation exists, the basicranial lines of *Pan paniscus* resemble those of the other pongids or *Macaca* much more closely than they do *Homo.* At least as concerns this parameter of the base, *Pan paniscus* shows few similarities to the markedly flexed crania of modern *Homo.*

Our observations concerning the similarities of the basicranial lines of *Pan paniscus* to those of juvenile and subadult members of other groups supports the findings of Coolidge (1933), Cramer (1977), and Fenart and Deblock (1973) regarding the general paedomorphic condition of its skull. It is interesting to note that *Pan paniscus* basicrania resemble not only those of juvenile or subadult *Pan troglodytes* but also the younger stages of *Pongo, Gorilla,* and *Macaca,* and even an occasional *Homo.* The basicrania of the younger stages of primates, particularly among the Pongidae, resemble each other quite closely, denoting an almost "general pattern" that exists before the final adult specializations have begun. By its resemblance to these younger juvenile and subadult stages *Pan paniscus* appears to approximate this somewhat generalized condition.

The retention of the "juvenile" appearance of the basicranial line in the older stages of *Pan paniscus* may be partly responsible for the lack of any marked developmental pattern within the species. Since both subadult and adult stages of *Pan paniscus* do not undergo any major changes, they appear largely similar to their own juvenile predecessors. *Pan paniscus* appears to show no adult or subadult specializations. It thus appears as if all major developmental change in this parameter of the cranium of *Pan paniscus* has ceased after its own juvenile period (stage 3) has been achieved.

Examination of one parameter of the skull is certainly not sufficient to enable us to offer comprehensive conclusions about either the general growth patterns or taxonomic relationships of *Pan paniscus.* Nevertheless, the cranial base is a unique region of the skull, and one that may be of considerable diagnostic importance in assessing the developmental and comparative morphology of *Pan paniscus.* Further study of this region will hopefully aid us in better understanding some of the distinctive features of this least known of the great apes.

ACKNOWLEDGMENTS. We would like to express our appreciation to Drs. D. Thys van den Audenaerde, M. Louette, and D. Meirte of the Musée de l'Afrique Centrale, Tervuren, Belgium, both for granting access to the *Pan paniscus* collection in their charge, and for their kindness and generosity during our visit. We are greatly indebted to Drs. E. S. Crelin, D. R. Pilbeam, I. Tattersall, E. Delson, T. Barka, and especially L. B. Laitman, without whose suggestions, support, and advice our research program would not be possible. We also wish to thank our colleague P. J. Gannon for his ongoing participation in our project and his expertise in photography and graphic illustration; and G. DeFalco for editorial assistance. This work was supported by Grant BNS 8025476 from the National Science Foundation.

References

Cameron, J., 1927, The main angle of cranial flexion (the nasion–pituitary–basion angle), *Am. J. Phys. Anthropol.* **10**:275–279.

Coolidge, H. J., 1933, *Pan paniscus:* Pygmy chimpanzee from south of the Congo River, *Am. J. Phys. Anthropol.* **18**(1):1–57.

Cousin, R. P., Fenart, R., and Deblock, R., 1981, Variation ontogeniques des angles basicraniens et faciaux: Etudes comparatives chez *Homo* et chez *Pan, Bull. Mem. Soc. Anthropol. Paris* **8**(13):189–212.

Cramer, D. L., 1974, Cranio-Facial Form in Two African Pongidae, with Special Reference to the Pygmy Chimpanzee, *Pan paniscus,* Ph.D. Dissertation, University of Chicago.

Cramer, D. L., 1977, Craniofacial morphology of *Pan paniscus, Contrib. Primatol.* **10**:1–64.

Deblock, R., and Fenart, R., 1973, Differences sexuelles sur crânes adultes chez *Pan paniscus, Bull Assoc. Anat. (Nancy)* **57**(157):299–306.

Deblock, R., and Fenart, R., 1977, Les angles de la base du crâne chez les chimpanzées, *Bull. Assoc. Anat. (Nancy)* **61**(173):183–188.

Fenart, R., and Deblock, R., 1973, *Pan paniscus* et *P. troglodytes* craniométrie: Étude comparative et ontogenique selon les méthodes classiques et vestibularies, *Mus. R. Afr. Cent. Tervuren, Belg. Ann. Ser. Octavo. Sci. Zool.* **204**:1–473.

Giles, E., 1956, Cranial allometry in the great apes, *Hum. Biol.* **28**:43–58.

Huxley, T. H., 1867, On two widely contrasted forms of the human cranium, *J. Anat. Physiol.* **1**:60–77.

Laitman, J. T., 1977, The Ontogenetic and Phylogenetic Development of the Upper Respiratory System and Basicranium in Man, Ph.D. Dissertation, Yale University.

Laitman, J. T., and Crelin, E. S., 1976, Postnatal development of the basicranium and vocal tract region in man, in: *Symposium on Development of the Basicranium* (J. F. Bosma, ed.), U.S. Government Printing Office, Washington, D. C., pp. 206–219.

Laitman, J. T., and Heimbuch, R. C., 1982, The basicranium of Plio-Pleistocene hominids as an indicator of their upper respiratory systems, *Am. J. Phys. Anthropol.* **59**(3):323–340.

Laitman, J. T., Heimbuch, R. C., and Crelin, E. S., 1978, Developmental change in a basicranial line and its relationship to the upper respiratory system in living primates, *Am. J. Anat.* **152**:467–483.

Laitman, J. T., Heimbuch, R. C., and Crelin, E. S. 1979, The basicranium of fossil hominids as an indicator of their upper respiratory systems, *Am. J. Phys. Anthropol.* **51**:15–34.

McHenry, H. M., and Corruccini, R. S., 1981, *Pan paniscus* and human evolution, *Am. J. Phys. Anthropol.* **54**:355–367.

Mosimann, J. E., 1970, Size allometry: Size and shape variables with characterizations of the lognormal and generalized gamma distributions, *J. Am. Stat. Assoc.* **65**:932–945.

Schulter, F. P., 1976, Studies of the basicranial axis: A review, *Am. J. Phys. Anthropol.* **44**:453–468.

Schultz, A. H., 1969, The skeleton of the chimpanzee, in: *The Chimpanzee,* Volume I (G. H. Bourne, ed.), S. Karger, Basel, pp. 50–103.

Schwarz, E., 1929, Das Varkomen des Schimpansen auf der linken Kongo-Ufer, *Rev. Zool. Bot. Afr.* **XVI**(4):425–426.

Scott, J. M., 1958, The cranial base, *Am. J. Phys. Anthropol.* **16**:319–348.

Sirianni, J. E., and Swindler, D. R., 1979, A review of postnatal craniofacial growth in Old World monkeys and apes, *Yearb. Phys. Anthropol.* **22**:80–104.

Sneath, P. H. A., 1967, Trend-surface analysis of transformation grids, *J. Zool. (London)* **151**:65–111.

Tatsuoka, M. M., 1971, *Multivariate Analysis: Techniques for Education and Physiological Research,* Wiley, New York.

CHAPTER 5

The Dentition of the Pygmy Chimpanzee, Pan paniscus

WARREN G. KINZEY

1. Introduction

Prior to 1974 the only studies on the dentition of *Pan paniscus* involved very small samples. Remane (1960) presented measurements of teeth of 4–11 adult individuals and later (1962) measurements of deciduous teeth of two maxillas and 3–4 mandibles. Conroy (1972) published indices based on measurements of maxillary teeth in three specimens of *Pan paniscus* to show that relative reduction of incisors and canine of *Ramapithecus* (FT1271-2) was greater than that in either species of chimpanzee. Vandebroek (1969) used *Pan paniscus* extensively to compare with other primates in his book on vertebrate evolution. He included several photos of teeth of *P. paniscus:* one a lingual view of a deciduous dentition (Vandebroek, 1969, p. 289), and lateral views of male and female permanent dentitions (pp. 388–389). He also presented a graph of average measurements of the three lower molars compared with those of other hominoids (p. 394), and a drawing of lingual views of the lower teeth showing morphological variation (p. 415). Almquist (1974) measured incisors and canines in 18 female and 14 male *P. paniscus* from Tervuren, and compared these measurements with other African pongids and cercopithecids. He found significant sexual dimorphism in the length, breadth, and height of both upper and lower permanent canines of *P. paniscus*.

In 1974 Johanson (1974a) presented metrical data on teeth from 140

WARREN G. KINZEY ● Department of Anthropology, City College of New York, City University of New York, New York, New York 10031.

skulls in the collection in Tervuren, Belgium. This is the most thorough study available of measurements of the teeth of *P. paniscus*. Subsequently, Johanson (1974*b*) presented morphological data on the same specimens.

In 1971 I examined the permanent dentitions and deciduous canine teeth of the collection of *Pan paniscus* in Tervuren. My individual measurements of buccolingual trigonid breadths of 99 lower first permanent molars in *P. paniscus* were published previously (Thomas, 1976, p. 82), as was my distribution of permanent maxillary canine lengths and breadths (Cramer, 1977, p. 18).

2. Materials and Methods

The results in this chapter are based on my own measurements and observations of permanent dentitions of *Pan paniscus* in Tervuren, and on previously published measurements and observations, primarily those of Johanson (1974*a,b*), on the deciduous and permanent dentitions. In addition, comparisons were made by Johanson (1974*b*) between *P. paniscus* and each of the three subspecies of *Pan troglodytes*.

The collection of *P. paniscus* in Tervuren included (as of 1971) 140 skulls collected in Zaire, predominantly by Georges Vandebroek; some of these specimens were wild-shot, but many were from animals used in medical laboratories there. The numbers of available individual teeth at each position are listed in Table I. In this table each position is counted once, whether its antimere is present or not. For example, there are 110 maxillas in which a first molar is present and fully erupted; however, occlusal wear and breakage of enamel reduce the number of measurable teeth to 108. There are 139 maxillas and 136 mandibles from 140 individuals (Table II). There are 28 maxillas and 28 mandibles with complete permanent dentitions on the right and/or left side.

Measurements of teeth were taken with dial calipers on the left side unless the left tooth was damaged or missing, in which case the antimere was measured.

I also measured the breadth of the palate in both *P. paniscus* and a series of *P. troglodytes schweinfurthii* from Tervuren and the American Museum of Natural History, New York, using a modification of the method of Lavelle (Lavelle *et al.*, 1970). In order to include maxillas in which teeth were missing or broken, measurements of dental arch width were taken between midpoints of corresponding left and right alveoli, rather than between midpoints of tooth crowns.

Table I. Distribution of Teeth in *Pan paniscus* Specimens from Tervuren

	Deciduous teeth					Permanent teeth							
	i1	i2	c	dm1	dm2	I1	I2	C	P3	P4	M1	M2	M3
Maxilla													
Male	19	19	23	23	23	21	20	20	28	25	40	29	18
Female	17	19	21	21	21	26	25	27	30	29	49	32	18
Unknown	17	18	24	25	27	7	2	0	6	5	21	7	0
Total	53	56	68	69	71	54	47	47	64	59	110	68	36
Mandible													
Male	17	21	23	22	23	26	24	26	27	27	40	31	21
Female	16	21	19	21	22	28	29	27	32	31	46	30	20
Unknown	16	18	20	21	22	4	4	0	4	4	25	6	0
Total	49	60	62	64	67	58	57	53	63	62	111	67	41

3. Results

3.1. Upper Molar Morphology

3.1.1. Hypocone Reduction

Pan paniscus has a greater percentage of reduction of the hypocone in M^1, M^2, and M^3 than in any subspecies of *P. troglodytes:* it is completely missing from 3% of M^2 and 1% of M^1; whereas it is always present on M^1 and M^2 in *P. troglodytes* (Johanson, 1974*b*). In M^3, it is fully developed in only 9% of *P. paniscus*, whereas, it is fully developed in 21% of *P. troglodytes*. As in the larger chimpanzee, the trend in *P. paniscus* is increasing reduction in the size of the hypocone from M^1 to M^3. The degree of reduction is greater from M^1 to M^3 in *P. paniscus* than in *P. troglodytes*. Figure 1 shows a maxilla of *P. paniscus* in which the hypocone is reduced on M^1 and is completely missing from M^2.

3.1.2. Metacone Reduction

Pan paniscus has the least reduction of the metacone on M^3 and M^2 compared with any subspecies of *P. troglodytes* (Johanson, 1974*b*).

3.1.3. Anterior Transverse Crest (Preprotocrista) Configuration

The anterior transverse crest is located between the paracone and the protocone. Korenhof (1960) defined three configurations, and

Table II. Distribution of Dentitions in *Pan paniscus* Specimens from Tervuren

		Complete dentitions					
		Deciduous teeth			Permanent teeth		
	Total specimens	One side	Both sides	Total	One side	Both sides	Total
Maxilla							
Male	52	6	7	13	7	7	14
Female	60	5	9	14	7	7	14
Unknown	27	3	9	12	0	0	0
Total	139	14	25	39	14	14	28
Mandible							
Male	52	1	13	14	5	8	13
Female	59	3	11	14	4	11	15
Unknown	25	1	14	15	0	0	0
Total	136	5	38	43	9	19	28

Johanson (1974*b*) presents the incidence of each for *P. paniscus* and each of the subspecies of *P. troglodytes*. On all three molars *P. paniscus* has the lowest frequency of type I and the highest frequency of type III at both the lingual and buccal ends of the crest (except the lingual end of M^3) compared with all subspecies of *P. troglodytes*. This means that in *P. paniscus* the crest has its lingual attachment closer to the protocone and farther from the mesial edge of the tooth, and its buccal attachment mesial to the paracone (rather than at the tip of the paracone); thus, the crest has a more transverse orientation in *P. paniscus* (and a more angled orientation—from mesiolingual to buccodistal) than in *P. troglodytes*. Figure 2 shows the transverse orientation of the anterior transverse crest in an M^1 of *P. paniscus* in which the buccal attachment is mesial to the paracone.

3.1.4. Morphology of the Crista Obliqua

The crista obliqua is generally a continuous crest from protocone to metacone in both species of chimpanzee. The only difference in *P. paniscus* is that in the M^2 (but not M^1 or M^3) the crest is more often missing than in *P. troglodytes*. It is missing from 18% of *P. paniscus,* and from 2–6% of various subspecies of *P. troglodytes* (Johanson, 1974*b*). In both species of chimpanzee the crista obliqua increases in frequency of its absence from dm^2 to M^3.

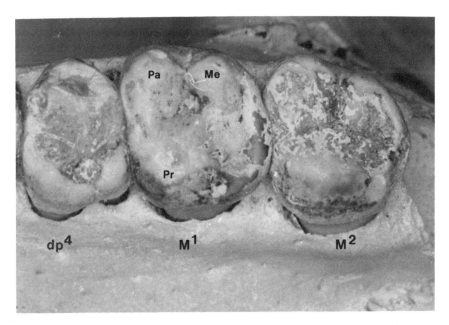

Figure 1. Upper first and second molar showing M¹ with reduced hypocone and M² lacking hypocone in *Pan paniscus* (Pa, paracone; Pr, protocone; Me, metacone).

3.1.5. Lingual Cingulum. Upper Molars

The lingual cingulum is most strongly expressed on M¹ and decreases distally to M³. There is no difference between *P. paniscus* and *P. troglodytes* in rate of decrease. On M¹ there is no significant difference between the species in expression of the cingulum, but the cingulum tends to extend more distally in *P. paniscus*. On M² the cingulum is significantly extended more distally on *P. paniscus* (see Fig. 3). On M³ this is the case also, except that *P. troglodytes verus* also has a well-developed cingulum distal to the protocone. Thus, overall, *P. paniscus* tends to have a better developed lingual cingulum on upper permanent molars than *P. troglodytes* (see Fig. 4). In addition, "the larger percentage of lingual cingula in the *paniscus* sample [of dm²] differs significantly from the other samples" of *P. troglodytes* (Johanson, 1974*b*, p. 128). See Fig. 5.

3.1.6. Buccal Cingulum. Upper Molars

Normally the buccal cingulum is poorly developed in higher primates. It occurs on only up to 9% of samples of *P. troglodytes* (Johanson, 1974*b*). There is no trace of a buccal cingulum on any upper molars in *P. paniscus*.

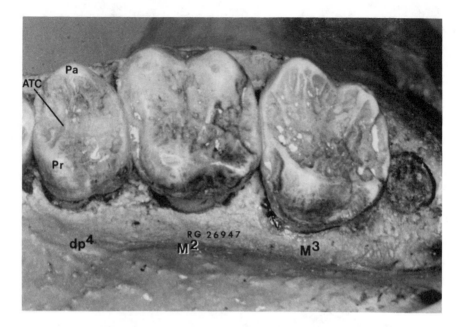

Figure 2. Upper first molar showing anterior transverse crest with buccal attachment mesial to the paracone in *Pan paniscus* (ATC, anterior transverse crest).

3.1.7. Distoconule (Supernumerary Cusp)

The distoconule, when it occurs, is an accessory cusp between the hypocone and the metacone. This cusp is completely absent from *Pan paniscus* molars. It occurs, however, on all subspecies of *P. troglodytes*, generally at low incidence, but as high as 40% of M^3 in *P. t. troglodytes* in the Powell-Cotton collection (Johanson, 1974*b*). It generally increases in frequency from M^1 to M^3.

3.2. Lower Molar Morphology

3.2.1. Paraconid on dm₂

The paraconid is present in 44–61% of *P. troglodytes* subspecies (Johanson, 1974*b*), but in only 32% ($^{18}/_{56}$) of *Pan paniscus*. Thus, *P. paniscus* has a relatively low incidence of the paraconid on dm_2.

3.2.2. Position of the Metaconid

Pan paniscus is significantly different from the three subspecies of *P. troglodytes* in having the metaconid positioned less frequently in the distal position (more frequently opposite the protoconid) in dm_2, M_1, M_2 (also in *P. t. schweinfurthii*), and M_3 (also in *P. t. schweinfurthii*) (Johanson, 1974*b*). See Fig. 6. This results in a greater relative distance (and a deeper groove) between the metaconid and the entoconid in *P. paniscus*. In both chimpanzee species there is a clear trend of increase in mesial positioning of the metaconid from dm_2 to M_3. (In M_3 it is actually most frequently in the mesial position.)

3.2.3. Position of the Hypoconulid

Pan paniscus has a more centrally placed (more lingual) hypoconulid on dm_2 and M_1 than the (more buccally placed) position in all subspecies of *P. troglodytes* (Johanson, 1974*b*). See Fig. 7. There is a trend of decreasing incidence of buccal position from M_1 to M_3 in all subspecies of *P. troglodytes*. In *P. paniscus*, however, M_2 has the highest incidence of the buccal position.

3.2.4. Tuberculum Sextum (Sixth Cusp)

The tuberculum sextum is an accessory cusp, which, when present, is located between the entoconid and the hypoconulid. *Pan paniscus* has a significantly lower incidence than all subspecies of *P. troglodytes* on dm_2, M_1, and M_2. On M_3 the decreased incidence of a tuberculum sextum in *Pan paniscus* is also found in *P. t. verus* (Johanson, 1974*b*). The incidence of the sixth cusp increases in frequency from M_1 to M_3 on *P. paniscus* and on all subspecies of *P. troglodytes* except *P. t. verus*, in which the lowest incidence is on M_3. The reduced incidence in *P. paniscus* is perhaps related to the more lingual location of the hypoconulid.

3.2.5. Groove Pattern

The Y-5 groove pattern on lower molars predominates in all chimpanzees and there is no significant difference between *P. paniscus* and *P. troglodytes* (Johanson, 1974*b*). The greatest variability is found on M_3 in both species.

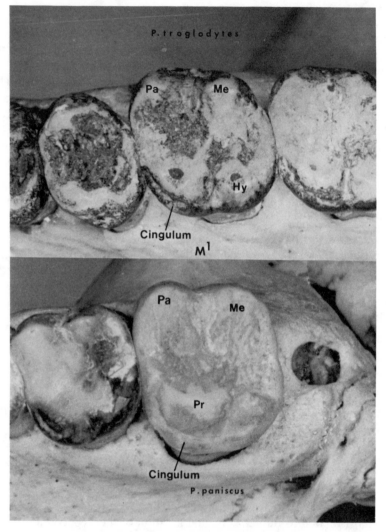

Figure 3. Upper first molar in (top) *Pan troglodytes* and (bottom) *P. paniscus*. Note greater extent of lingual cingulum in *P. paniscus* (Hy, hypocone).

3.3. Metrical Data

3.3.1. Deciduous Dentition. Size

a. Sex Differences. In *Pan paniscus* there is no sexual dimorphism in any mesiodistal, buccolingual, crown module (length + breadth/2), or

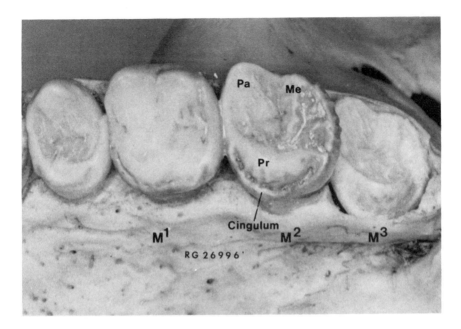

Figure 4. Upper last premolar and molars in *Pan paniscus* showing well-developed lingual cingulum.

crown area (length × breadth) dimension of either the maxillary or the mandibular deciduous dentitions (Johanson, 1974*b*). Further, there is no significant difference between males and females in crown height of deciduous maxillary or mandibular canines (Johanson, 1974*a*).

    b. Relative Size of Deciduous Teeth. On average the maxillary central incisors are both longer and wider than laterals, whereas the opposite is the case for the mandibular deciduous incisors. The upper canine is longer and narrower than the lower, on average. The second deciduous molar is on average both longer and wider than the first, in both upper and lower dentitions. Summaries of the dental measurements are published in Johanson (1974*a*).

    c. Interspecies Comparisons. In all mesiodistal, buccolingual, and crown area dimensions of both maxillary and mandibular teeth, *Pan paniscus* is significantly smaller ($p < 0.001$; except $p < 0.01$ for di_2) than *P. troglodytes*. The one exception is the lack of a significant difference between the two species in the buccolingual dimension of dm_1 (Johanson, 1974*b*).

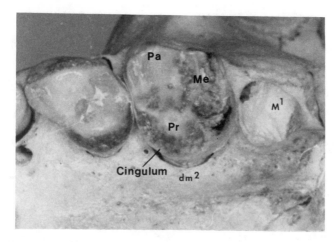

Figure 5. Upper second deciduous molar in *Pan paniscus* showing well-developed lingual cingulum.

3.3.2. Deciduous Dentition. Shape

a. Sex Differences. There are no sex differences in the shape index (mesiodistal length/buccolingual breadth × 100) of either *P. paniscus* or *P. troglodytes,* except in the dm² of *P. paniscus,* in which females have "a significantly $p < 0.05$ higher index reflecting a somewhat longer tooth among females. This difference could reflect greater tooth loss due to interproximal wear in males" (Johanson, 1974b, p. 187).

b. Interspecies Comparisons. There is no difference in shape index of dm¹ between *P. paniscus* and *P. troglodytes.* In all other deciduous teeth there is a difference in shape between the species on the average. In the mandibular dc and di₂ *P. paniscus* is relatively longer than *P. troglodytes;* in all other teeth, *P. paniscus* is relatively wider. It is of interest that in the di2 and the dc the maxillary teeth are relatively wider (or shorter) in *P. paniscus,* whereas the homologous mandibular teeth are relatively narrower (or longer) than in *P. troglodytes.* Although *P. paniscus* teeth are absolutely smaller than those of *P. troglodytes,* in the maxilla, mesiodistal differences between the two species are relatively greater than buccolingual differences, whereas in the mandible, buccolingual differences are relatively greater (Johanson, 1974b).

c. Ratio of Anterior to Posterior Teeth. When the combined length of mandibular deciduous incisors is divided by the combined lengths of mandibular deciduous molars, the ratio is significantly larger in *P. paniscus* than in *P. troglodytes* ($p < 0.001$), indicating a proportionately larger

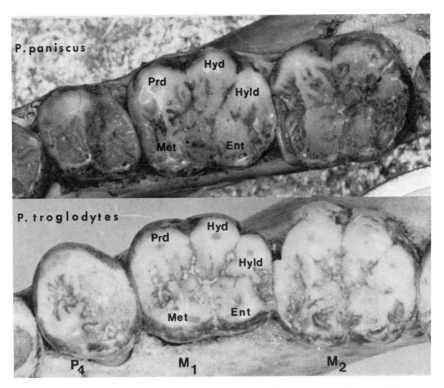

Figure 6. Lower first molar in (top) *Pan paniscus* and (bottom) *P. troglodytes*. Note that metaconid is positioned more mesially in *P. paniscus* Prd, protoconid; Hyd, hypoconid; Hyld, hypoconulid; Met, metaconid; ent, entoconid).

anterior tooth set in the pygmy chimpanzee (Johanson, 1974*b*). A similar ratio for the maxillary dentition is not significantly different between the two species.

3.3.3. Permanent Dentition. Size

a. Sex Differences. There are no sex differences in mesiodistal or buccolingual dimensions, crown, module, or crown area of any permanent teeth of *Pan paniscus*, except in the maxillary and mandibular canines. This is in marked contrast with *Pan troglodytes*, in which virtually every permanent tooth is significantly larger on average in the male than in the female (Ashton and Zuckerman, 1950; Johanson, 1974*b*). In *P. paniscus* the maxillary canine is 23% longer and 25% broader in the male than the

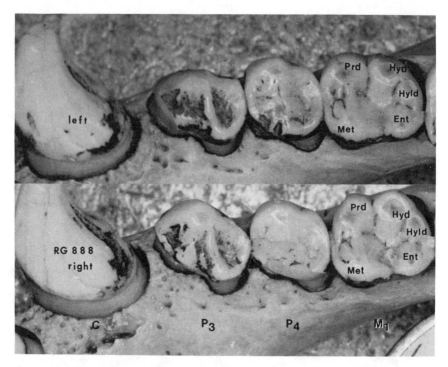

Figure 7. Right and left (reversed) sides of mandible of *Pan paniscus*. Note the centrally placed position of the hypoconulid on M_1.

female, and the mandibular canine is 17% longer and 21% broader in the male. Measurements of *P. paniscus* canine dimensions show almost no overlap between measurements of males and females (Fenart and Deblock, 1973; Kinzey, in Cramer, 1977, pp. 17–18).

Almquist (1974), on a sample of nine females and six males, and Johanson (1974a), on a sample of 18 females and 14 males, found males to have a 33.6% higher maxillary canine crown than females. Mandibular canine crown height is 23% greater in males than in females (Johanson, 1974a).

b. Relative Size of Permanent Teeth. On average the maxillary central incisors are both longer and wider than laterals; the two mandibular incisors are equally long, but the medial is slightly wider (labiolingually) than the lateral. The upper canine is longer and broader than the lower, on average. The anterior upper premolar is longer and broader on average than the posterior one; the anterior lower premolar is longer and narrower than the posterior one. On average the first two upper molars are equal in length and longer than M^3; all three are equal in breadth. The second lower molar on average is the longest and broadest of the three molars in the mandible. There is considerable individual variation, however (see Section 3; Table III). Summaries of measurements are in Johanson (1974*a*).

 c. Relative Size of Individual Molars. There is no difference between male and female in relative sizes of the molars, so the sexes have been combined. Although the average length of the first two upper molars is the same, when teeth in individual dentitions are compared the second molar is most frequently the largest tooth in the molar row, in both the maxilla and the mandible, and in both mesiodistal and buccolingual dimensions (Table III).

 d. Interspecies Comparisons. In all mesiodistal and buccolingual dimensions, crown module, and crown area of both maxillary and mandibular teeth the average measurement in *Pan paniscus* is significantly less ($p < 0.001$) than that of *P. troglodytes* (Johanson, 1974*b*). In *P. paniscus* the second molar is most frequently the largest tooth in the molar series. This is in marked contrast to the *P. troglodytes* maxilla in which M^1 is most frequently the longest molar (Schuman and Brace, 1954; Mahler, 1973; see Table IV) and M^1 most frequently the broadest (Mahler, 1973) or $M^1 = M^2$ in breadth (Schuman and Brace, 1954). In the mandible of *P. troglodytes* M_1 and M_2 are most frequently equally long (Schuman and Brace, 1954), or M_2 is most often longest (Mahler, 1973), and M_2 is most frequently the broadest of the mandibular molars.

Table III.   Relative Lengths of Permanent Molar Teeth in *Pan paniscus*

		Maxilla				Mandible			
		Length		Breadth		Length		Breadth	
		N	Percent	N	Percent	N	Percent	N	Percent
M1 largest {	M1 > M2 > M3	6	23	4	18	5	16	1	10
	M1 > M3 > M2	2		1		1		2	
M1 = M2 > M3		3	8	0	0	5	13	0	0
	M2 > M1 > M3	16		11		19		19	
M2 largest {	M2 > M3 > M1	4	66	6	67	5	68	6	90
	M2 > M1 = M3	3		1		1		3	
M2 = M3 > M1		0	0	3	11	0	0	0	0
M3 largest {	M3 > M2 > M1	1	3	0	4	1	3	0	0
	M3 > M1 > M2	0		1		0		0	
Total		35		27		37		31	

3.3.4. Permanent Dentition. Shape

a. Sex Differences. As in the case of the deciduous dentition, there are no sex differences in the shape index of upper teeth of *Pan paniscus* or *P. troglodytes*, except that in the M^1 of the latter the female tooth is relatively longer. "It would not be unreasonable to interpret this difference as a result of more extensive interproximal tooth loss in the males" (Johanson, 1974*b*, p. 229). In the mandibular dentition the only sex difference in shape index of *P. paniscus* occurs on M_2, where females have relatively longer molars. Again, this is probably due to greater interproximal wear in males. (In *P. troglodytes* the P_3 is relatively longer in males than in females).

b. Interspecies Comparisons. There is no difference in shape index between *Pan paniscus* and *P. troglodytes* in Upper C, P^3, M^1, M^2; lower P_3, P_4, and M_3 (Johanson, 1974*b*). The incisors of *P. paniscus* are relatively longer (mesiodistally) than those of *P. troglodytes*, both in the maxilla and the mandible, and also in the mandibular canines. Johanson (1974*b*, p. 252) suggests this may be a result of wear and/or "may be a consequence of a larger number of somewhat younger individuals in the [*paniscus*] sample" (Johanson, 1974*b*, p. 229). In upper P^4, M^3 and lower M_1 *P. troglodytes* has relatively longer teeth; the M_2 in *P. paniscus* is relatively longer.

c. Ratio of Anterior to Posterior Teeth. Unlike the finding for the deciduous teeth, there is no significant difference between the two species of chimpanzee in the ratio of length of incisors to length (or area) or molars, in either upper or lower dentitions (Johanson, 1974*b*).

Table IV. Relative Lengths of Upper Molars in Chimpanzees

	Pan paniscus[a]	Pan troglodytes			
		Male[b]	Female[b]	Male[c]	Female[c]
M^1 longest	23%	57%	74%	68%	75%
$M^1 = M^2$	8%	24%	22%	0%	0%
M^2 longest	66%	14%	2%	23%	20%
$M^2 = M^3$	0%	1%	1%	0%	0%
M^3 longest	3%	0%	0%	11%	5%

[a]This work.
[b]Data from Schuman and Brace (1954); N = 110 male + 85 female.
[c]Data from Mahler (1973); N = 286.

3.3.5. Permanent Dentition. Age Changes

Johanson (1974*b*) measured the permanent dentitions of old adults separately from those of young adults, in order to determine age changes due to wear. As expected, there are no differences between the two groups in buccolingual measurements, except in M^1, which is presumably due to sampling error. In mesiodistal dental dimensions of *P. paniscus* upper I^1, I^2, P^3, M^1, and M^2 and lower I_1, I_2, P_4, and M_1 all have significant reductions in the old adults compared with the unworn teeth of young adults. The male canines are significantly larger mesiodistally in the old adults, both in the maxilla and in the mandible. This may be of particular importance when comparing *P. troglodytes*, since the latter does not show a similar increase with age.

The ratio $\dfrac{\text{(anterior teeth lengths)}}{\text{(molar teeth lengths)}}$ is significantly less in the old adult *P. paniscus* compared with the adults with unworn teeth. This indicates significantly greater wear on the anterior teeth (both upper and lower incisors) than on the posterior teeth. Further, whereas there was no difference in ratios between *P. paniscus* and *P. troglodytes* in the unworn teeth of adults, in the teeth of old adults the ratios were significantly lower in *P. paniscus*, indicating relatively greater wear on incisors in the latter species than in *P. troglodytes*.

3.3.6. Shape of the Palate

Measurements between midpoints of left and right alveoli of all eight pairs of maxillary teeth were not significantly different between males and females of *P. paniscus*, so the sexes were combined. Measurements increase mesiodistally to M^1, indicating a posteriorly diverging arch from incisors through the first molar (Table V). This gives the arch a more "parabolic" appearance (Fig. 8) than that of *P. troglodytes*, in which the two rows of cheek teeth are essentially parallel (see Table V and Fig. 9).

3.4. Roots of Teeth

According to Monteil (1970), the premolars of *P. paniscus* have two roots, except the upper anterior premolar, which has three. Upper molars have three roots, and lower molars have two, although sometimes upper molars, especially the third, may have two roots or even one root. Monteil does not indicate the degree of variability in number of roots of *P. pan-*

Table V. Breadths of Palate between Midpoints of Corresponding Alveoli (Mean $\pm$ SE in mm)[a]

Species	Sex	N	I^1	I^2	C	P^3	P^4	M^1	M^2	M^3
Pan paniscus	Both	44	10.33 ± 0.10 <	24.78 ± 0.21 <	36.85 ± 0.34 <	39.26 ± 0.36 =	40.05 ± 0.35 <	41.02 ± 0.34	39.87 ± 0.35	38.35 ± 0.41
P. troglodytes	Male	18	12.65 ± 0.23 <	31.45 ± 0.60 <	49.71 ± 0.91 =	49.36 ± 0.68 =	48.42 ± 0.73 =	49.10 ± 0.75	47.99 ± 0.76	45.42 ± 0.79
schweinfurthii	Female	14	12.44 ± 0.44 <	29.57 ± 0.37 <	44.87 ± 0.53 <	45.14 ± 0.56 =	45.25 ± 0.62 =	46.14 ± 0.53	45.55 ± 0.42	43.60 ± 0.51

[a] =, No significant difference between adjacent breadths. <, Adjacent breadth is greater ($p < 0.05$).

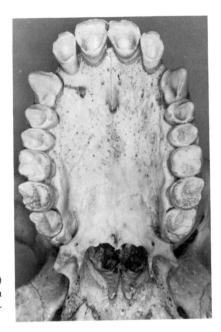

Figure 8. Maxilla of *Pan paniscus* (RG26996) showing increasing divergence of the teeth from I¹ to M¹, giving a somewhat "parabolic" appearance to the palatal arch.

iscus. These values are within the normal range given for *P. troglodytes* (Remane, 1960).

3.5. Sequence of Eruption

The most frequent order of eruption of the permanent teeth in *P. paniscus* was M1, I1, M2, I2, P3, P4, C, and M3, although there was some variation (see Table VI). The second incisor was found erupted before M2 in three mandibles and three maxillas, but M2 had erupted prior to I2 in four mandibles and eight maxillas.

The usual order of eruption of the permanent teeth in apes is: M1, I1, I2, M2, P3, P4, C, and M3 (Clements and Zuckerman, 1953; Schultz, 1956; Krogman, 1969). In none of the foregoing publications, however, was *Pan paniscus* considered. Of 86 common chimpanzees, Schultz (1935) found I2 to erupt prior to M2 in nine, but M2 to erupt prior in I2 in only one. There appears to be a marked difference between the two species of chimpanzee in the timing of eruption of the second permanent incisor.

The order of eruption of the premolars in *P. paniscus* was variable. In five mandibles and three maxillas I2 erupted before P3, whereas in two mandibles and five maxillas P3 erupted prior to I2. In five mandibles and

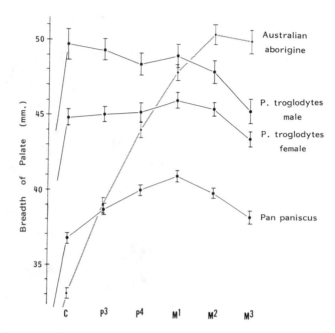

Figure 9. Breadth of the palate between midpoints of paired dental alveoli from canine to third molar in *Pan paniscus* and *P. troglodytes*. The dotted line is *Homo sapiens* (Australian aborigine), $N = 20$. Means and standard deviations.

four maxillas I2 erupted prior to P4, whereas in one mandible and four maxillas P4 erupted before I2 (see Table VI). Unlike *P. troglodytes*, in which the permanent canine frequently erupts after M3, especially in males (Krogman, 1930; Schultz, 1956), no jaw of *P. paniscus* was seen in which a canine erupted after the third molar.

In *P. paniscus* the deciduous canine is the last deciduous tooth to erupt and rarely erupts after the first permanent molar has erupted (Table VI). This is the same as in *P. troglodytes* (Krogman, 1930; Nissen and Riesen, 1945). Earlier stages of dental development were not available to determine the sequence of eruption of the deciduous teeth in *P. paniscus*.

The most unusual sequence of eruption in *P. paniscus* compared with that of the common chimpanzee was the tendency for the eruption of I2 to be delayed, as in the case of some monkeys (Schultz, 1956). It is certainly a trend away from that in the human, in which the eruption of M2 is delayed, with the result that I2 erupts well before M2. It is not clear why there is a delay in eruption of I2 in the pygmy chimpanzee, or whether it is actually an advancement in the eruption of M2. This needs to be

Table VI. Summary of Dental Eruption Stages in Jaws of *Pan paniscus*[a]

								Mandible	Maxilla
di1	di2		dm1	dm2				7	2
di1	di2	dc	dm1	dm2				11	20
di1	di2		dm1	dm2	M1			1	0
di1	di2	dc	dm1	dm2	M1			36	34
I1	di2	dc	dm1	dm2	M1			2	9
I1	I2	dc	dm1	dm2	M1			2	3
di1	I2	dc	dm1	dm2	M1			1	0
I1	di2	dc	dm1	dm2	M1	M2		4	3
I1	di2	dc	P3	dm2	M1	M2		2	1
I1	di2	dc	P3	P4	M1	M2		0	4
I1	di2	dc	dm1	P4	M1	M2		1	0
I1	I2	dc	P3	P4	M1	M2		7	7
I1	I2	dc	dm1	dm2	M1	M2		2	0
I1	I2	dc	P3	dm2	M1	M2		0	1
I1	I2	C	P3	P4	M1	M2		11	10
I1	I2	C	P3	P4	M1	M2	M3	44	45
Total								131	139

[a]The number of mandibles and maxillas is given for each stage of eruption.

determined by a longitudinal growth study. Figures 10 and 11 show, respectively, a mandible and a maxilla with M2 erupted, the second deciduous incisor retained, and the permanent second incisor not yet erupted.

4. Discussion

Differences in the dentition between the pygmy chimpanzee and the common chimpanzee will be discussed from two points of view: phylogeny and function.

4.1. Distribution of Advanced and Primitive Traits

It is clear (see Table VII) that the pygmy chimpanzee does not display either a majority of primitive hominoid traits in the dentition or a majority of advanced traits. Whereas Frisch (1965) could clearly distinguish species of gibbons with more conservative dental characteristics from those with more progressive dental features (for example, island forms were more progressive than mainland forms), there is no such clear distinction between the two species of chimpanzee. Each species has some advanced traits and some primitive retentions. Thus, the suggestion of Gould (1975) and others that the pygmy chimpanzee is merely a scaled-down model of the common chimpanzee is not reflected in the dental data.

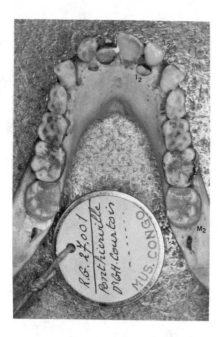

Figure 10. Mandible of *Pan paniscus* (RG27001) showing M_2 erupted before eruption of I_2.

The suggestion of Vandebroek (1959) and others that the pygmy chimpanzee represents the ape most similar to humans is also not generally borne out by the dental data. Although the pygmy chimpanzee and the human share some advanced traits, there are several traits in which *Pan paniscus* is less like *Homo sapiens* than is *P. troglodytes*. For example, the upper M^2 is most frequently the largest molar; in the sequence of eruption the permanent incisors normally erupt after the second molar. Although the arch of the palate appears more "parabolic" than that of the common chimpanzee, the *P. paniscus* maxillary arch increases in width posteriorly only to M1, not to M2 as in the human. There do not appear to be any shared derived dental features in both *P. paniscus* and *Homo sapiens*.

4.2. Functional Differences between Chimpanzee Species

Functional features are not independent of evolutionary features, but are discussed separately here for convenience. Functional differences are found in two general areas: anterior teeth, especially the incisors; and, posterior teeth, especially the anterior molars. (See Table VIII).

The incisors of *P. paniscus* appear to receive proportionately more wear than those of *P. troglodytes*. This is evident from (1) larger ratios

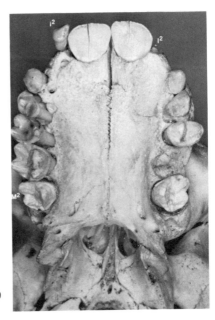

Figure 11. Maxilla of *Pan paniscus* (RG11528) showing M² erupted before eruption of I².

of deciduous lower incisors to deciduous lower molars in *P. paniscus* than in *P. troglodytes;* (2) maxillary and mandibular ratios of anterior/posterior teeth that are significantly larger in old adult *P. paniscus* than in unworn adult teeth of *P. paniscus;* and (3) the ratios of old adult *P. paniscus* that are significantly less than those of *P. troglodytes.* These differences in-

Table VII.   Traits in Dentition of *Pan paniscus* Compared with Those of *Pan troglodytes*

Advanced hominoid traits
1. Lower incidence of paraconid on dm₂
2. Buccal cingulum completely absent (upper molars)
3. Hypocone reduced and more often absent
4. Crista obliqua more often absent from M²
5. Metaconid more mesially oriented

Primitive hominoid traits
1. Upper molars lack distoconule
2. Lingual cingula (upper molars) better developed
3. Hypoconulid more lingually placed on M₁ and M₂
4. Tuberculum sextum less frequent
5. Less reduction of metacone on M² and M³

Table VIII. Dental Morphology of *Pan paniscus* Compared with
Pan troglodytes

Anterior dentition
1. Incisors receive proportionately more wear in *P. paniscus*
2. Canine teeth are proportionately larger in old adults
3. Arch of palate diverges more posteriorly (to M^1)

Posterior dentition
4. Anterior transverse crest (upper molars) is more transverse
5. Metaconid and entoconid are relatively more separated
6. Second molar is relatively the largest
7. There is no sexual dimorphism in dental measurements

---

dicate that there is proportionately more interproximal and incisal wear
in the pygmy chimpanzee than in the common chimpanzee.

Both maxillary and mandibular canine teeth of *P. paniscus* are larger
in old adults than in adults with unworn dentitions, indicating possible
selection for larger canines with age. This difference was not reported for
P. troglodytes.

The anterior transverse crest on upper molars has a more buccolingual orientation in *P. paniscus* than in *P. troglodytes*. This correlates with
the more mesially positioned metaconid in *P. paniscus* and the deeper
groove between protoconid and hypoconid into which the anterior transverse crest occludes. This would appear to provide a more efficient shearing mechanism than the corresponding features in *P. troglodytes*.

Badrian and Malenky (Chapter 11, this volume) report that pygmy
chimpanzees have a relatively high incidence of pith and leaf petioles in
their diet, as well as a large amount of immature leaves. Both the incidence
of folivorous materials in the diet and the manner of eating pith may bear
a significant relationship to dental features. They report that *Haumania
liebrechtsiana*, a canelike plant, is "the single most frequently eaten food
in the Lomako area" and its "soft juicy pith is enclosed in a tough fibrous
sheath, which is bitten open with the incisors." This kind of feeding,
which is frequent, could account for the high incidence of incisal wear.
Kay (1973) has shown a positive correlation between the expression of
dental design features associated with shearing and the incidence of plant
foods in the diet that are high in structural carbohydrate. Thus, the apparently more efficient shearing mechanism of upper and lower molars in
P. paniscus may reflect the relatively high incidence of pith and leaf
petioles in the diet. A comparison of measurements of molar shearing
blades in the two species of chimpanzee is needed to confirm this. Further

studies of both occlusal morphology and diet in free-ranging pygmy chimpanzees are required to determine the functional relationships between the diet and dentition.

ACKNOWLEDGMENTS. I should like to thank the Wenner-Gren Foundation for Anthropological Research and especially the Director of Research, Lita Osmundsen, for a generous grant (No. 2811), which made possible study of the collection of pygmy chimpanzee dentitions in Tervuren. I am also indebted to Dr. Max Poll, Curator of Vertebrates, Musée Royal de l'Afrique Centrale, Tervuren, Belgium, for permission to study the collection. I also thank the authorities of the Department of Mammalogy, American Museum of Natural History, New York, for permission to study specimens in their collection. I am grateful to Dr. R. L. Susman for inviting me to participate in the Symposium, Evolutionary Morphology and Behavior or *Pan paniscus,* and for providing some dental casts. My daughters, Andrea and Claudia, were very helpful in transcribing data. The completion of this study was supported in part by a grant from the Research Foundation of the City University of New York.

References

Almquist, A. J., 1974, Sexual differences in the anterior dentition in African primates, *Am. J. Phys. Anthropol.* **40:**359–367.

Ashton, E. H. and Zuckerman, S., 1950, Some quantitative dental characteristics of the chimpanzee, gorilla and orang-outang, *Philos. Trans. R. Soc.* **234B:**471–479.

Clements, E. M. B., and Zuckerman, S., 1953, The order of eruption of the permanent teeth in the Hominoidea, *Am. J. Phys. Anthropol.* **11:**313–332.

Conroy, G. C., 1972, Problems in the interpretation of *Ramapithecus:* with special reference to anterior tooth reduction, *Am. J. Phys. Anthropol.* **37:**41–48.

Cramer, D. L., 1977, Craniofacial morphology of *Pan paniscus, Contrib. Primatol.* **10:**1–64.

Fenart, R., and Deblock, R., 1973, *Pan paniscus* et *Pan troglodytes* craniométrie. Étude comparative et ontogénique selon les méthodes classiques et vestibulaires, *Mus. R. Afr. Centr. Tervuren Belg. Ann. Ser. Octavo Sci. Zool.* **204:**1–473.

Frisch, J. E., 1965, Trends in the evolution of the hominoid dentition, *Bibl. Primatol.* **3:**1–130.

Gould, J. S., 1975, Allometry in primates, with emphasis on scaling and the evolution of the brain, in: *Approaches to Primate Paleobiology* (F. S. Szalay, ed), S. Karger, Basel, pp. 244–292.

Johanson, D. C., 1974*a*, Some metric aspects of the permanent and deciduous dentition of the pygmy chimpanzee *(Pan paniscus), Am. J. Phys. Anthropol.* **41:**39–48.

Johanson, D. C., 1974*b*, An Ondontological Study of the Chimpanzee with Some Implications for Hominoid Evolution, Ph.D. Dissertation, University of Chicago, Chicago, Illinois.

Kay, R. F., 1973, Mastication, Molar Tooth Structure and Diet in Primates, Ph.D. Dissertation, Yale University, New Haven, Connecticut.

Korenhof, C. A. W., 1960, *Morphogenetical Aspects of the Human Upper Molar*, Uitgeversmaatschappy, Utrecht.

Krogman, W. M., 1930, Studies in growth changes in the skull and face of anthropoids. I. The eruption of the teeth in anthropoids and Old World apes, *Am. J. Phys. Anthropol.* **46**:303–313.

Krogman, W. M., 1969, Growth changes in skull, face, jaws, and teeth of the chimpanzee, in: *The Chimpanzee*, Volume 1 (G. H. Bourne, ed.), S. Karger, Basel, pp. 104–164.

Lavelle, C. L. B., Flinn, R. M., Foster, T. D., and Hamilton, M. C., 1970, An analysis into age changes of the human dental arch by a multivariate technique, *Am. J. Phys. Anthropol.* **33**:403–412.

Mahler, P., 1973, Metric Variation in the Pongid Dentition, Ph.D. Dissertation, University of Michigan, Ann Arbor, Michigan.

Monteil, G. A., 1970, Étude du deplacement ontogénique des dents du *Pan paniscus* dans les axes vestibulaires d'orientation, Thesis, Doctor of Dental Surgery, Faculty of Medicine, Paris.

Nissen, H. W., and Riesen, A. H., 1945, The deciduous dentition of chimpanzee, *Growth* **9**:265–274.

Remane, A., 1960, Zähne und Gebiss, in: *Primatologia*, Volume III (H. Hofer, A. H. Schultz, and D. Starck, eds.), S. Karger, Basel, pp. 637–846.

Remane, A., 1962, Masse und Proportionen des Milchgebisses der Hominoidea, *Bibl. Primatol.* **1**:229–238.

Schultz, A. H., 1935, Eruption and decay of the permanent teeth in primates, *Am. J. Phys. Anthropol.* **19**:489–581.

Schultz, A. H., 1956, Postembryonic age changes, in: *Primatologia*, Volume I (H. Hofer, A. H. Schultz, and D. Starck, eds.), S. Karger, Basel, pp. 887–964.

Schuman, E. L., and Brace, C. L., 1954, Metric and morphologic variations in the dentition of the Liberian chimpanzee; comparisons with anthropoid and human dentitions, *Hum. Biol.* **26**:239–268.

Thomas, D. H., 1976, *Figuring Anthropology*, Holt, Rinehart and Winston, New York.

Vandebroek, G., 1959, Notes ecologiques sur les anthropoides africains, *Ann. Soc. R. Zool. Belg.* **89**:203–211.

Vandebroek, G., 1969, *Evolution des Vertebres de leur origine à l'homme*, Masson, Paris.

CHAPTER 6

An Allometric Perspective on the Morphological and Evolutionary Relationships between Pygmy (Pan Paniscus) and Common (Pan troglodytes) Chimpanzees

BRIAN T. SHEA

1. Introduction

The pygmy chimpanzee *(Pan paniscus)*, or bonobo, has been the object of considerable primatological interest since its "discovery" in the late 1920s and early 1930s. Because pygmy and common chimpanzees differ in average adult size (although the amount of difference and degree of overlap are debatable), many have suggested that morphological differences between the species may be size-related, or *allometric*. For example, Corruccini and McHenry (1979) and McHenry and Corruccini, (1981) analyzed the cranium, postcranium, and dentition of *P. paniscus* and *P. troglodytes*, and concluded that many of the shape differences between the two species are due to allometry. Horn (1979) has made a similar suggestion. By contrast, Zihlman and others (e.g., Zihlman and Cramer, 1978; Zihlman *et al.*, 1978; Zihlman, 1979, 1981) have argued that allometry cannot be invoked to explain the morphological differences

BRIAN T. SHEA • Department of Anthropology, American Museum of Natural History, New York, New York 10024. *Present address:* Departments of Anthropology and Cell Biology and Anatomy, Northwestern University, Evanston, Illinois 60201.

between the chimpanzee species. In fact, Zihlman (1981, p. 6) suggests that we might view allometry as a "favorite anatomical fudge factor" which has obscured our understanding of the two chimpanzee species. In this paper and in other works (Shea, 1981*a*, 1982, 1983*a–d;* Coolidge and Shea, 1982), I have also analyzed the morphological differences between the chimpanzee species (and among the African pongids as a group) from an allometric perspective.

The morphological differences between pygmy and common chimpanzees have also been described using DeBeer's (1958) categories and processes of *heterochrony*. In his paper on *P. paniscus*, Coolidge (1933, p. 56) concluded:

> It is widely believed among scientists that the life cycle of a single individual in some ways recapitulates the evolution of that living organism. It is further thought that the less specialized, which usually means the more primitive or juvenile forms, approach more closely to an ancestral state in the evolution of a given animal than do the more highly developed adult forms. If this is true, the *paniscus, a true paedomorphic species,* which shows definitely juvenile characteristics in an adult state, is the most important of the chimpanzees in a study of the phylogeny and relationships of this high order of anthropoid apes. It may approach more closely to the common ancestor of chimpanzees and man than does any living chimpanzee hitherto discovered and described (italics added).

Susman (1980) and Latimer *et al.* (1981) have suggested that these comments were prompted by Louis Bolk's "fetalization" theories. I think we might attribute some influence to Haeckel, and perhaps even DeBeer himself, considering that the first edition of his book *Embryology and Evolution* was published in 1930. Gould's (1977) recent synthesis has placed DeBeer and heterochrony firmly within modern evolutionary biology.

As for other specific suggestions concerning the chimpanzees, Tuttle (1975) has claimed that the morphological differences between *P. paniscus* and *P. troglodytes* indicate that the former may have been derived from the latter via the process of "neoteny." MacKinnon (1978) argues that neoteny is the primary basis of the pygmy chimp's specializations. McHenry and Corruccini (1981) have also suggested that neoteny might have played an important role in the differentiation of the chimpanzee species. Gijzen (1975, p. 137) uses the term "infantilism" to describe the chimpanzee differences.

The labels "pygmy" and "dwarf" have also been used to describe the morphological differences between *P. paniscus* and *P. troglodytes.* Various definitions of these and related terms have been discussed by Marshall and Behrensmeyer (ms.), and the specific case of the chimpanzee species has been considered by Coolidge (1933), Frechkop (1935), Gijzen (1975), Zihlman and Cramer (1978), McHenry and Corruccini (1981), Latimer *et al.* (1981), Johnson (1981), Susman (1980), and Coolidge and Shea (1982). Kortlandt (1972), Horn (1979), Johnson (1981), and Susman and Jungers (1981) have considered the ecological evidence for claims that *P. paniscus* might represent a subgroup of chimpanzees that decreased in size due to isolation in Central African forests. Scenarios specifically relating behavioral and environmental difference to the morphological distinctions between pygmy and common chimpanzees have ranged from claims that *P. paniscus* is more specialized for arboreal arm swinging and may be convergent toward hylobatids (Coolidge, 1933; Frechkop, 1935: Roberts, 1974; Susman, 1980), to suggestions that they are adapted for a more terrestrial existence and locomotion (Horn, 1979; Johnson, 1981).

It is clear that the long and varied series of investigations into the meaning of the morphological differences between pygmy and common chimpanzees has resulted in an unusually confusing plethora of names, labels, and hypotheses. This has proven to be an appreciable hindrance to our understanding of the morphological and evolutionary relationships of the chimpanzees. For example, it is confusing to read in the same paper (Zihlman, 1979, p. 165) both that "common chimpanzees . . . can be viewed as larger and more robust versions of pygmy chimpanzees," and that "body build differs between pygmy and common chimpanzees but not allometrically." Notions of allometric and nonallometric differences need to be clarified here. In regard to claims that paedomorphosis or neoteny (or progenesis or hypermorphosis) has played an important role in chimpanzee evolution, neither rigorous morphological criteria necessary for adequate testing, nor the requisite developmental timing data (see Shea, 1983*d*), have been analyzed. The use of these labels in such a context can only produce confusion. From another perspective, the multiple uses of the terms "pygmy" and "dwarf" and the numerous distinct *types* of dwarfism (Gould, 1975*b;* Gijzen, 1975; Marshall and Corruccini, 1978; Marshall and Behrensmeyer, ms.; Shea, 1981*b;* Johnson, 1981, and comments), have also caused confusion.

Significant advances in our understanding of these issues will only come when labels are clearly defined, hypotheses are carefully formulated, and appropriate data are collected and used to test predictions. In this chapter, I consider the morphological differences between the chimpanzee species in relation to allometry. An attempt is made to clarify the different uses and meanings of allometry and the related categories of heterochrony, and to provide the data necessary to analyze these issues in some detail.

2. The Types and Meanings of Allometry

Allometry has been broadly defined as the study of size and its consequences (Gould, 1966). However, there are several different *types* of allometry, which require careful distinction; failure to properly distinguish among these different types is responsible for much of the confusion in allometric studies in general (Gould, 1975a), and the studies of the chimpanzees in particular. *Ontogenetic allometry* (growth allometry, relative growth) involves examination of changing proportions during growth of an individual or species. *Static adult intraspecific allometry* focuses on proportion changes among adults of a single species, as between males and females, for example. Finally, *interspecific allometry* refers to size-related change in shape among adults of different species (evolutionary allometry is thus a special case of interspecific allometry). These three types of allometry are sometimes broken down into "growth allometry" and "size allometry" (Simpson *et al.*, 1960; Sprent, 1972). It is worth pointing out at this juncture a claim that I have detailed elsewhere (Shea, 1981a, 1983a): the usefulness of static adult intraspecific allometry is very limited. Attempts to infer growth phenomena from such adult data are clearly problematic (Cock, 1966); similarly, extension or truncation of the curve of static adult allometry in order to infer the results of selection for size increase or decrease is also biologically infeasible (Lande, 1979; Atchley *et al.*, 1981; Cheverud, 1982; Shea, 1983a).

A more important distinction than the different types of allometry involves the different *meanings* or *uses* of allometry. These have been reviewed by Gould (1966, 1975a) and Shea (1981a, 1982), among others. We may conclude that shape differences between or among adults of different species are size-related, or allometric, if they result from the simple extension or truncation of common growth allometries (or ontogenetic trajectories of size/shape change) to new adult sizes. In such cases, selection may act to increase or decrease overall body size, and certain shape changes are correlated consequences of extension or truncation of the ancestral ontogenetic patterns of shape change. This meaning or use of allometry is referred to as *ontogenetic scaling* (Gould, 1975a; Shea, 1981a, 1982). On the other hand, certain shape differences between or among adults of different species are size-required, or allometric, if they result from selection to maintain functional equivalence at different body sizes. In such cases, selection has presumably acted to change overall body size (or, perhaps, duration of ontogeny), but shifts in ancestral growth allometries are also required in order to maintain behavioral or physiological equivalence at the new sizes. This meaning or use of allometry is referred to as *biomechanical scaling* (Shea, 1981a, 1982). Ontogenetic growth patterns of the adults comprising such a biomechanically scaled series rarely coincide, perhaps because growth allometries are ultimately restrictive (Gould, 1966), or perhaps because ontogenetic size ranges are usually not great enough to require the broad patterns of biomechanical alterations observed among different species (McMahon, 1975b). Von Bertalanffy and Pirozynski (1952) note that the "balance of organs" necessary for proper functioning of species varying in size but belonging to a certain "bauplan" often requires changes in developmental patterns.

These different meanings or uses of allometry suggest that we need to ultimately refine our questions and hypotheses past the point of asking "are pygmy chimpanzees simply a scaled-down version of common chimpanzees" (Zihlman and Cramer, 1978, p. 87), or hypothesizing that "*P. paniscus* and *P. troglodytes* are allometrically scaled versions of one another" (McHenry and Corruccini, 1981, p. 355). This is because the "tests" for ontogenetic scaling and biomechanical scaling are different. In short, we test for ontogenetic scaling by comparison of growth allometries to see if they remain unchanged with size increase and decrease (Shea, 1981a). (Ontogenetic data themsleves must be examined—growth allometries can never be reliably inferred from static adult intraspecific data, or static interspecific curves.) By contrast, we test for biomechanical scaling by determining if observed interspecific shape changes correspond with those expected on the basis of certain theoretical biomechanical principles. For example, McMahon's (1973, 1975a,b, 1980) model of elastic similarity scaling yields expected allometric exponents for changes in body and bone shape (among other factors) as size is altered. We test such predictions of biomechanical scaling primarily by examining observed and predicted *interspecific* coefficients of shape change, and not by examining growth allometries (although we might also examine growth patterns to determine if the same biomechanical principles are at work throughout ontogeny). Of course, in addition to the empirical testing of proposed biomechanical explanations, we may also question the theoretical basis of the specific predictions [in the case of McMahon's models, see Alexander et al. (1979) and Maloiy et al. (1979)].

The distinction between ontogenetic scaling and biomechanical scaling may be more fully appreciated by referring to Fig. 1. This is a plot of arm and leg length among subadult and adult hylobatids (gibbons and siamangs). The points A and B represent adult gibbons of different body size; point C is for the adult siamang. Provided that the slope of the gibbon line differs from 1.0 (i.e., the relation is allometric), the shape differences between A and B are allometric and result from ontogenetic scaling, because they represent extension of a common growth allometry to larger sizes among closely related species. I will present theoretical and empir-

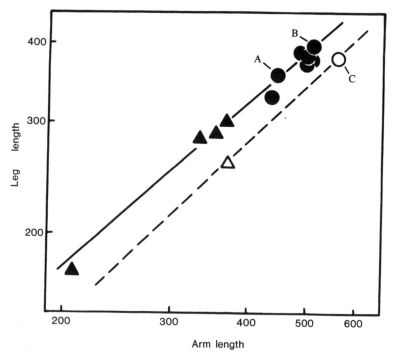

Figure 1. A plot of arm length versus leg length in gibbons and siamangs. Triangles represent subadults; circles represent mean adult values for different species. Closed symbols are gibbons; open symbols are siamangs. The shift of the siamang growth allometry relative to that of the gibbon illustrates that the lower intermembral index of gibbons is present throughout ontogeny. Proportion differences between A and B are size-related due to ontogenetic scaling; the differences between the gibbons as a group (A and B) and the siamang (C) may also be size-related, but in this case is a result of biomechanical scaling. (From Lumer, 1939.)

ical arguments below, however, that might support the claim that the shape differences between A and B as a group, and C, are also due to allometry, but in this case biomechanical scaling, in order to maintain functional equivalence at different body sizes. That both sets of comparisons and shape changes can be labeled as "allometric" may seem counterintuitive and confusing, but it is precisely the failure to understand such different types and meanings of allometry that is responsible for so much imprecision in studies of size and shape.

The previous discussions allow a hierarchial outline for a complete allometric investigation. Given two species, A and B, we can assess allometric influences, and more clearly comprehend nonallometric changes, between average adults of A and B by the following procedure [Shea (1982), modified from Huxley (1932), p. 204]:

1. Determine if there are significant differences in shape as well as in weight and overall size between A and B. If there are no significant size or weight differences, then whatever proportion differences exist cannot be size related or the result of allometry in any sense of that word.

2. Determine if the interspecific proportion differences between A and B are the result of ontogenetic scaling. Comparison of growth allometries in A and B for the proportions concerned provide the relevant test and "criterion of subtraction." If proportion differences result from the extension or truncation of a common growth allometry, then we may conclude that they are allometric due to ontogenetic scaling.

3. For those proportion differences demonstrated not to be ontogenetically scaled via step 2, determine if the interspecific differences between A and B are the result of biomechanical scaling. Testing and criteria of subtraction here are based on theoretical expectations, which may differ with the body region being investigated (teeth, limbs, etc.). If observed proportion differences between A and B correspond with predicted trends, then we may conclude that they are allometric due to biomechanical scaling.

4. For those proportion differences between A and B that cannot be explained by ontogenetic scaling (step 2) or biomechanical scaling (step 3), we may conclude that the interspecific differences are *nonallometric,* and are the result of selection for altered proportions unrelated to size differences. An alternative, nonallometric, adaptive or "nonadaptive" explanation is required.

This step-by-step application of the allometric approach can be best illustrated by way of a specific example. In his comparison of *P. paniscus* and *P. troglodytes,* Coolidge (1933) noted that pygmy chimpanzees have a narrower, more gracile scapula than do common chimpanzees (see Fig. 2). A number of authors have stressed this interspecific difference, noted a convergence toward scapula form in hylobatids, and suggested that it might reflect significant locomotor differences between the chimpanzee species (e.g., Coolidge, 1933; Frechkop, 1935; Roberts, 1974; Horn, 1979; Susman, 1980; McHenry and Corruccini, 1981). Susman *et al.* (1980) have suggested that the long, narrow scapula of *P. paniscus* provides evidence for a greater degree of arboreal leaping, diving, and armswinging in the smaller chimpanzee. These claims may all be true. However, they are

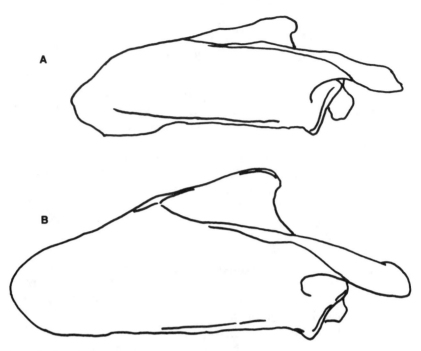

Figure 2. Scapula form in (A) *Pan paniscus* and (B) *Pan troglodytes* adults. The shape differences have been claimed to indicate greater arboreality in the pygmy chimpanzee. (From Coolidge, 1933.)

based on a comparison of one or two adults, and we do not know to what extent scapulae of similar sizes differ or correspond in shape between the species. Do adult *P. paniscus* scapulae differ significantly in shape from those of (subadult) *P. troglodytes* of equal size? In other words, we first need to investigate the potential effects of ontogenetic allometry in producing the interspecific shape differences. Only then can we discuss the morphological differences in scapula form with the confounding effects of body size controlled for. Of course, we may still seek correlations between behavioral (i.e., locomotor) and anatomical differences, whether we compare young and old *P. troglodytes* or adult pygmy and common chimpanzees, but this is also something we need to know for a full assessment. I have chosen the example of the scapula because it *may* be one of the features in which the chimpanzees differ most [although see Zihlman and Cramer (1978)], it has attracted much attention and speculation, and yet no systematic investigation from an allometric perspective has been completed on it.

3. Allometry and Heterochrony

Analysis of ontogenetic allometries is also a fundamental aspect of studies of heterochrony (DeBeer, 1958; Gould, 1977; Alberch *et al.*, 1979; Shea, 1983*c,d*). Heterochrony is defined as the displacement of a feature in descendants relative to the time at which it appeared in ancestors (Gould, 1977). Since hypotheses of heterochronic change have been forwarded to account for the morphological differences between the chimpanzee species, we need a rigorous and consistent approach to test these suggestions. Allometry, or the study of size/shape relations, provides one basic component of the data we use to assess heterochronic change. Another basic component involves *timing* of development events, such as length of ontogeny in time, and onset and cessation of growth. Alberch *et al.* (1979) have provided a methodology that can be used to analyze morphological differences and transformations in terms of a basic set of parameters controlling development. Their model builds on Gould's (1977) revision of DeBeer (1958). I have analyzed allometry and heterochrony in the African apes as a whole in detail elsewhere (Shea, 1983*d*).

The basic categories and processes of heterochrony are presented using Gould's (1977) "clock models" of size, shape, and age in Fig. 3. *Paedomorphosis* is defined as the development in descendants of a morphology that resembles subadult stages in ancestors. *Peramorphosis* occurs when ancestral adult morphologies are transcended, or "passed beyond," in descendants (Alberch *et al.*, 1979). The *results* of paedomorphosis and peramorphosis are produced via different developmental *processes* (Gould, 1977). In Fig. 3, paedomorphosis may result from neoteny, rate hypomorphosis, or time hypomorphosis. Similarly, peramorphosis may result from acceleration, size hypermorphosis, or age hypermorphosis. In the case of neoteny and acceleration, ancestral size/shape relations are "dissociated" or uncoupled; significant departures from ontogenetic scaling occur in these cases. The processes of rate and time hypo- and hypermorphosis (Shea, 1983*d*) are characterized by ontogenetic scaling of proportions. Significant extension or truncation of either time of maturity or ultimate size attained are the primary shifts in these cases, and paedomorphosis or peramorphosis results. Underlying trajectories of size/shape change in the respective ontogenies remain unaltered, and recapitulation or "reverse recapitulation" occur (Gould, 1977). This brief outline pro-

PAEDOMORPHOSIS   PERAMORPHOSIS

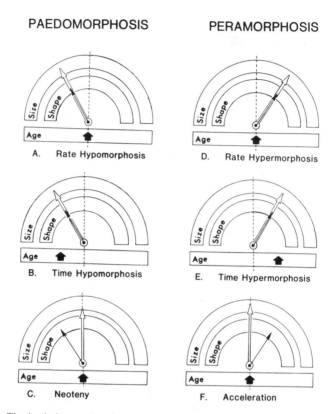

A. Rate Hypomorphosis   D. Rate Hypermorphosis

B. Time Hypomorphosis   E. Time Hypermorphosis

C. Neoteny   F. Acceleration

Figure 3. The basic heterochronic processes in terms of size, shape, and age. The clock depicts changes relative to an ancestral condition where size, shape, and age are aligned along the dashed line. Paedomorphic or peramorphic morphological results may be produced by different processes. In the case of (rate or time) hypomorphosis or hypermorphosis, ancestral ontogenetic allometries are extended or truncated; length of ontogeny in time may or may not be altered. With neoteny and acceleration, ancestral allometries (size/shape) are dissociated. See text for additional discussion. (Modified from Gould, 1977.)

vides a quantitative test for the different categories and processes of heterochony.

4. Proportion and Size Differences between *Pan paniscus* and *Pan troglodytes*

Differences in size and shape between *P. paniscus* and *P. troglodytes* pervade the cranium, dentition, and postcranium. A very brief summary of some previous studies suggests that, compared to adult common chimpanzees, adult *P. paniscus* are characterized by: a more gracile skull with smaller splanchnocranial to neurocranial proportions (Coolidge, 1933; Weidenreich, 1941; Fenart and Deblock, 1973; Cramer, 1977; Shea, 1982), smaller teeth (Kinzey, 1971; Johanson, 1974; Shea, 1982), less sexual dimorphism (Fenart and Deblock, 1972; Cramer, 1977; Zihlman and Cramer, 1978; Shea, 1982), smaller clavicles (Zihlman and Cramer, 1978), a relatively narrower scapula (Coolidge, 1933; Frechkop, 1935; Roberts, 1974), a narrower chest girth (Shea, 1981*a*, 1982; Coolidge and Shea, 1982), a shorter "head-to-fork" length (Coolidge and Shea, 1982), different arm-to-leg proportions and a lower intermembral index (Shea, 1981*a;* Coolidge and Shea, 1982), a smaller pelvis (Zihlman and Cramer, 1978), and metacarpals and phalanges that differ in a number of respects (Susman, 1979).

As noted previously, the first step in determining whether such shape differences are allometric is to firmly establish significant differences in overall size and weight. In his original comparison of a single adult *P. paniscus* and several *P. troglodytes*, Coolidge (1933) suggested that the "pygmy" chimpanzee was approximately one-half the size of its larger relative. This estimate was based on an unusually small (but fully adult) female specimen from the American Museum of Natural History in New York. The figure of one-half was frequently cited in the subsequent literature, particularly in Hill's (1967, 1969) reviews of the chimpanzees. Recently, a number of authors have noted that the size differences between *P. paniscus* and *P. troglodytes* are not as great as has been frequently claimed (e.g., Vandebroek, 1969; Zihlman and Cramer, 1978; Horn, 1979; McHenry and Corruccini, 1981; Coolidge and Shea, 1982). Coolidge and Shea (1982) report a wild-shot pygmy chimpanzee weighed in the field at 61 kg.

I have plotted much of the available body weight data for subadult and adult *P. paniscus* in Fig. 4. Mean values for *P. troglodytes* as given by Grether and Yerkes (1940) are also included. Although there is clearly a great deal of overlap, it is apparent that throughout growth and at adulthood, *P. paniscus* is a smaller species. The insert in Fig. 4 provides additional information on adult chimpanzee weights. Again, overlap is strong, but *P. troglodytes* are significantly larger animals. I believe that the following weights given by Pilbeam and Gould (1974) are a fairly accurate statement of the *mean* weight differences between pgymy and common chimpanzees: *P. paniscus*, 35 kg (male, 38 kg; female, 32 kg); *P. troglodytes*, 45 kg (male, 49 kg; female, 41 kg).

Whether these size differences merit the label "pygmy" for *P. paniscus* is another issue, and perhaps one of less biological than semantic import (see Marshall and Behrensmeyer, ms.). A selected series of cranial,

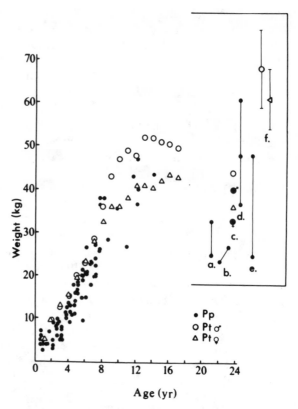

Figure 4. A plot of growth in body weight against time for pygmy and common chimpanzees. Data for *P. troglodytes* are mean values from Grether and Yerkes (1940). Mixed cross-sectional data points (including some longitudinal segments) for *P. paniscus* are from Benirschke *et al.* (1980), Neugebauer (1980), and zoo weights obtained by this author from U. S. and European institutions. The insert at right (set properly on the axes) represents adult values from various sources: (a) Lokalema, an aged female from Yerkes; drop due to sickness; (b) Linda (Benirschke *et al.*, 1980); also sick weights; (c) interspecific adult means given by Jungers (1984); (d) three individual wild-shot weights from Coolidge and Shea (1982); (e) range of adult body weights given by Zihlman and Cramer (1978); (f) mean and standard deviation for *P. troglodytes* reared in captivity as given by Smith *et al.* (1975).

dental, and postcranial dimensions in Table I give the percentage size differences between adult chimpanzees. These data are used to construct a schematic diagram of chimpanzee proportions in Fig. 5. *Pan paniscus* is clearly a smaller species "above the waist" than *P. troglodytes*. It is only in absolute or relative hindlimb length that pygmy chimpanzees are not diminutive.

Table I.   Means (Sexes Pooled) of Adult Craniodental and Postcranial
Dimensions in *Pan paniscus* and *Pan troglodytes*[a]

Dimensions, mm	Pan paniscus	Pan troglodytes	Significance	Percent difference[b]
Craniodental				
Basicranial length	84.8 (41)	98.0 (66)	[c]	0.87
Palate length	58.8 (41)	74.8 (66)	[c]	0.79
Facial height	68.0 (41)	83.5 (66)	[c]	0.81
Skull length	163.0 (41)	195.0 (66)	[c]	0.84
Bizygomatic width	112.8 (41)	126.8 (66)	[c]	0.89
Bimaxillary width	76.8 (41)	91.3 (66)	[c]	0.84
Canine length	10.0 (41)	13.0 (66)	[c]	0.77
M² length	8.6 (41)	10.0 (66)	[c]	0.86
Postcranial				
Head-to-fork	748.1 (8)	812.7 (45)	[d]	0.92
Height	1153.2 (8)	1201.2 (50)	[e]	0.96
Girth	805.4 (7)[f]	901.3 (54)[g]	[c]	0.89
Span	1588.1 (8)	1779.1 (48)	[d]	0.89
Clavicle[h]	105.0 (21)	122.0 (18)	[d]	0.86
Arm	556.7 (12)	570.5 (82)	[e]	0.98
Leg	544.7 (10)	540.6 (82)	NS	1.01
Humerus	289.4 (12)	297.9 (82)	[d]	0.97
Radius	267.3 (12)	272.6 (82)	[d]	0.98
Femur	297.7 (10)	293.4 (82)	NS	1.01
Tibia	246.2 (12)	247.2 (82)	NS	1.00

[a]Sample sizes are given in parentheses.
[b]*Pan paniscus* mean divided by *P. troglodytes* mean.
[c]Mean of mean values taken from Shea (1982); significance not tested.
[d]Significant at 0.05 or less in a two-tailed test.
[e]Significant difference at 0.05 in a one-tailed test.
[f]From Coolidge and Shea (1982).
[g]From Ashton (1954).
[h]Data from Zihlman and Cramer (1978).

Several workers have implied or suggested that there may be no mean weight difference between *P. paniscus* and *P. troglodytes* (Zihlman, 1979, 1981; Zihlman and Cramer, 1978). As Coolidge and Shea (1982) pointed out, this question will only be fully resolved when large samples of body weights from the wild, especially for the wide-ranging *P. troglodytes,* are available. In fact, Jungers and Susman (Chapter 7, this volume) conclude that the eastern subspecies of common chimpanzee *(P. t. schweinfurthii)*

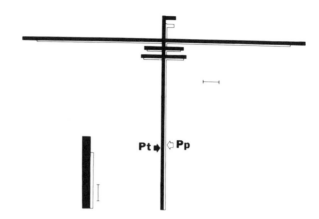

Figure 5. A schematic diagram of body proportions in adult *Pan paniscus* and *Pan trog-lodytes* (sexes pooled). The scale at upper right is 10 cm. The insert at lower left is for mean weight, with a scale of 10 kg [weights from Pilbeam and Gould (1974)]. Dimensions used for the linear contrast are basicranial skull length, span length, body height, clavicle length (from Zihlman and Cramer, 1978), and chest diameter. The arrows indicate the location of the crotch, as determined by subtracting head-to-fork length from total body height. This figure illustrates the *P. paniscus* is smaller than *P. troglodytes* in all dimensions plotted here, except for length of the hindlimbs.

does not differ significantly in weight from *P. paniscus*,while *P. t. trog-lodytes* is significantly larger than the pygmy chimpanzee. Nevertheless, Fig. 4 provides strong support for the claim that *P. paniscus* is smaller on average than *P. troglodytes* as a whole. The criticism that these weights are for captive animals is not valid in this case. First of all, the Grether and Yerkes (1940) weights used in Fig. 4 are substantially lower than those from other studies (e.g., Smith *et al.,* 1975). More important, however, the *P. paniscus* weights in Fig. 4 are also for captive animals, so those who would reject the growth pattern and size differences between the species must explain why captivity has affected the two species differently.

The data presented in Fig. 5 and Table I allow consideration of this size issue from another perspective. If we accept the claim that there are no significant weight differences between the chimpanzee species, and given that *P. paniscus* is clearly significantly smaller in most linear skeletal and external body dimensions, then it follows that *P. paniscus* must be a stockier, more robust creature. (Perhaps *P. paniscus* should be known as the "pudgy" chimpanzee!). In fact, such a claim runs directly contrary to a half-century of qualitative and quantitative morphological assessment, which has stressed the gracility and slenderness of *P. paniscus* (e.g.,

Coolidge, 1933; Frechkop, 1935; Hill, 1967, 1969; Gijzen, 1975; Horn, 1979; Zihlman, 1979; Johnson, 1981).

5. Postcranial Allometry

A number of studies have examined poscranial anatomy in *P. paniscus* and *P. troglodytes* (e.g., Schultz, 1954; Zihlman and Cramer, 1978; Zihlman, 1979; Corruccini and McHenry, 1979; Susman, 1979; McHenry and Corruccini, 1981). However, studies focusing on ontogenetic data and relative growth are fewer (Shea, 1981*a*, 1982; Coolidge and Shea, 1982). The latter studies have combined data on skeletal limb lengths with external body dimensions identical to those taken on many wild-shot *P. t. troglodytes* in the early part of this century (Ashton, 1954; Shea, 1981*a*).

Examination of growth allometries for postcranial dimensions reveals a number of examples where *P. paniscus* and *P. troglodytes* are ontogenetically scaled. Several of these are illustrated in Figs. 6 and 8. A plot of chest girth and arm span against head-to-fork length demonstrates that adult *P. paniscus* have the proportions observed in subadult *P. troglodytes*

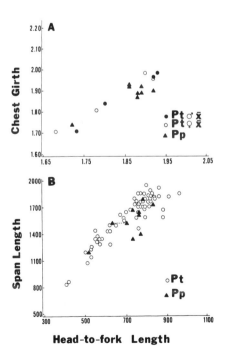

Figure 6. A plot of (A) chest girth and (B) span length against head-to-fork length in pygmy and common chimpanzees. Individual data points for *P. paniscus* are from Coolidge and Shea (1982); means in part A are from Ashton (1954). The dashed line connecting two values of *P. paniscus* in part B represents two possible values of "head-to-fork" for Coolidge's (1933) adult female. Both span length and chest girth appear to be ontogenetically scaled in the chimpanzees; adult pygmy chimps have the proportions observed in subadult common chimps.

(Fig. 6). This is also true for plots of humerus versus radius length and femur versus tibia length [Fig. 8; see also Shea 1981a, 1982)]. Another way of expressing this relationship is to note that the proportions observed in adult common chimpanzees are those predicted if the growth patterns of the pygmy chimpanzee are simply extended to larger terminal sizes. Jungers and Susman (Chapter 7, this volume) have replicated these results on ontogenetic scaling of intralimb proportions for chimpanzees using slightly different measurements than Shea (1981a, 1982). This replication and confirmation—after all, the heart of the scientific method—is most welcome. Jungers and Susman also present new data on limb girdles and joint surfaces. Some dimensions exhibit ontogenetic scaling and others do not.

A clear departure from ontogenetic scaling of postcranial proportions is observed for the hindlimbs [or for a dimension that includes hindlimb length, such as total body height; see Coolidge and Shea (1982)]. A plot of arm length against leg length (Fig. 7) reveals a downward shift or transposition of the *P. paniscus* growth allometry (or, conversely, an upward shift of the *P. troglodytes* growth allometry). The lower inter-membral index characteristic of adult *P. paniscus* is present throughout growth (Shea, 1981a). A plot of humerus versus femur length reveals a similar shifting, while a plot of radius versus tibia seems to suggest on-togenetic scaling (although a slight shifting is possible; Fig. 8). Therefore, pygmy chipmanzees get their lower intermembral index primarily from the proportions of their proximal limb segments, and not the distal ones. Comparison of absolute sizes also indicates an anomalous situation for the hindlimb (see Table I). Whereas adult *P. paniscus* are significantly smaller than adult common chimpanzees in almost all dimensions, their hindlimbs (and their femora in particular) are not significantly smaller (see also Zihlman and Cramer, 1978). The *relatively* longer hindlimbs of *P. paniscus* (or, conversely, the relatively shorter hindlimbs of *P. troglo-dytes*) are clearly not allometric in the sense of simply being the result of ontogenetic scaling.

I have previously argued (Shea, 1981a; Coolidge and Shea, 1982) that the limb proportion differences betewen *P. paniscus* and *P. troglodytes* are the result of biomechanical scaling, and therefore they are allometric. This argument was based on analysis of climbing behavior presented by Cartmill (1974; but see Bock and Winkler, 1978) and applied by Jungers (1977, 1978, 1979) to limb proportions in prosimians. The crux of the argument is that *relative* shortening of the hindlimbs is required to main-tain functional equivalence in climbing behavior as body size increases. Rotational torques about the foothold point during vertical climbing in-crease with the animal's weight and the distance of the axis of weight

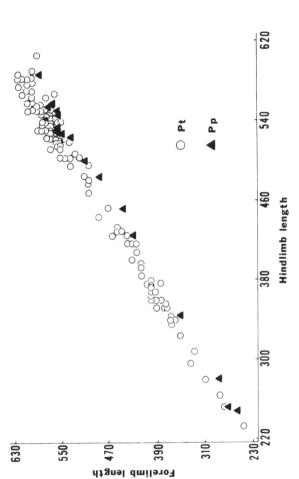

Figure 7. A plot of forelimb length (humerus + radius) against hindlimb length (femur + tibia) in the chimpanzees. Mixed cross-sectional data; each point represents a different individual, from infant to adult. A slight but clear shift or transposition of the growth allometries demonstrates that the higher intermembral index of *P. troglodytes* is present throughout ontogeny. Common chimpanzees have relatively short legs for their forelimb length compared to pygmy chimpanzees. See text for additional discussion.

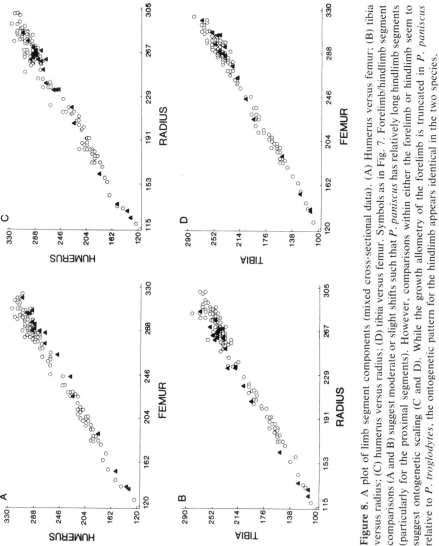

Figure 8. A plot of limb segment components (mixed cross-sectional data). (A) Humerus versus femur; (B) tibia versus radius; (C) humerus versus radius; (D) tibia versus femur. Symbols as in Fig. 7. Forelimb/hindlimb segment comparisons (A and B) suggest moderate or slight shifts such that *P. paniscus* has relatively long hindlimb segments (particularly for the proximal segments). However, comparisons within either the forelimb or hindlimb seem to suggest ontogenetic scaling (C and D). While the growth allometry of the forelimb is truncated in *P. paniscus* relative to *P. troglodytes*, the ontogenetic pattern for the hindlimb appears identical in the two species.

bearing from the foothold point (Cartmill, 1974; Jungers, 1978). Therefore, a decrease in leg length would proportionately reduce such rotational torques—this would be of biomechanical advantage whether the forelimbs and/or the hindlimbs provide the major propulsive force during climbing. The differences in relative hindlimb length between the chimpanzees is clarified when gorillas are included in the analysis (Shea, 1981a). Strong negative allometry of hindlimb length is the cause of the progressively increasing intermembral index among the African apes. It is because we can also apply this argument to a gibbon-siamang comparison (Jungers, 1984) that the differences in arm/leg proportions referred to in Fig. 1 might be explained as the result of allometry (i.e., biomechanical scaling).

Shea (1981a) and Aiello (1981) have suggested different explanations for why hindlimb shortening might be biomechanically required as size increased in a *terrestrial* environment. I have noted that these proportion shifts would be of advantage in forms that frequently adopt postures of truncal erectness (Shea, 1981a), as the African apes do. Aiello (1981), extending an argument made by Stern (1976), has suggested that an increase in the intermembral index with increasing body size would result in a proportionately greater amount of body weight being supported by the forelimbs. This, she argues, would be of biomechanical advantage during terrestrial quadrupedalism. Both of the above arguments have more relevance to a gorilla-chimpanzee contrast rather than a *P. paniscus-P. troglodytes* comparison, however.

Other scaling factors are potentially related to differences in locomotor patterns between the chimpanzee species, although additional data on captive and natural locomotion are badly needed (Susman *et al.*, 1980). A tendency toward increased arboreal behavior (such as swinging or leaping) might be expected in the smaller species of a related series. Falls from the aboreal environment become increasingly dangerous as size increases (Cartmill and Milton, 1977); thus young pongids are much more agile in the trees than are adults. Fleagle and Mittermeier's (1980) study of South American monkeys of various body sizes provides additional empirical evidence. Furthermore, Taylor *et al.* (1972) have shown that the *relative* difference between energy required for vertical and horizontal locomotion increases as body size becomes larger. Thus, we might expect the series of the African apes to display an increasing tendency toward terrestrial locomotion, and a decreasing proclivity for arboreal swinging and leaping, as body size increases. This prediction agrees with observed locomotor behavior (e.g., Schaller, 1963; Susman *et al.*, 1980).

According to the procedure outlined above, if one rejects the present argument based on biomechanical scaling, then an alternative adaptive explanation for the limb proportions and relatively elongate hindlimbs of

P. paniscus is required. Zihlman (1980, 1981) has indeed suggested that these differences might reflect a greater dependence on bipedal locomotion in the pygmy chimpanzee, although there are no good behavioral or morphological observations to strongly support this claim at the present time (Susman and Jungers, 1981). Of course, if the *P. paniscus* condition is primitive, an explanation of the form differences in terms of allometry (biomechanical scaling) need only account for the relative hindlimb shortening in *P. troglodytes*. Relative proportion differences could then predispose the species to slightly different behavioral patterns (thus the body proportions of the gibbon may predispose it to bipedal behavior, but it seems unlikely that these proportions were specifically selected for reasons having to do with bipedalism). Therefore, one could argue that the behavior/anatomy argument made by Zihlman in regard to limb proportion differences and the present explanation based on biomechanical scaling are not mutually exclusive.

Whether these morphological differences are due to biomechanical scaling or nonallometric factors, the relative differences in hindlimb length provide one of the most important shape distinctions between the chimpanzee species. We might as well refer to the pygmy chimpanzee as the "long-legged" chimpanzee; or, if *P. troglodytes* is derived from a form with pygmy chimpanzee-like proportions, the common chimpanzees (and especially the gorillas) are "short-legged" apes (see also Aiello, 1981). It would seem that a decrease in relative hindlimb length was, along with marked size increase, a fundamental factor in the evolution of the African apes (Shea, 1981a). Developmentally, this trend of relative hindlimb shortening could have been produced by a minor alteration in limb growth patterns. A simple shift such that hindlimb growth began at somewhat larger overall body size (and at a more advanced stage) in descendant forms would produce the observed downward "transpositioning" in bivariate plots. This alteration has been labeled "post-displacement" by Alberch *et al.* (1979). Jungers (1978) has reviewed some of the genetic evidence relevant to hindlimb shortening.

Body proportion differences between *P. paniscus* and *P. troglodytes* might also be considered in light of McMahon's (1973, 1975a,b, 1980) model of "elastic similarity scaling." McMahon proposes that body proportions should alter as size increases so that larger forms are elastically similar to smaller species, i.e., they maintain equal resistance to buckling or bending under their own weight. Additional elaboration can be found in McMahon's articles, as well as in applications or rebuttals (e.g., Alexander *et al.*, 1979; Maloiy *et al.*, 1979; Shea, 1981a), but it is important to note that the model sets up theoretical expectations of biomechanical scaling. The two chimpanzee species represent much too small a sample and far too restricted a weight range to provide a good test for McMahon's model. However, I have noted (Shea, 1981*a*) that the African pongids as a group fit the expectations of McMahon's model quite well. Body height increases less quickly, and chest girth more quickly, than does overall weight in the three species *(P. paniscus, P. troglodytes,* and *G. gorilla).* This statement also holds for a more restricted comparison of just the two chimpanzee forms. Again, the strongest departure from the prediction of elastic similarity scaling is in the strong negative allometry of hindlimb length.

6. Cranial Allometry

Detailed discussions of cranial allometry in the African apes are given in Shea (1982, 1983*b*). I will concentrate more specifically here on the relationship between the two chimpanzee species. A number of detailed craniometric studies have compared craniofacial growth and morphology in *P. paniscus* and *P. troglodytes* (e.g., Weidenreich, 1941; Fenart and Deblock, 1972, 1973; Cramer, 1977; Shea, 1982). These have clearly demonstrated some significant differences in overall size and shape. The pygmy chimpanzee has a more gracile and rounded skull, with less development of the "facial mass," less marked sexual dimorphism, and a foramen magnum positioned more centrally than in *P. troglodytes.* My own craniometric studies (Shea, 1982, 1983*b*) have focused on comparison of ontogenetic allometries in the two species. This approach not only allows an adequate assessment of the role of size differences in affecting craniofacial proportions, but it also permits analyses based on processes of transformation rather than static adult morphology or simple description of growth changes. As Alberch and Alberch (1981, p. 263) have noted, "morphology must be viewed as the end result of a dynamic ontogenetic process, rather than as a static entity, if we wish to interpret morphological transformations in a phylogenetic context."

The growth allometries of the craniofacial region in the two species fall along a single continuum for most major dimensions considered (Shea, 1982). Examples are provided in Fig. 9. In these cases, adult pygmy chimpanzees have the cranial proportions observed in subadult *P. troglodytes;* the simple extension of a common growth trajectory to larger sizes accounts for the shape differences between the adults of the respective species. These ontogenetic studies greatly clarify the differences in skull shape between *P. paniscus* and *P. troglodytes.* A simple positive or negative (depending on whether small size is primitive or derived in this case) shift in ultimate size attained accounts for many of the multiple

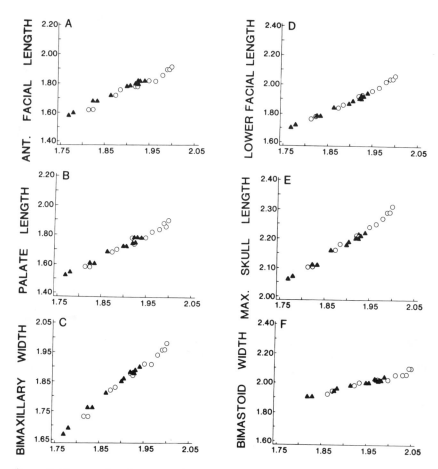

Figure 9. Six examples of ontogenetic scaling of cranial proportions in *P. paniscus* and *P. troglodytes*. The *x* axis in all cases equals basicranial length (basion–nasion). For these comparisons, adult pygmy chimpanzees have the proportions observed in subadult common chimpanzees of comparable skull sizes. See text for discussion.

differences in cranial shape. Again, it is of interest that many growth allometries of the gorilla skull also follow this pattern (Shea, 1982, 1983*b*, 1984). The results reported here are supported by the conclusions of Ramboux (1981), who has completed a similar comparative growth study of chimpanzees and gorillas.

A visual assessment of the size/shape relations in cranial form in *P. paniscus* and *P. troglodytes* is provided in Fig. 10. These are cranial "maps" based on coordinate point locations as described in Shea (1982).

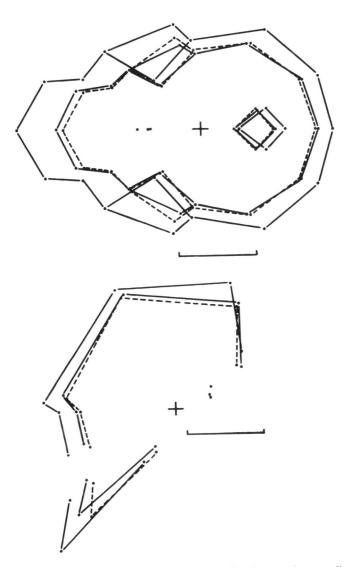

Figure 10. Cranial maps in lateral and basal perspective drawn using coordinate point locations as detailed in Shea (1982). Dashed outline is for adult male *P. paniscus*; smaller and larger solid outlines represent subadult and adult male *P. troglodytes*, respectively. The overall size and shape similarities between adult pygmy chimpanzees and subadult common chimpanzees is clearly indicated. Some of the cranial length/width differences metrically illustrated in Fig. 9 may also be discerned here.

The three outlines from basal and lateral perspective are for subadult and adult male *P. troglodytes* and adult male *P. paniscus*. The overall similarity in size and shape between adult pygmy chimpanzees and subadult common chimpanzees is striking. This figure illustrates in a more composite fashion what the bivariate regressions reveal quantitatively—pygmy and common chimpanzees of comparable skull size are very similar in overall shape.

Not all of the 55 cranial dimensions are ontogenetically scaled as in Fig. 9. As Rensch (1948) has noted, such a total concordance of growth allometries would hardly be expected. Those growth allometries of the skull that differ between *P. paniscus* and *P. troglodytes* suggest an interesting conclusion. Several of these growth allometries are illustrated in Fig. 11. A principal-components analysis of variables "regression-adjusted" against the bivariate growth regressions of *Pan paniscus* summarizes variation in cranial form resulting from divergent growth patterns in the chimpanzees (Fig. 12). Those variables most responsible for the

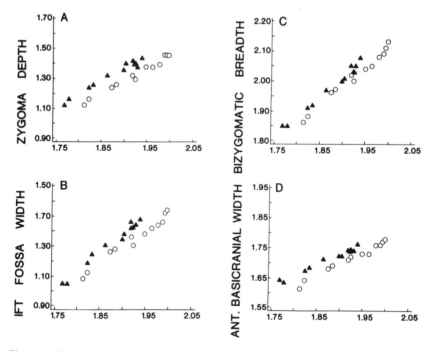

Figure 11. Four examples of bivariate shifts of ontogenetic allometries in *P. paniscus* and *P. troglodytes*. The *x* axis in all cases equals basicranial length (basion–nasion). At comparable skull sizes, pygmy chimpanzees exhibit relatively broader dimensions in these examples. See text for discussion.

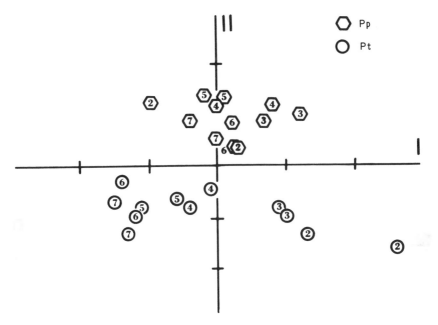

Figure 12. Results of a principal-components analysis on the covariance matrix of values of 55 cranial dimensions "regression-adjusted" relative to values predicted by the *P. paniscus* growth allometries. Component I accounts for 44.3% of the total variance, and it appears to summarize aspects of slope deviations between the species' growth allometries. Component II accounts for 27.1% of the total variance, effects a separation of the two chimpanzee species, and summarizes shape variation due to shifts or transpositions of the bivariate regressions. This analysis clearly demonstrates that not all growth allometries are ontogenetically scaled in the two species. See Shea (1982) for additional discussion.

species separation along axis II yield vertical shifts or "transpositions" in the bivariate plots (Shea, 1982). In all cases where the growth allometries of the pygmy chimpanzee diverge from those of common chimpanzees, it is in the direction of a *relatively* larger or more robust cranium (the sole exception is for dimensions of the foramen magnum or cranial vault). At a common basicranial length (where we actually compare chimpanzees of different *ages* but the same size), pygmy chimpanzees have greater cranial widths, larger infratemporal fossae, and a larger zygomatic root. Common chimpanzee crania are less robust in these features than we would expect in pygmy chimpanzees ontogentically scaled to their sizes. This implies specific selection for altered proportions, particularly because the majority of facial and cranial dimensions exhibit no significant shifts in their growth patterns. In almost all cases where the chimpanzee growth allometries differ in position, the gorillas diverge in the direction of *P. troglodytes*, only to a more exaggerated degree. Thus, allometric control leads us to exactly the opposite conclusion than that normally made concerning cranial morphology in the African ages—the pygmy chimpanzees are *relatively* the most robust, while the gorillas are *relatively* the most gracile. It is the huge difference in adult body size that obscures this finding. It should be remembered that these conclusions are relative to the expectations of ontogenetic scaling; as in the traditional usage, an interspecific adult trend of positive allometry implies relative increase (regardless of the changes in underlying growth allometries), and interspecific negative allometry implies relative decrease in size of the feature being considered.

The inclusion of *Pongo* in these analyses should prove an interesting addition, but I would predict that for those growth allometries where *P. paniscus* and *P. troglodytes* differ, the pygmy chimpanzees tend to differ in the direction of the orangutan, while *P. troglodytes* and *Gorilla* diverge in the opposite direction. This is based on the apparent finding that *Pongo* has a relatively shorter, broader, and deeper skull than the African apes (Krogman, 1931a,b,c; Dean and Wood, 1981). This is not to suggest that cranial morphology in *P. paniscus* is particularly orang-like, but it might lend some support to the notion that skull form in pygmy chimpanzees is more generalized than in the larger African apes. It would also provide support for claims that the common ancestor of the large-bodied hominoids resembled about equally *P. paniscus, Pongo,* and *Australopithecus* (Pilbeam, 1979), and that the larger African apes represent a highly derived series (Wolpoff, 1982). This possibility requires additional investigation.

The scaling of brain size in the chimpanzees has been examined by Pilbeam and Gould (1974) and Shea (1983a). Common chimpanzees have absolutely larger brains (Cramer, 1977) and larger bodies, but their brain size scales at approximately the one-third power of body weight in a comparison of the two species. This exponent is frequently observed in closely related species differing primarily in body size (Jerison, 1973, 1979; Gould, 1975a; Pilbeam and Gould, 1974; Lande, 1979; Shea, 1983a). I have argued elsewhere (Shea, 1983a) that we would expect lower than "normal" interspecific scaling coefficients (i.e., the two-thirds power of weight) when size increase has been produced via terminal additions (Rensch 1959), or an increase in size affecting primarily *postnatal* growth periods. This is true of the size differences in the chimpanzee species, as the previous growth and size plots demonstrate. Differences in absolute brain size or encephalization quotients should not be taken as evidence of significant adaptive distinctions between pygmy and common chimpanzees, but rather they are indicative of the maintenance of functional equivalence with size change (Jerison, 1979).

7. Dental Allometry

Absolute differences in the size of the deciduous and permanent dentition between *P. troglodytes* and *P. paniscus* have been detailed by Remane (1962), Johanson (1974), Shea (1982), and Kinzey (1971). The only previous study that has focused on scaling of the dentition in the chimpanzees is that of Pilbeam and Gould (1974), and that investigation is based on species means of body weight and postcanine tooth area. I have analyzed dental allometry in the chimpanzees (in addition to gorillas) in greater detail in Shea (1981*b*, 1982, 1983*b*).

Two issues are of direct relevance here. First, given the overall size differences between the chimpanzee species, are the differences in tooth size the result of allometry? And second, does the interspecific scaling relationship provide any insights into hypotheses of evolutionary transformations, specifically processes of dwarfing? In the first case, we must again distinguish between different meanings of allometric scaling, although we have no direct case of ontogenetic allometry, since teeth do not grow continuously like many other body structures. Therefore, I have used the curve of static adult allometry to set up an *intra*specific criterion of subraction for "ontogenetic" scaling. The dangers of this approach have been outlined elsewhere (Cock, 1966; Shea, 1983*a*), and additional investigations of true ontogenetic approaches to dental scaling are required (e.g., Cochard, 1981).

A substantial debate has arisen over the issue of biochemical scaling of the dentition. Primary contributions include those of Kay (1975*a,b*, 1978), Gould (1974, 1975*b*), Pilbeam and Gould (1974, 1975), Pirie (1978), Goldstein *et al.* (1978), Creighton (1980), Corruccini and Henderson (1978), Marshall and Corruccini (1978), Hylander (1975), Wood (1979*a,b*), Wood and Stack (1980), Gingerich (1977), Gingerich *et al.* (1982), Harvey *et al.* (1978), Wolfpoff (1978), and Smith (1981). A more complete discussion is provided in Shea (1982), but much of the present debate focuses on the original proposition of Gould (1974, 1975*b;* Pilbeam and Gould, 1974, 1975) that positive allometry of the postcanine dentition is expected in order to maintain functional equivalence (i.e., interspecific positive allometry is an example of biomechanical scaling). In fact, although problems with samples, statistical approaches, and control for dietary differences exist,

Table II. Regression Analysis of Some Dental Dimensions in *Pan paniscus* (Pp) and *Pan troglodytes* (Pt)

Measurement y against basicranial length x in logarithms	*Pan* slope[a] (Pp/Pt)	*Pan troglodytes* slope[b] Pt	Significance of difference[c]
Temporal fossa area	2.16 (1.75–2.57)	2.06 (1.16–2.96)	NS
	2.95 (0.73)	3.89 (0.53)	
Incisal chord	1.05 (0.89–1.21)	0.48 (0.21–0.75)	0.01
	1.30 (0.81)	1.09 (0.44)	
Incisor length	0.64 (0.38–0.89)	—[d]	—
	1.27 (0.50)		
Canine area	3.28 (2.60–3.98)	2.20 (0.74–3.66)	0.01
	4.62 (0.71)	5.79 (0.38)	
Canine length	1.60 (1.25–1.99)	0.40 (0.38–1.17)	0.01
	2.32 (0.68)	1.00 (0.40)	
Postcanine length	0.68 (0.51–0.85)	—[d]	—
	1.05 (0.66)		
Premolar area	1.31 (0.96–1.66)	—[d]	—
	2.11 (0.62)		
Molar area	1.45 (1.13–1.77)	—[d]	—
	2.10 (0.69)		
Postcanine area	1.36 (1.05–1.67)	—[d]	—
	1.97 (0.69)		
Total area	1.68 (1.32–2.04)	0.76 (0.19–1.32)	0.01
	2.26 (0.74)	2.13 (0.36)	

[a]Sample of *P. paniscus* and *P. troglodytes* adults. Top value of each pair is the least-squares slope (95% confidence interval); bottom value is the reduced major axis slope (correlation coefficient).
[b]Sample of *P. troglodytes* adults. Top and bottom values as in preceding column.
[c]Test for least-square slope differences.
[d]Nonsignificant correlation coefficient.

the bulk of the evidence suggests that *isometry* of tooth size provides the appropriate interspecific criterion of subtraction (Shea, 1982).

Results of bivariate regression analyses for various dental dimensions are given in Table II. Regression slopes were fitted to the *P. troglodytes* adult scatter, and then to the joint (*P. paniscus* plus *P. troglodytes*) adult scatter. A significant difference in the slopes theoretically indicates shape differences unrelated to simple size change. However, all of the correlation coefficients for the within-*P. troglodytes* scatter are quite low, and many are not significantly different from zero. The low correlations produce extreme differences in least squares and reduced major axis slopes for the same intraspecific scatter. Visual examination of the adult scatters (see Fig. 13) suggests no clear pattern differences between the two chimpanzee species. It thus appears likely that the differences in tooth size

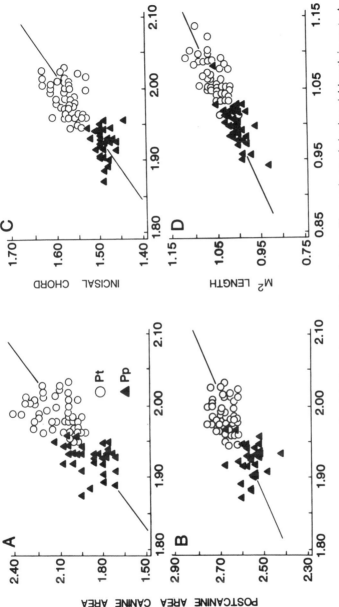

Figure 13. Plots of dental dimensions in adult *P. paniscus* and *P. troglodytes*. The *x* axis equals basicranial length in parts A–C; in D the *x* axis equals M² width. The regression line is fitted to both scatters simultaneously. No clear pattern of divergence in the scatters of the two chimpanzees emerges here (see also Table II). Dental size differences appear to correspond to differences in overall skull size.

are largely accounted for by the overall skull size differences [this is clearly not the case when we compare chimpanzees and gorillas, however (Shea, 1982, 1983*b*)]. This finding accords with recent claims that the diets in the pygmy and common chimpanzees are very similar (e.g., Horn, 1979; Badrian *et al.*, 1981). Specific arguments about crown shape (e.g., McHenry and Corruccini, 1980, 1981; Kinzey, Chapter 5, this volume) are a separate issue, of course, and require additional investigation.

In his paper on dental scaling, Gould (1975*b* suggested that lineages charcterized by "dwarfing" exhibited different (i.e., lower) coefficients of interspecific allometry than standard interspecific series. In other words, negative allometry of the postcanine dentition yielded *relatively larger* teeth in dwarfs. Gould further suggested that this scaling trend might be utilized to recognize dwarfed forms in neontological and paleontological series. Marshall and Corruccini (1978) examined dental allometry in dwarfed marsupial lineages. In addition to their empirical findings, they stressed that different types of dwarfing might be characterized by differing scaling exponents. Rapid, "ontogenetic" dwarfing, or truncation of ancestral growth patterns, might yield negative allometry, while long-term selection for smaller forms might produce allometric coefficients that coincide with the expectations of biomechanical scaling (isometry or positive allometry). Marshall and Corruccini's (1978) distinction between "phenotypic" dwarfing and "genotypic" dwarfing is thus analogous in some ways to the distinction between ontogenetic and biomechanical scaling.

These arguments have potential application to hypotheses of chimpanzee evolution. For example, Johnson (1981, p. 364) writes "If selection had merely eliminated the large end of the size range of the ancestral bonobo population, both the body-size differences between bonobos and common chimpanzees and the reduction in bonobo sexual dimorphism could have developed, and dental size differences should correspond to those observed by Marshall and Corruccini (1978) for within-species allometric trends." In response to Johnson's (1981) argument, both Susman and Jungers (1981) and I (Shea, 1981*b*) present data relevant to the question of dwarfing and dental allometry. First, however, it should be noted that Johnson's statement suggests that selection would have eliminated most of the (large) male ancestral population and few of the (small) females. This seems biologically unlikely. Furthermore, the use of strictly adult data to infer size changes is problematic (Shea, 1983*a*); increase or decrease into *new size ranges* (as in the present case) requires ontogenetic modification, even if it is merely extension or truncation of ancestral growth patterns. The argument based on disappearance of a portion of a previously greater size range also cannot be supported, given the demonstrated shifts in ontogenetic allometries [small adult *P. troglodytes* do not have the lower intermembral indices of *P. paniscus,* for example (Shea, 1981*a*)]. Finally, if we are to believe that pygmy chimpanzees represent what is left of an ancestral population that extended to larger body sizes, is *P. troglodytes* a remnant of this population that had the lower end of the range erased by negative selection?

My results (Shea, 1981*b*, 1982) indicate that tooth size scales with either negative allometry or isometry (depending on the dimension and line-fitting technique utilized) relative to basicranial length in the chimpanzees. Mean adult tooth surface areas and body weights as given by Pilbeam and Gould (1974) yield a coefficient greater than 0.66, but one that is not significantly different than isometry when confidence intervals are included. Susman and Jungers (1981) present data to suggest that *P. paniscus* "possess teeth smaller than would be expected on the basis of their body size alone." However, this assessment is made relative to interspecific regressions for postcanine area against body weight in hominoids, cercopithecoids, and ceboids. Hypotheses of dwarfism in *P. paniscus* relate only to the species pair *P. troglodytes/P. paniscus.* This is the appropriate level at which to carry out the analysis, and it appears that pygmy and common chimpanzees have a proportionately similar-sized postcanine tooth area *for their respective skull sizes.* Nevertheless, overall skull (and tooth) size is relatively smaller in *Pan paniscus* (see Section 9). This supports the conclusions of Susman and Jungers (1981).

Finally, the most recent analysis of dental scaling in dwarfing lineages indicates that Gould's (1975*b*) original prediction may not be viable in any case (Prothero and Sereno, 1982; see also Johnson, 1981) or at least that it is in need of quantitative support. Prothero and Sereno (1982) find that dental scaling does not depart significantly from isometry in the dwarfing lineages they examined. It appears that dental scaling can provide us no special insights into questions about the direction of size change during chimpanzee evolution. These must be addressed by use of (presently nonexistent) fossil evidence and the determination of derived features (Susman, 1980).

8. Heterochrony and Morphology

A brief introduction to DeBeer's (1958) categories and processes of heterochrony was given in Section 3. More detailed analyses are presented in Gould (1977), Alberch *et al.* (1979), and Shea (1983*d*). The size and shape information provided in allometric analyses can serve as the central metric with which to judge heterochronic hypotheses. Again, it is important to stress that the processes of neoteny (which yields paedomorphosis) and acceleration (which yields peramorphosis) require dissociation of ancestral allometries, and thus a clear departure from ontogenetic scaling. Such alterations indicate selection for specific morphological configurations (Rensch, 1948), presumably in response to external, environmental factors. By contrast, the processes of rate and time hypo- and hypermorphosis are characterized by the extension of truncation of a common growth pattern. In these cases, selection may have acted to alter either length of ontogeny in time, or rate of body-weight growth and ultimate size attained [see Shea (1983*d*) for a more detailed analysis of rate and time hypo- and hypermorphosis].

Certain of the postcranial and cranial examples illustrated in the above figures are clearly ontogenetically scaled. Assuming the *P. paniscus* pattern is primitive, the morphology observed in adult *P. troglodytes* is peramorphic, for these are characterized by "shapes beyond" the ancestral configuration (Alberch *et al.*, 1979). Conversely, if the *P. paniscus* pattern is derived, the morphology exhibited by the adult pygmy chimpanzee is paedomorphic, for it resembles the sizes and shapes characterizing subadult common chimpanzees. In the first case, we observe recapitulation; in the second, "reverse recapitulation" (Gould, 1977). An allometric plot provides no reliable information on time or age, however, and therefore we cannot distinguish between what I have labeled rate hypo- or hypermorphosis, and time hypo- and hypermorphosis. In both cases, ontogenetic allometries are simply extended or truncated in descendants, but with time hypo- or hypermorphosis, the length of ontogeny *in time* has been shortened or lengthened, respectively, while ancestral rates of body-weight growth and ontogenetic allometries remain unaltered. In rate hypo- and hypermorphosis, the rate of body-weight growth in time is decreased or increased, respectively, while length of ontogeny in time and ontogenetic allometries remain unaltered (see Fig. 3). We require data on ontogenetic allometries, patterns of growth in body weight, and length of development in time in order to distinguish among these different processes (Shea, 1983*d*). As Gould (1977) has stressed, similar morphological *results* may be produced by fundamentally different *processes,* and these may reflect different interactions among development, ecology, and natural selection.

Data on patterns of growth in body weight have been presented for the chimpanzees in Fig. 4. Data on developmental chronologies are more difficult to come by, particularly in the case of the pygmy chimpanzee. Gijzen (1975) has implied that pygmy chimpanzees may mature more quickly than common chimpanzees. Thus, if pygmy chimpanzees are derived, we could say that many of their morphological features result from the process of *time hypomorphosis* (progenesis in DeBeer's and Gould's terminology). We might then seek an ecological correlation for their precocious maturation, which yields their paedomorphic morphology. On the other hand, if pygmy chimpanzees are generalized in their growth patterns, the morphology of common chimpanzees would be peramorphic via the process of *time hypermorphosis*. Gould (1977) suggests a number of possible relationships between ecological factors and prolonged development or delayed maturity. However, Fenart and Deblock (1973, p. 19) have "visually" implied in a figure that pygmy chimpanzees mature *more slowly* than common chimpanzees by placing M_3^3 eruption in *P. paniscus* at a later time on a dental eruption scheme (see also Rode, 1941). R. Fenart (personal communication) has stressed that this was inadvertant, and was not meant to imply a longer growth period in *P. paniscus*. In fact, there are no good data to indicate that length of ontogeny in time differs significantly at all between the two chimpanzee species (Shea, 1982, 1983*d*). They appear to grow for about the same length of time, and become sexually mature at about the same age (although firm longitudinal studies of animals born in captivity will be required to confirm this). As indicated in Fig. 4, pygmy chimpanzees simply grow somewhat more slowly than *P. troglodytes*, not more slowly in terms of the coefficients of growth allometry (which are mostly very similar in the two species), but more slowly in terms of total weight gain or size increase in a given time.

9. Interspecific Dissociations

An overall assessment of proportion differences between the chimpanzees reveals an interesting combination of ontogenetic scaling and ontogenetic dissociations. That is, ontogenetic scaling *within* the major body regions appears quite pervasive (for example, skull—see Fig. 9; trunk/forelimb—see Fig. 6; hindlimb—see Fig. 8), but comparisons among these regions reveal dissociations, or uncoupling of the ontogenetic patterns between the species (Shea, 1983*c*). The dissociation of a hindlimb versus trunk or forelimb regression, and its potential functional bases, has been discussed in Section 5. A plot of skull size versus trunk or forelimb dimensions indicates a strong deviation of adult *Pan paniscus* from the ontogenetic allometry of *Pan troglodytes*. In other words, at a given trunk length (measured here by the head-to-fork distance), pygmy chimpanzees have relatively small crania. This is even more true of a comparison between skull size and hindlimb length.

Absolute dimensions support this conclusion and provide the following results. Average skull size in *Pan paniscus* falls roughly below dental age 4 and 5 of *Pan troglodytes*, while the trunk, hindlimbs, and, most likely, body weight in pygmy chimpanzees fall between dental age 5 and 6 (young adults) of the common chimpanzees (Shea, 1983c). Given the lack of significant differences in hindlimb length, this indicates a "gradient" of size/shape differences in the chimpanzees—strong in the craniodental region, moderate in the trunk and forelimb region, and nonexistent for the hindlimbs. [Thus, as several workers have noted (McHenry and Corruccini, 1981; Zihlman, 1982), the skull and teeth in *Pan paniscus* appear disproportionately small in a general comparison between adults of the two species.] Because of the strong ontogenetic scaling within each of these regions, however, a *differential* paedomorphosis or peramorphosis (depending on the direction of size change) is observed between these chimpanzee species. The strongly paedomorphic skull of *P. paniscus* results from the heterochronic process of *neoteny*, since the (presumptive) ancestral growth trajectory of skull vs. body is dissociated and rates of shape change are retarded in the pygmy chimpanzee (i.e., decrease in regression slope value).

The most intriguing question, however, is why the skull of *P. paniscus* is so markedly reduced in size (with concomitant juvenilization of shape) relative to the body as a whole. I have tentatively concluded (Shea, 1983c) that the neoteny and paedomorphosis of the skull in the pygmy chimpanzee is related to sexual dimorphism in both morphology and behavior. Schultz (1962) demonstrated that the greater the percentage change in the size of a structure during growth, the greater the degree of sexual dimorphism in that structure, arguing (Schultz, 1969, p. 203) that . . .

> "sexual dimorphism is least pronounced in the skulls of such species—as well as in such cranial features—as change least during growth and vice versa. . . . It appears quite likely that this may turn out to represent a generally valid rule, according to which the more infantile a bodily feature remains the less marked will become its sexual distinction in adults and the smaller the postnatal changes of a species are the less conspicuous remain its sex differences."

The retardation of cranial shape change and the truncation to smaller overall size in *P. paniscus* clearly yields a skull with reduced sexual dimorphism (Deblock and Fenart, 1972; Fenart and Deblock, 1972, 1973, 1974; Cramer and Zihlman, 1978; Cramer, 1977; Johanson, 1974; Almquist, 1974; Shea, 1982).

Recent behavioral field work may provide a link with the morphological results reported here. Field studies of *P. paniscus* by Japanese primatologists have indicated that pygmy chimpanzees are characterized by high male/female affinity and less sexual differentiation in social structure than is characteristic of *P. troglodytes* (Kuroda, 1979; Kano, 1980).

Interestingly, in a later paper, Kuroda (1980, p. 181) compares the behavior of adult *P. paniscus* to subadult *P. troglodytes:*

"Coolidge (1933) pointed out that the pygmy chimp preserved the paedomorphic traits in its morphology even after maturity. Similar preservation of infantile traits after maturity may be plausibly assumed in its behavioral patterns. Among the higher primates there is seen a tendency that the period of maturity is prolonged and the adult preserves more infantile behavior patterns (Bruner, 1976). The pygmy chimps copulate frequently in the ventro-ventral posture, and the GG (genito-genital) rubbing is also in the ventro-ventral; in primates in general, this posture is often observed in immatures' copulation and plays but rare in the adults. The plays among the adults of the pygmy chimps and their frequent food-sharing behavior may be interpreted as part of the remnant of the immature's behavior patterns. The little social differentiation by sex in the pygmy chimps can also be interpreted as one of the pedomorphism."

Although some sexual differentiation (particularly of adult canine size) exists in *P. paniscus*, this is significantly less marked than in the common chimpanzees (see above). Therefore, I suggest that the heterochronic shift yielding paedomorphic skull form and reduced sexual dimorphism in the dentofacial complex in *P. paniscus* may be directly related to the evolution of the social and behavioral differences noted here. The face and teeth are presumably differentially affected (i.e., more paedomorphic than the rest of the body) due to their prominent role in sociosexual recognition and differentiation. The reverse case would hold for a hypothetical *P. paniscus* to *P. troglodytes* transition, of course. Shea (1983c), provides a more detailed discussion of these dissociations between pygmy and common chimpanzees. This differential size and shape change among the body regions is at the root of much of the confusion and debate over the size differences between these chimpanzee species.

10. Conclusions

In general, the evidence suggests that the primary proportion differences of the postcranial and craniodental regions in the pygmy and common chimpanzees are the result of allometry—either ontogenetic scaling or biomechanical scaling, as noted. I do not believe that this is using allometry as an obfuscating "fudge factor." It is simply the conclusion I come to after rigorous application of the different levels of allometric analysis. Furthermore, I think we can all agree that "allometric" differences are of no greater (or lesser) intrinsic biological interest than those proportion differences unrelated to size per se.

Integration of analyses of relative growth and heterochrony reveals that many of the morphological distinctions between the chimpanzee species are the result of rate hypomorphosis or hypermorphosis (Shea, 1983c). In such cases, ontogenetic scaling of proportions occurs, and shape differences result from selection for a change in rate of growth of the body as a whole, or of specific components of the body. Elsewhere (Shea, 1983c), I have shown that the paedomorphic skull of *P. paniscus* results from neoteny, or decreased rates of shape change in the skull relative to overall body size.

The conclusion that many shape differences in a comparison of *P. troglodytes* and *P. paniscus* result from ontogenetic scaling provides no particular insight into the important question of primitive and derived status. The frequency of neoteny, rate hypomorphosis, and time hypomorphosis (progenesis) argues against the simple assumption of "ontogenetic priority" (Nelson, 1978; Gould, 1977; Alberch *et al.*, 1979). As Bretsky (1979, p. 148) has noted, "recognizing morphological differences between species as correlated with size differences neither gives an unambiguous delineation of the polarity of ancestor–descendant relationship between two sister species nor eliminates the possibility of both species being descended from an unknown common ancestor." In addition, just as no individual species will exhibit either a totally primitive or derived set of character states relative to another closely related form, no species will likely represent in its entirety a paedomorphic or peramorphic version of its ancestor (Fink, 1982; Shea, 1983d). Many of the most vexing questions about chimpanzee evolution and morphology must await the discovery of relevant fossils. In this light, the extremely poor fossil record of the African apes is unfortunate, for, as has become apparent in the past few years, a cogent theory of human origins requires a concomitant understanding of the evolution of the great apes (Zihlman, 1979; Wolpoff, 1982).

ACKNOWLEDGMENTS. The research reported on here is part of a Ph.D. thesis from Duke University completed in 1982. For help in that endeavor, I thank R. F. Kay, M. Cartmill, W. Hylander, E. Simons, K. Glander, and E. Mertz. The assistance of Lester Barton at the Powell-Cotton Museum (Birchington, England) and Dr. Thys van den Audenaerde at the Central African Museum (Tervuren, Belgium) greatly facilitated my research. Financial support was provided by the National Science Foundation, Duke University Graduate School, the Society of Sigma Xi, and the Southern Regional Educational Board.

I would also like to express my gratitude to R. L. Susman for inviting me to participate in the Atlanta IPS symposium on *Pan paniscus*, and to contribute to this volume. This chapter is publication #7 from the Richard Lounsbery Laboratory of Physical Anthropology at the American Museum of Natural History in New York. The financial support of a Lounsbery Postdoctoral Research Fellowship is gratefully acknowledged. I would also like to express my appreciation to Pat Bramwell, who typed the manuscript.

References

Aiello, L. C., 1981, Locomotion in the Miocene Hominoidea, in: *Aspects of Human Evolution* (C. B. Stringer, ed.), Taylor and Francis, London, pp. 63–97.

Alberch, P., and Alberch, J., 1981, Heterochronic mechanisms of morphological diversification and evolutionary change in the neotropical salamander, *Bolitoglossa occidentalis*, *J. Morpho.* **161**:249–264.

Alberch, P., Gould, S. J., Oster, G. F., and Wake, D. B., 1979, Size and shape in ontogeny and phylogeny, *Paleobiology* **5**:296–317.

Alexander, R. M., Jayes, A. S., Maloiy, G. M. O., and Wathuta, E. M., 1979, Allometry of the limb bones of mammals from shrews *(Sorex)* to elephants *(Loxodonta)*, *J. Zool. Lond.* **189**:305–314.

Almquist, A. J., 1974, Sexual differences in the anterior dentition in African primates, *Amer. J. Phys. Anthropol.* **40**:359–367.

Ashton, E. H., 1954, Age changes in some bodily dimensions in apes, *Proc. Zool. Soc. Lond.* **124**:587–594.

Atchley, W., Rutledge, J. J., and Cowley, E., 1981, Genetic components of size and shape. II. Multivariate covariance patterns in the rat and mouse skull, *Evolution* **35**:1037–1055.

Badrian, N., Badrian, A., and Susman, R. L., 1981, Preliminary observations on the feeding behavior of *Pan paniscus* in the Lomako Forest of Central Zaire, *Primates* **22**:173–181.

Benirschke, K. M., Bogart, H., and Adams, F., 1980, The status of the pygmy chimpanzee in the U.S.A., *Int. Zoo Yearb.* **20**:71–76.

Bock, W. J., and Winkler, H., 1978, Mechanical analysis of the external forces on climbing mammals, *Zoomorphologie* **91**:49–61.

Bretsky, S., 1979, Recognition of ancestor-descendant relationships in invertebrate paleontology, in: *Phylogenetic Analysis and Paleontology* (J. Cracraft and N. Eldredge, eds.), Columbia University Press, New York, pp. 113–164.

Bruner, J. S., 1976, Nature and uses of immaturity, in: *Play* (J. S. Bruner, A. Jolly, and K. Sylva, eds.), Penguin Books, Harmondsworth, Middlesex, pp. 28–64.

Cartmill, M., 1974, Pads and claws in arboreal locomotion, in: *Primate Locomotion* (F. A. Jenkins, Jr., ed.), Academic Press, New York.

Cartmill, M., and Milton, K., 1977, The lorisiform wrist joint and the evolution of "brachiating" adaptations in the Hominoidea, *Am. J. Phys. Anthropol.* **47**:249–272.

Cheverud, J. M., 1982, Relationships among ontogenetic, static, and evolutionary allometry, *Am. J. Phys. Anthropol.* **59**:139–149.

Cochard, L. R., 1981, Ontogenetic and intersexual dental allometry in the rhesus monkey *(Macaca mulatta)*, *Am. J. Phys. Anthropol.* **54**:210 (Abstract).

Cock, A. G., 1966, Genetical aspects of metrical growth and form in animals, *Q. Rev. Biol.* **41**:131–190.

Coolidge, H. J., 1933, *Pan paniscus:* Pygmy chimpanzee from south of the Congo River, *Am. J. Phys. Anthropol.* **18**(1):1–57.

Coolidge, H. J., and Shea, B. T., 1982, External body dimensions of *Pan paniscus* and *Pan troglodytes* chimpanzees, *Primates* **23**:245–251.

Corruccini, R. S., and Henderson A. M., 1978, Multivariate dental allometry in primates, *Am. J. Phys. Anthropol.* **48**:203–208.

Corruccini, R. S., and McHenry, H. M., 1979, Morphological affinities of *Pan paniscus*, *Science* **204**:1341–1343.

Cramer, D. L., 1977, Craniofacial morphology of *Pan paniscus*, *Contrib. Primatol* **10**:1–64.

Cramer, D. L., and Zihlman, A. L., 1978, Sexual dimorphism in the pygmy chimpanzee, *Pan paniscus*, in: *Recent Advances in Primatory*, Volume 3 (D. J. Chivers and K. A. Joysey, eds.), Academic Press, London, pp. 487–490.

Creighton, G. K., 1980, Static allometry of mammalian teeth and the correlation of tooth size and body size in contemporary mammals, *J. Zool. Lond.* **191**:435–443.

Dean, J. C., and Wood, B. A., 1981, Metrical analysis of the basicranium of extant hominoids and *Australopithecus*, *Am. J. Phys. Anthropol.* **54**:63–71.

DeBeer, G. R., 1958, *Embryos and Ancestors*, rev. ed., Clarendon Press, Oxford.

Deblock, R., and R. Fenart, 1972, Differences specifiques craniennes, entre *Pan troglodytes* et Pan paniscus, *Sciences* **III(2)**:162–167.

Fenart, R., and Deblock, R., 1972, Sexual difference in adult skulls of *Pan paniscus*, in: *Medical Primatology*, 1972, part I, pp. 342–348.

Fenart, R., and Deblock, R., 1973, *Pan paniscus* et Pan troglodytes craniométrie: Étude comparative et ontogénique selon les méthodes classiques et vestibulaires. Tome I, *Mus. R. Afr. Centr. Tervuren Belg. Ann. Octavo Ser. Sci. Zool.* **204**:1–473.

Fenart, R., and R. Deblock, 1974, Sexual differences in adult skulls in *Pan troglodytes*, *J. Hum. Evol.* **3**:123–133.

Fink, W. L., 1982, The conceptual relationship between ontogeny and phylogeny, *Paleobiol* **4**:7–76.

Fleagle, J. G., and Mittermeier, R. A., 1980, Locomotor behavior, body size, and comparative ecology of seven Surinam monkeys, *Am. J. Phys. Anthropol.* **52**:301–314.

Frechkop, S., 1935, Notes sur les mammifères. XVII; A propos du chimpanzé de la rive gauche du Congo, *Mus. R. Hist. Nat. Belg. Bull.* **11**:1–41.

Gijzen, A., 1975, Studbook of *Pan paniscus*, Schwarz, 1929, *Acta Zool. Pathol. Antverp.* **61**:119–164.

Gingerich, P. D. 1977, Correlation of tooth size and body size in living hominoid primates with a note on relative brain size in *Aegyptopithecus* and *Proconsul*, *Am. J. Phys. Anthropol.* **47**:395–398.

Gingerich, P. D. Smith, B. H., and Rosenberg, K., 1982, Scaling in the dentition of primates and prediction of body weight from tooth size in fossils, *Am. J. Phys. Anthropol.* **58**:81–100.

Goldstein, S., Post, D., and Melnick, D., 1978, An analysis of Cercopithecoid odontometrics, *Am. J. Phys. Anthropol.* **49**:517–532.

Gould, S. J., 1966, Allometry and size in ontogeny and phylogeny, *Biol. Rev.* **41**:587–640.

Gould, S. J., 1974, The nonscience of human nature, *Nat. Hist.* **83(4)**:21–25.

Gould, S. J. 1975a, Allometry in primates, with emphasis on scaling and the evolution of the brain, *Contrib. Primatol.* **5**:244–292.

Gould, S. J., 1975b, On the scaling of tooth size in mammals, *Am. Zool.* **15**:351–362.

Gould, S. J., 1977, *Ontogeny and Phylogeny*, Harvard University Press, Cambridge, Massachusetts.

Grether, W. F., and Yerkes, R. A., 1940, Weight norms and relations for chimpanzees, *Am. J. Phys. Anthropol.* **27**:181–197.

Harvey, P. H., Kavanaugh, M., and Clutton-Brock, T. H., 1978, Sexual dimorphism in primate teeth, *J. Zool. Lond.* **186**:475–485.

Hill, W. C. O., 1967, The taxonomy of the genus *Pan*, in: *Neue Ergebnisse der Primatologie* (D. Starck, R. Schneider, and H. Kuhn, eds.), Fischer, Stuttgart, pp. 47–54.

Hill, W. C. O., 1969, The nomenclature, taxonomy, and distribution of chimpanzees, in: *The Chimpanzee*, Volume 1 (G. H. Bourne, ed.), Karger, Basel, pp. 22–49.

Horn, A. D., 1979, The taxonomic status of the bonobo chimpanzee, *Am. J. Phys. Anthropol.* **51**:273–282.

Huxley, J. S., 1932, *Problems of Relative Growth*, MacVeagh, London.

Hylander, W. L., 1975, Incisor size and diet in Anthropoids with special reference to Cercopithecidae, *Science* **189**:1095–1098.

Jerison, H. J., 1973 *Evolution of the Brain and Intelligence*, Academic Press, New York.

Jerison, H. J., 1979, The evolution of diversity in brain size, in: *Development and Evolution of Brain Size: Behavioral Implications* (M. E. Hahn, C. Jensen, and B. C. Dudek, eds.), Academic Press, New York, pp. 29–57.

Johanson, D. C., 1974, Some metric aspects of the permanent and deciduous dentition of the pygmy chimpanzee *(Pan paniscus)*, *Am. J. Phys. Anthropol.* **41**:39–48.

Johnson, S. C., 1981, Bonobos: Generalized hominid prototypes or specialized insular dwarfs? *Curr. Anthropol.* **22**:363–375.

Jungers, W. L., 1977, Hindlimb and pelvic adaptations to vertical climbing and clinging in *Megaladapis*, a giant subfossil prosimian from Madagascar, *Yearb. Phys. Anthropol.* **20**:508–525.

Jungers, W. L., 1978, The functional significance of skeletal allometry in *Megaladapis* in comparison to living prosimians, *Am. J. Phys. Anthropol.* **19**:303–314.

Jungers, W. L., 1979, Locomotion, limb proportions and skeletal allometry in lemurs and lorises, *Folia Primatol.* **32**:8–28.

Jungers, W. L., 1984, Scaling of the hominoid locomotor skeleton with special reference to lesser apes, in: *The Lesser Apes: Evolutionary and Behavioral Biology* (H. Preuschoft, D. Chivers, W. Brockelman, and N. Creel eds.), Edinburgh University Press, Edinburgh (in press).

Kano, T., 1982, The social group of pygmy chimpanzees *(Pan paniscus)* of Wamba, *Primates* **23**:171–188.

Kay, R. F., 1975a, The functional adaptations of primate molar teeth, *Am. J. Phys. Anthropol.* **43**:195–216.

Kay, R. F., 1975b, Allometry and early hominids, *Science* **189**:63.

Kay, R. F., 1978, Molar structure and diet in extant Cercopithecidae, in: *Development, Function and Evolution of Teeth* (P. M. Butler and K. A. Josey, eds.), Academic Press, London, pp. 309–339.

Kinzey, W. G., 1971, Evolution of the human canine tooth, *Am. Anthropol.* **73**:680–694.

Kortlandt, A., 1972, *New Perspectives on Ape and Human Evolution*, Stichting voor Psychobiologies, Amsterdam.

Krogman, W. M., 1931a, Studies in growth changes in the skull and face of anthropoids, III, Growth changes in the skull and face of the gorilla, *Am. J. Anat.* **47**:89–115.

Krogman, W. M., 1931b, Studies in growth changes in the skull and face of anthropoids, IV, Growth changes in the skull and face of the chimpanzee, *Am. J. Anat.* **47**:325–342.

Krogman, W. M., 1931c, Studies in growth changes in the skull and face of anthropoids, V, Growth changes in the skull and face of the orang-utan, *Am. J. Anat.* **47**:343–365.

Kuroda, S., 1979, Grouping of the pygmy chimpanzees, *Primates* **20**:161–183.

Kuroda, S., 1980, Social behavior of the pygmy chimpanzees, *Primates* **21**:181–197.

Lande, R., 1979, Quantitative genetic analysis of multivariate evolution, applied to brain:body allometry, *Evolution* **33**:402–416.

Latimer, B. M., White, T. D. Kimbel, W. H., and Johanson, D. C., 1981, The pygmy chimpanzee is not a living missing link in human evolution, *J. Hum. Evol.* **10**:475–488.

Leutenegger, W., and Kelley, J. T., 1977, Relationship of sexual dimorphism in canine size and body size to social, behavioral and ecological correlates in anthropoid primates, *Primates* **13**:365–369.

Lumer, H., 1939, Relative growth of limb bones of anthropoid apes, *Hum Biol* **11**:379–392.

MacKinnon, J., 1978, *The Ape Within Us*, Holt, Rinehart and Winston, New York.

Maloiy, G. M. O., Alexander, R. M., Njan, R., and Jayes, A. S., 1979, Allometry of the legs of running birds, *J. Zool. Lond.* **187**:161–167.

Marshall, L. G., and Behrensmeyer, A. K., ms., A review of evolutionary dwarfism.

Marshall, L. G., and Corruccini, R. S., 1978, Variability, evolutionary rates, and allometry in dwarfing lineages, *Paleobiology* **4**:101–119.

McHenry, H. M., and Corruccini, R. S., 1980, Late Tertiary hominoids and hominid origins, *Nature* **285**:397–298.

McHenry, H. M., and Corruccini, R. S., 1981, *Pan paniscus* and human evolution, *Am. J. Phys. Anthropol.* **54**:355–367.

McMahon, T. A., 1973, Size and shape in biology, *Science* **197**:1201–1204.

McMahon, T. A., 1975a, Allometry and biomechanics: Limb bones of adult ungulates, *Am. Nat.* **109**:547–563.

McMahon, T. A., 1975b, Using body to understand the structural design of animals: Quadrupedal locomotion, *J. Appl. Physiol.* **31**:619–637.

McMahon, T. A., 1980, Scaling physiological time, *Lec. Math Life Sci.* **13**:131–163.

Nelson, G., 1978, Ontogeny, phylogeny, paleontology and the biogenetic law, *Syst. Zool.* **27**:324–345.

Neugebauer, W., 1980, The status and management of the pygmy chimpanzee in European zoos, *Int. Zoo Yearb.* **20**:64–70.

Pilbeam, D. R., 1979, Recent finds and interpretations of Miocene hominoids, *Annu. Rev. Anthropol.* **8**:333–352.

Pilbeam, D. R., and Gould, S. J., 1974, Size and scaling in human evolution, *Science* **186**:892–901.

Pilbeam, D. R., and Gould, S. J., 1975, Allometry and early hominids, *Science* **189**:64.

Pirie, P. L., 1978, Allometric scaling in the postcanine dentition with reference to primate diets, *Primates* **19**:583–591.

Prothero, D. R., and Sereno, P. C., 1982, Allometry and paleoecology of medial Miocene dwarf rhinoceroses from the Texas Gulf Coastal Plain, *Paleobiol.* **8**:16 to 30.

Ramboux, A., 1981, Croissance du massif cranio-facial chez *Pan paniscus, Pan troglodytes* et *Gorilla gorilla, Bull. Soc. Belge Anthropol. Prehist.* **92**:175–190.

Remane, A., 1962, Masse und proportionen des Milchgebisses der Hominoidea, *Bibl. Primatol.* **1**:229–238.

Rensch, B., 1948, Histological changes correlated with evolutionary changes of body size, *Evolution* **2**:218–230.

Rensch, B., 1959, *Evolution above the Species Level*, Columbia University Press, New York.

Roberts, D., 1974, Structure and function of the primate scapula, in: *Primate Locomotion* (F. A. Jenkins, Jr., ed.), Academic Press, New York, pp. 171–200.

Rode, P., 1941, Etude d'un chimpanzé pygmée adolescent *(Pan satyrus paniscus,* Schwarz), *Mammalia* **5**:50–68.

Schaller, G. B., 1963, *The Mountain Gorilla: Ecology and Behavior*, University of Chicago Press, Chicago.

Schultz, A. H., 1954, Bemerkungen zur Variabilitat und Systematik der Schimpansen, *Saugetierkd. Mitt.* **2**:159–163.

Schultz, A. H., 1962, Metric age changes and sex differences in primate skulls, *Zeit. Morphol. Anthropol.* **52**:239–255.

Schultz, A. H., 1969, *The Life of Primates*, Universe Books, New York.

Shea, B. T., 1981*a*, Relative growth of the limbs and trunk in the African apes, *Am. J. Phys. Anthropol.* **56**:179–202.

Shea, B. T., 1981*b*, Comment on: Bonobos: Generalized hominid prototypes or specialized insular dwarfs?, by S. C. Johnsons, *Curr. Anthropol.* **22**:368–369.

Shea, B. T., 1982, Growth and size allometry in the African Pongidae: Cranial and postcranial analyses, Ph.D. Thesis, Duke University.

Shea, B. T., 1983*a*, Phyletic size change and brain/body scaling: A consideration based on the African pongids and other primates, *Int. J. Primatol.* **4**:33–62.

Shea, B. T., 1983*b*, Size and diet in the evolution of African ape craniodental form, *Folia Primatol.* **40**:32–68.

Shea, B. T., 1983*c*, Paedomorphosis and neoteny in the pygmy chimpanzee, *Science* **222**:521–522.

Shea, B. T., 1983*d*, Allometry and heterochrony in the African apes, *Amer. J. Phys. Anthropol.* **62**:275–289.

Shea, B. T., 1984, Ontogenetic allometry and scaling: A discussion based on growth and form of the skull in African apes, in: *Size and Scaling in Primate Biology* (W. L. Jungers, ed.), Plenum Press, New York, pp. 175–205.

Simpson, G. G., 1953, *The Major Features of Evolution*, Columbia University Press, New York.

Simpson, G. G., Roe, A., and Lewontin, R. C., 1960, *Quantitative Zoology*, 2nd ed., Harcourt, World and Brace, New York.

Smith, R. J., 1981, Interspecific scaling of maxillary canine size and shape in female primates: Relationships to social structure and diet, *J. Hum. Evol.* **10**:165–173.

Smith, A. H., Butler, T. M., and Pace, N., 1975, Weight growth of colony-reared chimpanzees, *Folia Primatol.* **24**:29–59.

Sprent, P., 1972, The mathematics of size and shape, *Biometrics* **28**:23–37.

Stern, J. T., Jr., 1976, Before bipedality, *Yearb. Phys. Anthropol.* **19**:59–68.

Susman, R. L., 1979, Comparative and functional morphology of hominoid fingers, *Am. J. Phys. Anthropl.* **50**:215–236.

Susman, R. L., 1980, Acrobatic pygmy chimpanzees, *Nat. Hist.* **89**(9):33–39.

Susman, R. L., and Jungers, W. L., 1981, Comment on: Bonobos: Generalized homonid prototypes or specialized insular dwarfs?, by S. C. Johnson, *Cur. Anthropol.* **22**:369–370.

Susman, R. L., Badrian, N., and Badrian, A., 1980, Locomotor behavior of *Pan paniscus* in Zaire, *Am. J. Phys. Anthropol.* **53**:69–80.

Taylor, C. R., Caldwell, S. L., and Rowntree, V. J., 1972, Running up and down hills: Some consequences of size, *Science* **178**:1096–1097.

Tuttle, R. H., 1975, Parallelism, brachiation and hominoid phylogeny, in: *Phylogeny of the Primates* (W. P. Luckett and F. S. Szalay, eds.), Plenum Press, New York, pp. 447–480.

Vandebroek, G., 1969, *Evolution des Vertebres de leur origine à l'homme*, Masson, Paris.

Von Bertalanffy, L., and Pirozynski, W. J., 1952, Ontogenetic and evolutionary allometry, *Evolution* **6**:387–392.

Weidenreich, F., 1941, The brain and its role in the phylogenetic transformation of the human skull, *Trans. Am. Philos. Soc.* **31**:321–442.

Wolpoff, M., 1978, Some aspects of canine size in the Australopithecines, *J. Hum. Evol.* **7**:115–126.

Wolpoff, M., 1982, *Ramapithecus* and hominid origins, *Cur. Anthropol.* **23**:501–522.

Wood, B. A., 1979*a*, An analysis of tooth and body size relationships in five primate taxa, *Folia Primatol.* **31**:187–211.

Wood, B. A., 1979*b*, Models for assessing relative canine size in fossil hominids, *J. Hum. Evol.* **8**:493–502.

Wood, B. A., and Stack, C. G., 1980, Does allometry explain the differences between "gracile" and "robust" australopithecines, *Am. J. Phys. Anthropol.* **52**:55–62.

Zihlman, A. L., 1979, Pygmy chimpanzee morphology and the interpretation of early hominids, *S. Afr. J. Sci.* **75**:165–168.

Zihlman, A. L., 1980, Locomotor behavior in pygmy and common chimpanzees, *Am. J. Phys. Anthropol.* **52**:295.

Zihlman, A. L., 1981, The pygmy chimpanzee: Living model for the ape-human ancestor?, *L. S. B. Leakey Found. News* **1981**(19):6–7.

Zihlman, A. L., 1982, *The Human Evolution Coloring Book*, Barnes and Noble, New York.

Zihlman, A. L., and Cramer, D. L., 1978, Skeletal differences between pygmy *(Pan paniscus)* and common *(Pan troglodytes)* chimpanzees, *Folia Primatol.* **29**:86–94.

Zihlman, A. L., Cronin, J. E., Cramer, D. L., and Sarich, V. M., 1978, Pygmy chimpanzee as a possible prototype for the common ancestor of humans, chimpanzees, and gorillas, *Nature* **275**:744–746.

CHAPTER 7

Body Size and Skeletal Allometry in African Apes

WILLIAM L. JUNGERS AND RANDALL L. SUSMAN

1. Introduction

The African apes represent a group of closely related primate taxa that differ substantially in adult body size. The close phylogenetic affinity of pygmy chimpanzees *(Pan paniscus)*, common chimpanzees *(Pan troglodytes)*, and gorillas *(Gorilla gorilla)* is affirmed by both molecular and morphological data. Most biomolecular studies to date, however, have been unable to resolve the chimpanzee-gorilla-human trichotomy into a definitive chimpanzee-gorilla clade that would have humans as a sister group (e.g., Sarich, 1968, and this volume; Zihlman *et al.*, 1978; Bruce and Ayala, 1979; Goodman, 1982). The evolutionary tree for humans and pongids based on cleavage maps of mitochondrial DNA (Ferris *et al.*, 1981; Templeton, 1983) is a notable exception in this regard; these data first group pygmy and common chimpanzees into a phyletic unit that is linked next to gorillas. Humans are then joined to the African ape clade before the orangutan *(Pongo pygmaeus)*. A variety of shared, unique features of the karyotypes of African apes lend strong credence to this phylogeny (Mai, 1983). Morphological analyses of the teeth and locomotor skeletons of pongids and humans [summarized in Ciochon (1983)] also corroborate the Ferris *et al.* branching sequence. Those anatomical details of the hands, especially the digits, linked functionally to knuckle-walking are perhaps the best examples of shared, derived structures common to

WILLIAM L. JUNGERS AND RANDALL L. SUSMAN • Department of Anatomical Sciences, School of Medicine, State University of New York at Stony Brook, Stony Brook, New York 11794.

all African apes (Tuttle, 1967, 1969, 1975; Susman, 1979). In fact, the overall similarity between chimpanzees and gorillas has been regarded by some authorities as so overwhelming and so obvious as to warrant only one genus, *Pan*, to accommodate all three species of African apes (e.g, Mayr, 1950; Simpson, 1963; Tuttle, 1967; Goodall and Groves, 1977; Szalay and Delson, 1979). Although we endorse the evidence favoring an especially intimate genealogical relationship among pygmy and common chimpanzees and gorillas, we follow the more conservative and traditional course of leaving the gorilla in a genus of its own, *Gorilla* (cf. Oxnard, 1978; Dixon, 1981; C. P. Groves, personal communication in Maple and Hoff, 1982).

Adult individuals of African apes range in body weight from less than 30 kg (Rahm, 1967; Zihlman and Cramer, 1978) to over 200 kg (Cousins, 1972; Groves and Stott, 1979). This range is based on body weights of wild-collected animals, and the upper limit of this range would increase considerably if captive records were included (Schaller, 1963; Cousins, 1972). External body measurements taken on numerous adult individuals further attest to the great differences in overall size between the smallest (female) representatives of *Pan* and the largest (male) representatives of *Gorilla* (Coolidge, 1933, 1936; Grizimek, 1956; Wood, 1979; Shea, 1981; Coolidge and Shea, 1982).

In many ways, therefore, the African apes conform very closely to what Davis (1962) has described as the "ideal situation" to study size-related (allometric) phenomena: "comparisons between closely related organisms with similar habits and behavior, but differing substantially in size, for then minor individual variations and differences in technique are minimized" (Davis, 1962, p. 505). The potential insights to be gained from this approach to African ape morphology and evolution have been recognized for some time. Giles (1956), for example, explored the interspecific size-related differences in the crania of great apes, and postulated an underlying similarity in the growth trajectories of gorilla and chimpanzee crania that might account for their shape differences as adults. Later, Groves (1970a) likened the gorilla to a "very large-sized chimpanzee," and Pilbeam and Gould (1974) and Gould (1975) suggested that the taxonomic series of pygmy to common chimpanzee to gorilla could be viewed as the same creature expressed at different adult body sizes. In similar fashion, Zihlman et al. (1978) characterized the African apes as stepwise "size variants in a single morphotypic series" from pygmy chimpanzees to gorillas. More recently, Shea (1981, 1983a,b) has carefully analyzed selected body proportions, relative brain size, and cranial form in African apes in an explicit allometric context; among other significant findings, his results emphasize the need for more precise phrasing of allometric questions (e.g., ontogenetic versus static adult allometry). A related allometric issue is the question of whether or not many of the differences seen in pygmy and common chimpanzees are primarily size-related. Although their respective theoretical approaches, data bases, and specific conclusions differ greatly, both McHenry and Corruccini (1981, also cf. Horn, 1979) and Shea (1981, and this volume) would answer this question in the affirmative. By contrast, Zihlman and colleagues (Zihlman, 1979; Zihlman and Cramer, 1978; Zihlman *et al.*, 1978) contend that pygmy chimpanzees are not merely smaller or scaled-down versions of common chimpanzees, and, therefore, most differences between these two congeners cannot be explained as allometric in nature.

If one wishes to address the pygmy–common chimpanzee question in meaningful allometric terms (regardless of which type of allometry is deemed most appropriate), two separate issues require prior attention. First, the two taxa being compared must be shown to differ in overall size (Huxley, 1932); without systematic size discrepancies, nonallometric explanations must be sought to explain differences in pattern. This point, as obvious as it appears to be, is most relevant in this case because there is no current consensus as to how much (if any) size separates pygmy and common chimpanzees (Zihlman and Cramer, 1978; McHenry and Corruccini, 1981; Shea, 1982, and this volume; Jungers, 1984a). Second, one must be very specific about which subspecies of common chimpanzee is being compared to *Pan paniscus*; to lump all specimens into a *Pan troglodytes* composite assumes that *P. troglodytes troglodytes, P. t. schweinfurthii,* and *P. t. verus* are essentially indistinguishable in size and form. Whereas Schwartz (1929) argued that *P. t. troglodytes* is probably more closely related to *P. paniscus* than is *P. t. schweinfurthii,** Coolidge (1933) noted that on the basis of external appearance *P. paniscus* seemed more closely allied to *P. t. schweinfurthii*. The possibility that *Pan paniscus* and *P. t. schweinfurthii* are very similar in overall size (Zihlman and Cramer, 1978) makes a contrast of these two groups especially significant. Pooling the sexes of a given subspecies or species of *Pan* (or the inclusion of subadults in adult samples) will almost certainly further confuse differences in adult size and size-related aspects of morphology.

We believe that much remains to be discovered about size-related differences and similarities among the African apes. The primary focus of this contribution is on size and scaling of the locomotor skeleton in *Pan* and *Gorilla*. Before proceeding to specific allometric analyses, we

---

*Schwarz (1929, p. 426) states that, "Es ist bemerkswert, dass seine Verwandtschaftsbeziehungen auf *P. s. satyrus* [= *troglodytes troglodytes*] weisen und nicht auf *P. s. schweinfurthi* [= *troglodytes schweinfurthii*] vom oberen Kongo."

hope to clarify the issue of overall size differences among the various taxa. We have compiled published values and museum records of adult body weights (by sex) for *Pan paniscus*, *P. t. troglodytes*, *P. t. schweinfurthii*, *P. t. verus*, *Gorilla gorilla gorilla*, *G. g. beringei*, and *G. g. graueri*. Due to the numerous problems inherent in using captive body weights (Rothenfluh, 1976), and because our skeletal samples are all from wild-shot individuals, we have limited our comparisons of body weight to values that were collected in the field. Following the review of African ape body weights, other aspects of size (linear variables) are also considered by comparing selected skeletal dimensions in *Pan* and *Gorilla* (summary data are presented by taxa and sex and compared between taxa by sex). This step permits a quick check for consistency between the size series established by a measure of overall size (body weight in this case) and the size order suggested by linear measurements of the postcranial skeleton. For example, if body weight is found to differ insignificantly between any two groups, is the same true for lengths of long bones or articular dimensions? If not, a difference in pattern unrelated to size seems highly probable. This type of comparison has been referred to as "narrow allometry" (R. J. Smith, 1980) of similarly sized animals.

   We have employed *interspecific* (adult) allometric analysis to investigate size-related variation in numerous elements of the postcranial skeleton of African apes. One set of variables includes 13 dimensions of long bones, pectoral and pelvic girdles, and the vertebral column. Scale relationships in certain of these bony elements can be evaluated against a set of *a priori* expectations derived from both theroetical and empirical generalizations (e.g., interlimb proportions or vertebral allometry). For other parts of the locomotor skeleton, the resulting allometric patterns are best regarded as descriptive empirical relationships that may or may not be amenable to functional interpretation. A second set of variables includes ten measures of joint dimensions. Interspecific analysis of articular allometry permits a direct test of the biomechanically derived null hypothesis of geometric similarity of animal joints (Alexander, 1980, 1981). At the same time, specific differences in scaling within a single joint (e.g., radial head diameter versus trochlear width) may reflect subtle differences in the way loads are borne by the joints of African apes in comparison to other anthropoid primates, and thereby help explain disparate morphological configurations.

   Two very different aspects of *intraspecific* allometry, static adult and ontogenetic, are also considered in this study. Static scaling of several postcranial variables in a group of adult pygmy chimpanzees (27–61 kg) is compared to the growth allometry in an infant to nearly adult series of pygmy chimpanzee skeletons (5–37 kg). Both types of intraspecific allometric analysis have been utilized by other workers to address the ques-

tion of whether or not pygmy and common chimpanzees are scaled versions of one another [compare Corruccini and McHenry (1979) and McHenry
and Corruccini (1981) to Shea (1981, 1983a, and this volume)]. One could,
in theory, also analyze many of the differences among African apes as a
whole from either or both intraspecific allometric perspectives (Gould,
1975; Shea, 1981, 1983b). Both approaches hope to illustrate the probable
evolutionary trajectories taken by the African ape lineage as body size
increased (or decreased). Several *assumptions* are implicit in this type of
analysis. For example, because this last question is really one of "evolutionary allometry," the pygmy chimpanzee-common chimpanzee-gorilla
series is clearly serving as a model of ancestor-descendent relationships.
This is obviously not true in the strictest sense (cf. Cock, 1966; Fink,
1982; Shea, 1983c), but this assumption has been shown to be a very
useful one in regard to the evolution of the African apes (Shea, 1983a–c).

Using *either* the static *or* ontogenetic allometric relationship as a basis
for extrapolating to larger or smaller body size in any phyletic series (i.e.,
assuming selection for an increase or decrease in size) is risky business.
Just as it is only in very special cases that ontogenetic and static allometries will be the same (Cock, 1966; Shea, 1981; Cheverud, 1982), the
belief that "selection on body size alone would extrapolate static phenotypic (or ontogenetic) lines of allometry is incorrect in theory and may
be seriously misleading in practice" (Lande, 1984; also cf. Cock, 1966;
Cheverud, 1982; Atchley, 1983). This caveat follows from the proof by
Lande (1979, 1984) that the slope of the evolutionary trajectory when
selection acts soley on body size is equal to

$$\gamma_{xy} h_y \sigma_y / h_x \sigma_x \qquad (1)$$

where σ_y and σ_x refer to the standard deviations (log-transformed) of some
variable Y of interest and body size X, h_y and h_x are the square roots of
the heritabilities of these variables, and γ_{xy} is the additive *genetic* correlation between the two variables. Because the static intraspecific allometric slope is equal to the *phenotypic* regression coefficient,

$$\rho_{xy} \sigma_y / \sigma_x \qquad (2)$$

where ρ_{xy} is the *phenotypic* correlation, (2) will resemble (1) only when
the heritabilities of X and Y are similar and/or when the phenotypic correlation is approximately equal to the genetic correlation. Precisely because these two types of correlation need not be and frequently are not
similar [cf. references in Cheverud (1982) and Lande (1984)], one cannot
automatically substitute the phenotypic regression for the genetic one.
The danger in doing so is nicely illustrated by Atchley's (1983) example

of the effects of size selection on rat skulls and jaws. He showed that both the magnitude and the direction of evolutionary change based on extrapolation of phenotypic allometries would be very misleading in the majority of the characters in his study; in fact, in almost one third of the cases, the phenotypic regression predicted changes in the *opposite direction* to that indicated by the genetic regression. Lande (1984) argues that similar reservations obtain about the relationship between ontogenetic and evolutionary allometries. Cheverud (1982) speaks directly to this last issue and notes that extrapolation of ontogenetic allometry to interspecific (evolutionary) allometry implies that the slope and length of ontogenetic vectors are not genetically correlated, the possibility of which he regards as "highly problematical."

Despite these cautionary notes as to the advisability of blindly extrapolating phenotypic ontogenetic vectors into new size ranges to predict or explain the evolutionary pathway of size-related changes, the considerable merits of comparing empirically determined ontogenetic trends in African apes have been demonstrated previously by Shea (1981, 1983a,b) with respect to selected cranial and bodily proportions and relative brain size. Without the necessary (but nonexistent) genetic information, overlap or nonoverlap in phenotypic ontogenetic trajectories of pygmy chimpanzees, common chimpanzees, and gorillas cannot constitute unequivocal proof or disproof of ontogenetic extrapolation, but it can still be regarded as strong circumstantial evidence in favor of such an interpretation. Accordingly, we have also studied the ontogenetic allometries of the limbs and bony girdles of African apes in order to describe and compare the proportional changes related to growth in these regions. If only in phenotypic terms, we can still evaluate the possibilities that gorillas might be characterized correctly as "overgrown chimpanzees" and that pygmy chimpanzees seem to be ontogenetically reduced versions of common chimpanzees. In addition, this comparative approach allows us to address the questions of if and to what degree proportional differences seen in adult African apes are also expressed in infants, juveniles, and subadults.

2. Materials and Methods

2.1. Sources of Body Weights

Only wild-collected records of adult body weights are included in our comparisons (by sex) of species and subspecies of African apes. Based on the catalogued records of the Tervuren Museum, the unpublished field notes of Vandebroek (University of Louvain-la-Neuve), and the recent report by Coolidge and Shea (1982), 13 body weights were compiled for *Pan paniscus* (seven for males, six for females). Due to very minor discrepancies between Vandebroek's notes and the Tervuren records, Vandebroek's values have been given priority in this study. Body weight records are more extensive for *P. troglodytes schweinfurthii*. Based on values published by Rahm (1967) and Wrangham and Smuts (1980), on the records of the Tervuren Museum and the U. S. National Museum, and on the field notes of Vandebroek, 34 adult body weights were included (15 males, 19 females). By contrast, the available data on *P. t. troglodytes* are relatively meager. Only six reliable records of wild-shot animals could be located in the records at the Powell-Cotton Museum (three males, three females). A single male *P. t. verus* body weight value was taken from the records of the U. S. National Museum.

Body weight values for adult *Gorilla gorilla gorilla* are compiled from the Powell-Cotton Museum records [including figures published by Cousins (1972)], from Grzimek (1956), and from notes reported by L. G. Smith to C. P. Groves (personal communication) for a total of 17 individuals (14 males, three females). Fewer reliable records exist for *G. g. beringei*. Only seven adult individuals were included in our mountain gorilla totals (six male, one female), all from the Virungas. The source of these body weight records are Gyldenstolpe (1928), Akeley (1929), Hoier (1955), and Mills (1960). *Gorilla g. graueri* body weight records are fewer still, with only five individuals represented by trustworthy values (four males, one female) from Gregory and Raven (1937), Frechkop (1944), L. G. Smith via C. P. Groves (personal communication), and Vandebroek's field notes. Body weights for the ontogenetic series of infant to subadult, wild-shot *P. paniscus* specimens are also from the Tervuren Museum records and Vandebroek's field notes.

2.2. Skeletal Sample

A total of 275 African ape skeletons were measured in this study. Of this total, 142 were adults as judged by fusion of epiphyses of the long bones; several "dental adults" were excluded from the sample because epiphyseal fusion was incomplete. The final tally included 20 adult *Pan paniscus* (ten males, ten females), 22 *P. t. schweinfurthii* (12 males, ten females), 40 *P. t. troglodytes* (20 males, 20 females), six *P. t. verus* (four males, two females), 39 *Gorilla gorilla gorilla* (21 males, 18 females), and 15 *G. g. beringei* (seven males, eight females). Another 133 immature individuals were measured for the ontogenetic aspect of this study. Due to relatively small samples and an appreciable number of sexually indeterminate specimens, sexes were pooled for this part of the analysis: 30

P. paniscus, 39 *P. t. schweinfurthii,* 32 *P. t. troglodytes,* and 32 *G. g. gorilla.*

2.3. Analytical Methods

2.3.1. Body Weight as the Size Variable

The selection of an appropriate size variable is critically important for the analysis of scaling trends because all conclusions concerning isometry and allometry are valid only in reference to this specified variable (Mosimann and James, 1979; R. J. Smith, 1980, 1981*a;* Jungers and German, 1981; Steudel, 1981*a;* Holloway and Post, 1982). In this study we argue for the primacy of body weight as the size variable for a variety of reasons. Most theoretical and empirical scale effects in animal locomotion are a primary function of body weight (McMahon, 1975; Pedley, 1977; Heglund, 1984) and it is clear that mass must be overcome or resisted when an animal moves (cf. Schmidt-Nielsen, 1975, 1977; Currey, 1977). Theoretical models, therefore, dictate that body weight be employed in several of the interspecific analyses of African ape allometry in order to test the predictions of these models. For example, interlimb proportions should follow specified scaling trends with increasing body weight in clawless mammals that climb large vertical supports (Cartmill, 1974; Jungers, 1978, 1979). Interspecific articular allometry must also be evaluated against biomechanical expectations founded on joint reaction forces and body weight differences among species (Alexander, 1980). Substitute measures for body weight (e.g., skeletal trunk length, femur length) usually scale nongeometrically with weight or mass in anthropoid primates (R. J. Smith, 1981*a;* Steudel, 1981*a*), and can lead, therefore, to quite erroneous interpretations of skeletal allometry and proportionality (e.g., Biegert and Maurer, 1972; Mobb and Wood, 1977; Aiello, 1981). In sum, we agree with the option of Linstedt and Calder (1981, p. 2): "Body mass, more than any other single descriptive feature, is the primary determinant of ecological opportunities, as well as of the physiological and morphological requirements of an animal." The utility of body weight as the size variable in both intraspecific static and ontogenetic allometry is discussed in Lande (1979) and Jungers and Fleagle (1980), respectively.

2.3.2. The Bivariate Power Function and Regression Models

The popularity of Huxley's power formula for allometric relationships

$$Y = \beta X^k \tag{3}$$

is no accident. Although this equation is mathematically simple, we agree with Linstedt and Calder (1981, p. 2) that it remains "the most useful descriptive tool for understanding the evolution of size." An empirical fit of data to a straight line on logarithimic coordinates may be the ultimate argument for its use in many studies (R. J. Smith, 1980; Lande, 1984); however, theoretical justifications for its application come from many sources. Huxley (1924, 1932) himself based this equation and its log-transformed counterpart

$$\log Y = k \log X + \log \beta \qquad (4)$$

firmly within the theoretical context of multiplicative growth (also cf. Katz, 1980). The great value of this bivariate approach in a statistical consideration of quantitative genetics and evolutionary allometry has been demonstrated elegantly by Lande (1979). In various biomechanical analyses of size-related locomotor phenomena, predictive bivariate power functions have been derived from both theoretical considerations and empirical observations (e.g., McMahon, 1973, 1975, 1977; Prange, 1977; Taylor, 1977; Alexander, 1980). The legitimate place of the bivariate power function in modern biology can also be traced to simplifying assumptions in the dimensional analysis of biological similarity (Lambert and Teissier, 1927; Günther, 1975; Economos, 1982).

The pros and cons of which regression technique is the most appropriate one in allometric studies have been reviewed extensively elsewhere (Kuhry and Marcus, 1977; Jungers, 1978, 1984a). Reasonable arguments can be made in favor of both model I (least squares) and model II (e.g., major axis) regression methods. When the correlation between a pair of variables is reasonably high, both regression models tend to produce almost identical results and have little, if any, effect on interpretations. In those cases where the choice of regression technique determines one's conclusions, the logical strength (and biological importance) of the size relationship itself should be questioned. Both least squares and major axis methods are used throughout the present study.

2.3.3. Skeletal Variables, Sample Composition, and Level of Allometric Analysis

Instead of regressing mean values of osteometric valuables on mean body weights for each taxonomic group in a given interspecific analysis of scaling (e.g., Gingerich et al., 1982; Jungers, 1984a), we have elected to include only individual skeletal specimens with associated body weights. Although such specimens are not abundant in museum collections, it was reasoned that their use should minimize the guesswork and inevitable noise inherent in combining skeletons drawn from one source with body weights taken from several others. We have assembled data on 27 such adult African ape skeletons, including 12 pygmy chimpanzees (six males, six females) ranging from 27 to 61 kg, nine common chimpanzees (four males, five females) ranging from 31.4 to 70 kg, and six gorillas (four males, two females) ranging from 68.2 to 187 kg.

Thirteen linear measurements were taken on the long bones, bony girdles, and vertebral column of each specimen:

1. Maximum humerus length
2. Maximum radius length
3. Forelimb length (1 + 2)
4. Maximum femur length
5. Maximum tibia length
6. Hindlimb length (4 + 5)
7. Scapula breadth along the spinal axis
8. Maximum clavicle length
9. Maximum ilium length (from center of acetabulum)
10. Maximum pubis length (from center of acetabulum)
11. Ischium length (from center of acetabulum along axis of the shank)
12. Total lumbar length (ribless definition)
13. Total thoracic length (rib-bearing definition)

An additional ten dimensions of articular surfaces were measured for the interspecific analysis of joint allometry:

1. Maximum breadth of the glenoid fossa
2. Humeral head diameter (perpendicular to greater tubercle)
3. Articular width of distal humerus (both capitulum and trochlea)
4. Trochlear width (ventral aspect)
5. Maximum diameter of radial head
6. Acetabulum height
7. Femoral head diameter (anteroposterior plane)
8. Width of medial femoral condyle (posterior aspect)
9. Width of lateral femoral condyle (posterior aspect)
10. Width of the lumbosacral articulation

Intraspecific static allometry was analyzed for an adult size series of *Pan paniscus* ranging from 27 to 61 kg. The sample was comprised of 12 skeletons (six of each sex), all with associated body weights. Only a subset of the original 13 linear variables were considered: humerus length, femur length, clavicle length and ilium length. The corresponding ontogenetic sample of pygmy chimpanzees included nine skeletons with associated weights ranging from 5 to 37 kg; seven of the nine specimens were males. Linear variables considered include: length of the humeral diaphysis (variable 1 in Fig. 1), length of the femoral diaphysis (variable 3 in Fig. 1), clavicle length (variable 6), and ilium length minus the crest (variable 8).

For the second ontogenetic aspect of the analysis, selected skeletal dimensions were plotted against each other (log transformed) in order to compare the growth trajectories in four taxonomic groups: *P. paniscus, P. t. schweinfurthii, P. t. troglodytes,* and *G. g. gorilla.* Eight linear variables (Fig. 1) were plotted pairwise against each other in several combinations: radial diaphysis (variable 2) versus humeral diaphysis (variable 1); tibial diaphysis (variable 4) versus femoral diaphysis (variable 3); femoral diaphysis versus humeral diaphysis; clavicle length (variable 6) versus scapula breadth (variable 5); and pubis length (variable 7) versus ilium length (variable 8). These same eight measurements were taken on several adults of each group and the means of each sex were plotted as terminal points in the corresponding ontogenetic series. The exponents (slopes) and constants (intercepts) of each bivariate pair were calculated for each group using a major axis fit to the log-transformed X and Y coordinates. The results of these mixed cross-sectional analyses (cf. Cock, 1966) can then be compared to the analogous analyses of Lumer (1939) and Shea (1981). In those bivariate comparisons where intergroup overlap was especially extensive, Tsutakawa and Hewett's (1977) "quick test" for comparing two samples was enlisted to determine if the pooled dis-

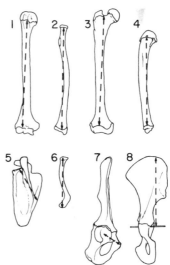

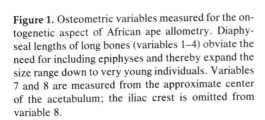

Figure 1. Osteometric variables measured for the ontogenetic aspect of African ape allometry. Diaphyseal lengths of long bones (variables 1–4) obviate the need for including epiphyses and thereby expand the size range down to very young individuals. Variables 7 and 8 are measured from the approximate center of the acetabulum; the iliac crest is omitted from variable 8.

tribution was truly random or if some consistent difference between the taxa did in fact exist.

3. Results and Discussion

3.1. Body Size of African Apes

3.1.1. Body Weights

Table I summarizes our compilations of body weight information on wild-collected pygmy and common chimpanzees and gorillas. Adult male gorillas are on the average five times larger than the average female *P. paniscus* or female *P. t. schweinfurthii*. At one extreme is a 26.4-kg female *P. t. schweinfurthii* and a 27-kg female *P. paniscus;* at the other is a 218.2-kg male *G. g. gorilla.** Clutton-Brock and Harvey (1979) have suggested that an ecological shift to terrestrial folivory in gorillas forms the selective basis for the large body size of this group.

Our findings support Groves' (1970*b*) suggestion that *G. g. graueri* is probably the largest subspecies of gorilla. If the 219 kg value of a male reported by Gatti (1932) were included in our sample, the average figure for *graueri* males would increase to 184 kg.† Male *G. g. gorilla* seem to be larger on the average than are male *G. g. beringei*, although the differences are not statistically significant. Our average values for male gorillas, especially *G. g. gorilla*, differ from those originally reported by Groves (1970*b*) for several reasons. We did not include the male mountain gorilla reported by Galloway *et al.* (1959), because it was most probably a subadult at 120.5 kg; known subadult males of *G. g. graueri* at 124.5 kg (Vandebroek's field notes) and *G. g. gorilla* at 121 kg (Powell-Cotton Museum) support this conclusion. We also excluded several *estimates* of male body weights made by various authors. For example, values for males credited to Barns (1922) (450 lbs), Willoughby (1950) (over 500 lb), Merfield (in Cousins, 1972) (574–588 lbs) are all pure guesses. In addition,

---

*We have not included one western lowland gorilla male value of 280 kg reported by L. G. Smith to C. P. Groves (personal communication) because it weighs in at more than four standard deviations from the mean of 14 other male *Gorilla gorilla gorilla*. Although it would elevate the mean for males of this subspecies by only 6 kg, we question the accuracy of this report; it is more than 60 kg greater than any other reliable weight of any wild-shot gorilla and approaches the records of obese animals in captivity (Cousins, 1972).

†According to Groves (personal communication), Gatti was "prone to hyperbole," an assessment supported by the discrepancy between his measurements and those of Willoughby on the same animal (Cousins, 1972).

Table I. Body Weights of African Apes (kg)

Group		Males	Females	Source
Pan paniscus	$\bar{x}$	45.0[a]	33.2	Records of the Tervuren
	SD	8.4	4.2	Museum; Vandebroek's
	Range	37–61	27–38	field notes; Coolidge
	N	7	6	and Shea (1982)
Pan troglodytes	$\bar{x}$	43.0[b]	33.2	Rahm (1967); Wrangham
schweinfurthii	SD	7.2	5.3	and Smuts (1980);
	Range	33.6–61	26.4–46	Tervuren Museum;
	N	15	19	Vandebroek's field
				notes; U. S. National
				Museum records
Pan troglodytes	$\bar{x}$	60	47.4	Records of the Powell-
troglodytes	Range	50–70	42.3–50	Cotton Museum
	N	3	3	
Gorilla gorilla	$\bar{x}$	169.5	71.5	Powell-Cotton Museum
gorilla	SD	25.8	—	(including Cousins,
	Range	132.0–218.2	68.2–74.3	1972); Grzimek (1956);
	N	14	3	L. G. Smith (personal
				communication via Colin
				Groves)
Gorilla gorilla	$\bar{x}$	159.2	97.7	Gyldenstolpe (1982);
beringei	SD	23.4	—	Akeley (1929); Mills
(Virunga only)	Range	133.6–190.9	—	(1960); Hoier (1955)
	N	6	1	
Gorilla gorilla	$\bar{x}$	175.2	80.0	Gregory and Raven (1937);
graueri	Range	150–208.6	—	Frechkop (1944); L. G.
	N	4	1	Smith (via Colin
				Groves); Vandebroek's
				field notes

[a]Difference between male-female means significant at $p < 0.01$.
[b]Difference between male-female means significant at $p < 0.001$.

because both Bradley (in Cousins, 1972) and M. L. J. Akeley (1929) report a weight of 380 lb for the "Old Man of Karisimbi," we have used this value rather than the 360 lb figure cited originally by C. Akeley (1923). Finally, Groves (personal communication) has recently informed us that duplicate body weights of two male *G. g. gorilla* were inadvertently incorporated into his 1970 paper on gorilla population systematics; his mean values for males of this subspecies were based on six rather than 32 records as initially reported. Reliable body weights of wild-collected female gorillas are too few to make conclusive comparisons. Values range from a low 68.2 kg for a female *G. g. gorilla* to 97.7 kg for a female *G. g. beringei;* the single value for a female *G. g. graueri* lies in between these extremes at 80.0 kg. A 2:1 body weight ratio for male versus female gorillas appears to be normal.

The issue of body weights and therefore overall size differences among pygmy chimpanzees and subspecies of common chimpanzees is considerably more complicated than previously recognized. Conventional wisdom argues that *Pan paniscus* is significantly smaller and/or less robust than the common chimpanzees (Coolidge, 1933; Napier and Napier, 1967; Hill, 1969; Pilbeam and Gould, 1974; Cousins, 1978; Zihlman, 1979; Johnson, 1981; Shea, 1983c, this volume). Some workers, however, have questioned the appropriateness of labels such as "dwarf" or "pygmy" in reference to *P. paniscus,* pointing out that the purported size difference between pygmy and common chimpanzees is either quite small, inconsistent, or nonexistent (Schultz, 1969; Zihlman and Cramer, 1978; Horn, 1979; McHenry and Corruccini, 1981; Susman and Jungers, 1981). The summary data on body weights presented in Table I suggests that part of the debate about relative size in *Pan* is probably due to a failure to indicate which subspecies of *Pan troglodytes* one is comparing to *Pan paniscus.* Another potential complication is the inclusion of subadult values with adult ones, especially for *Pan paniscus* (e.g., Zihlman and Cramer, 1978; Corruccini and McHenry, 1979). Given the present data base, no significant difference in adult body weight exists between the same sexes of *P. paniscus* and *P. t. schweinfurthii;* not only are the mean values similar, but the ranges for each taxa are almost identical. The fact, perhaps historical irony, that the first pygmy chimpanzee skeleton described by Coolidge (1933) is the smallest adult specimen of that species among all museum collections has no doubt fostered the misunderstanding about pygmy chimpanzee body size. *Pan t. troglodytes,* however, most probably is *much* heavier on the average than is either *P. paniscus* or *P. t. schweinfurthii* if the six specimens at the Powell-Cotton Museum are truly representative of this subspecies. The sole *P. t. verus* male body weight at our disposal (46.4 kg) is closer to *P. t. schweinfurthii* than to *P. t. troglodytes.* These findings have very important implications for questions of size and scaling in the genus *Pan.* Without average body size differences between them, morphological differences between *P. paniscus* and *P. t. schweinfurthii* cannot be explained in simple allometric terms. On the other hand, given the significant differences between *P. t. troglodytes* and *P. paniscus* (or *P. t. schweinfurthii)* in adult body size, allometric analysis (ontogenetic or intraspecific) represents a very appropriate avenue of investigation into morphological (presumably size-related) differences between taxa. Moreover, the differences observed among subspecies of *Pan troglodytes* may turn out to be as important and instructive to our understanding of chimpanzee evolution as are those differences seen between pygmy chimpanzees and the various subspecies of common chimpanzee.

3.1.2. Linear Dimensions

As did body weight, measures of stature of wild-collected gorillas suggest that male *G. g. graueri* are the largest members of the genus:

Stature of male *graueri:* 176.8 cm (N = 3) (Coolidge, 1936)
Stature of male *beringei:* 170.4 ± 13.4 cm (N = 8–9) (Groves and Stott, 1979)
Stature of male *gorilla:* 166.4 ± 10.5 cm (N = 15) (Coolidge, 1936; Grzimek, 1956; Cousins, 1972)

Differences between stature in *G. g. beringei* and *G. g. gorilla* are in the opposite direction than those indicated by body weight, but the differences are again not statistically significant ($p > 0.4$). "Span" of the outstretched upper limbs is slightly greater in male *G. g. gorilla* ($\bar{x}$ = 236.8 ± 21.9 cm; N = 10) than in *G. g. beringei* males ($\bar{x}$ = 220.2 ± 16.9 cm; N = 10), a difference that almost attains statistical significance ($0.05 < p < 0.10$).

Differences between the lengths of the extremities in the same sexes of *G. g. gorilla* and *G. g. beringei* (Table II) are all insignificant. These data indicate that several of the traditional "characters distinguishing the mountain gorilla from the lowland gorilla" proposed by Schultz (1934) must now be viewed with extreme skepticism. For example, the intermembral indices of these two subspecies are almost identical at approximately 116. If anything, the intermembral index of mountain gorillas may be somewhat greater than that of the lowland gorilla; Goodall and Groves (1977) report values from 117.0 to 122.6 (cf. Groves and Stott, 1979) for the Tsundura troop of mountain gorillas. As Groves and Stott (1979, p. 167) also note: "one of the 'best' features of the mountain gorilla, the short humerus, has turned out to be not so cogent after all." The brachial index also differs insignificantly between these two subspecies (approximately 80 on the average). Future study will determine whether or not the other postcranial features that Schultz argued could be used to distinguish the mountain gorilla are also invalid.

Table II. Comparisons of Long Bone Dimensions in *Gorilla gorilla gorilla* and *Gorilla gorilla beringei*

Variable		Gorilla g. gorilla		Gorilla g. beringei[a]	
		Males	Females	Males	Females
Forelimb length	N	21	18	7	8
	$\bar{x}$, mm	796.5[b]	666.7[c]	788.3	646.6
	SD	39.7	25.1	32.9	20.3
Hindlimb length	N	21	18	7	8
	$\bar{x}$, mm	689.3[b]	572.2[c]	676.7	557.2
	SD	31.4	18.2	23.4	20.6
Humerus length	N	21	18	7	8
	$\bar{x}$, mm	441.8[b]	369.7[c]	434.6	357.8
	SD	25.2	16.7	18.4	10.4
Radius length	N	21	18	7	8
	$\bar{x}$, mm	354.8[b]	297.0[c]	353.7	288.9
	SD	16.2	10.4	17.1	11.5
Femur length	N	21	18	7	8
	$\bar{x}$, mm	376.8[b]	313.7[b]	375.4	307.6
	SD	17.9	9.9	13.4	12.1
Tibia length	N	21	18	7	8
	$\bar{x}$, mm	312.5[c]	258.5[c]	301.4	249.6
	SD	14.9	11.0	10.9	8.7

[a]Only specimens from the Virunga volcanoes included.
[b]No significant difference between means of same sexes of different subspecies.
[c]$0.05 < p < 0.1$ for same-sex comparisons of the different subspecies.

Despite apparently substantial differences in adult body weight, stature in *P. paniscus* and *P. t. troglodytes* differs insignificantly:

Pan paniscus males: 119.0 ± 4.9 cm ($N = 4$) (Coolidge and Shea, 1982)
     females: 111.7 ± 7.7 cm ($N = 4$) (Coolidge and Shea, 1982)
P. t. troglodytes males: 120.4 ± 6.4 cm ($N = 10$) (Wood, 1979)
     females: 116.6 ± 4.6 cm ($N = 16$) (Wood, 1979).

These data together suggest than *P. t. troglodytes* is a much more robust, stockier animal than is *P. paniscus*. Although we could not locate comparable data on stature in *P. t. schweinfurthii*, Allen (1925) does report "total length" for 11 individuals (seven males, four females). If, as it would appear, this measurement is roughly equivalent to the "crown to anus" variable published by Coolidge and Shea (1982) on pygmy chimpanzees, *P. t. schweinfurthii* would appear to have somewhat longer trunks than *P. paniscus*.

> *Pan paniscus* males: 77.6 cm (73.0–82.8; $N = 4$)
>   females: 73.8 cm (70.0–76.0; $N = 4$)
> *Pan t. schweinfurthii* males: 83.4 cm (77.0–92.5; $N = 7$)
>   females: 78.3 cm (70.0–85.0; $N = 4$)

Differences in measurement technique might account for some of the differences observed in this variable. If these differences are real, and in view of insignificant differences in body weight between these two taxa, pygmy chimpanzees are probably shorter and stockier creatures than *P. t. schweinfuthii*. Visual impressions gained by one of us (RLS) from studying pygmy chimpanzees in the wild (Lomako Forest) suggest that adults (particularly mature males) are indeed robust animals, perhaps considerably more robust than previously appreciated.

The summary data presented in Table III on linear dimensions of the locomotor skeleton in *P. paniscus, P. t. schweinfurthii,* and *P. t. troglodytes* allow us to clarify the conclusions of a number of previous studies (Coolidge, 1933; Schultz, 1969; Zihlman and Cramer, 1978; Corruccini and McHenry, 1979; McHenry and Corruccini, 1981; Shea, 1981).

No significant differences exist between the average hindlimb lengths of same sex individuals of the three different taxa. Given the overall greater body size of *P. t. troglodytes,* therefore, this subspecies possesses relatively short hindlimbs. Male pygmy chimpanzees are characterized by significantly shorter forelimbs than the males of either of the common chimpanzees, a relationship that extends to the comparison of female *P. paniscus* and female *P. t. troglodytes*. The forelimbs of female *P. t. schweinfurthii,* however, are *not* significantly longer than those of *P. paniscus*. The existence of significant sexual dimorphism in forelimb length in *P. t. schweinfurthii* but the absence of sexual dimorphism in this measure in *P. paniscus* may account for such a discrepancy.

Our results confirm the conclusions of other workers that the clavicles of pygmy chimpanzees are significantly shorter than those of common chimpanzees. Although clavicle length is sexually dimorphic in both subspecies of common chimpanzees, no significant difference exists between the sexes in pygmy chimpanzees. In view of what has been established about clavicular function during locomotion in higher primates (Jenkins

Table III. Comparisons of Skeletal Dimensions in *Pan paniscus*, *Pan troglodytes schweinfurthii*, and *Pan troglodytes troglodytes*

Variable		*Pan paniscus* (A)		*Pan t. schwein- furthii* (B)		*Pan t. troglo- dytes* (C)		Significance of differences between means								
		Males	Females	Males	Females	Males	Females	A♂/B♂	A♂/C♂	B♂/C♂	A♀/B♀	A♀/C♀	B♀/C♀	A♂/A♀	B♂/B♀	C♂/C♀
Forelimb length	N	9	9	12	10	20	20	<0.05	<0.001	NS	NS	<0.05	NS	NS	<0.05	<0.002
	$\bar{x}$, mm	552.9	546.3	589.6	551.1	593.1	567.0									
	SD	20.8	27.8	33.9	45.4	27.4	21.5									
Hindlimb length	N	10	10	12	10	20	20	NS	NS	NS	NS	NS	NS	NS	NS	<0.02
	$\bar{x}$, mm	544.5	529.9	559.2	526.8	557.7	540.0									
	SD	26.0	23.0	35.1	44.2	22.4	23.7									
Clavicle length	N	9	9	9	9	16	16	<0.001	<0.001	NS	<0.001	<0.001	NS	NS	<0.005	<0.05
	$\bar{x}$, mm	107.5	103.8	133.0	120.4	131.5	126.1									
	SD	4.0	5.0	9.3	7.4	7.7	7.2									
Scapula breadth	N	9	9	11	10	16	16	<0.01	<0.001	<0.05	NS	NS	NS	NS	<0.05	<0.001
	$\bar{x}$, mm	98.4	96.8	105.9	99.1	111.3	100.1									
	SD	3.8	7.4	6.9	5.6	6.5	5.4									
Ilium length	N	10	10	11	10	16	16	<0.05	<0.001	NS	<0.05	<0.002	NS	NS	NS	<0.001
	$\bar{x}$, mm	175.6	172.0	185.5	183.6	192.4	181.2									
	SD	10.1	5.4	10.9	13.6	7.7	6.9									

Measurement	Stat															
Pubis length	N	10	10	11	10	16	16									
	x̄, mm	67.0	69.1	74.8	72.6	74.8	73.4	<0.002	<0.001	NS	NS	<0.01	NS	NS	NS	NS
	SD	4.6	3.4	5.0	5.8	5.2	4.0									
Ischium length	N	10	10	11	10	16	16									
	x̄, mm	79.5	73.3	82.7	77.8	82.4	77.5	NS	NS	NS	NS	<0.002	NS	<0.01	<0.05	<0.001
	SD	5.3	3.3	3.8	6.1	4.2	2.7									
Glenoid cavity breadth	N	9	9	10	10	15	14									
	x̄, mm	22.6	22.1	24.9	23.1	25.3	23.2	<0.005	<0.005	NS	NS	NS	NS	NS	<0.02	<0.005
	SD	1.5	2.0	1.3	1.7	2.0	1.2									
Articular width of distal humerus	N	9	9	10	10	15	14									
	x̄, mm	42.9	39.2	46.0	42.2	47.6	44.4	<0.01	<0.001	NS	<0.05	<0.001	<0.05	<0.01	<0.01	<0.001
	SD	1.9	2.9	2.5	2.9	2.1	2.1									
Radial head diameter	N	9	9	10	10	15	15									
	x̄, mm	23.4	22.1	24.9	23.2	26.5	24.1	<0.02	<0.001	<0.01	NS	<0.002	NS	NS	<0.02	<0.001
	SD	0.9	1.7	1.4	1.5	1.2	1.1									
Femoral head diameter	N	10	10	10	10	15	14									
	x̄, mm	31.5	30.5	34.1	31.7	34.9	32.0	<0.05	<0.001	NS	NS	NS	NS	NS	<0.05	<0.001
	SD	2.2	2.2	2.7	2.3	2.0	1.4									
Lumbo-sacral articular width	N	9	10	10	10	15	14									
	x̄, mm	31.5	31.6	37.4	36.6	37.2	34.5	<0.001	<0.001	NS	<0.01	<0.05	NS	NS	NS	<0.02
	SD	3.0	3.6	1.7	3.4	2.7	2.9									

et al., 1978), the mechanical significance of the shorter clavicles in *P. paniscus* is not entirely clear. This shortness does probably correlate with Shea's (1981) finding that smaller African apes are less "barrel-chested" than larger ones, but the longer clavicle in *P. t. schweinfurthii* than in *P. paniscus* (despite similar body size) complicates this inference.

The supposedly narrow scapula of *P. paniscus* is evident in males only. Compared to females of the subspecies of chimpanzee, the scapular breadth (spinal axis) of female pygmy chimpanzees differs insignificantly. The same pattern holds for the breadth of the glenoid cavity. Although other measurements are clearly needed to characterize accurately the shape of the pongid scapula, prior speculations about especially narrow scapulae in pygmy chimpanzees somehow being convergent on the scapulae of hylobatids (and, therefore, of locomotor significance) are clearly premature (also cf. Shea, this volume). In addition to a more precise understanding of scapular size and shape in African apes, we desperately need *quantitative* locomotor data on the subspecies of *Pan troglodytes* that can be compared in meaningful fashion to the growing body of such information on *Pan paniscus* (Susman *et al.*, 1980; Susman, this volume).

Whereas the ilium is significantly shorter in *P. paniscus* than in both *P. t. troglodytes* and *P. t. schweinfurthii*, the length of the ischium differs insignificantly with the exception of the female *P. paniscus*—female *P. t. troglodytes* comparison (shorter in *P. paniscus*). Pubis length is also shorter in pygmy chimpanzees, except in the contrast of female *P. paniscus* to female *P. t. schweinfurthii*. It is only in male *P. paniscus*, therefore, that the ratio of pubis to ischium is especially low; *P. paniscus* females actually possess one of the highest values for this index. Ischium length is one of the rare linear dimensions that is sexually dimorphic in pygmy chimpanzees (as it is in the other chimpanzees).

The articular dimensions of male *P. paniscus* tend to be smaller than those of male *P. t. troglodytes* and *P. t. schweinfurthii*. Differences among females are less clearcut. For example, glenoid breadth and femoral head diameter are not significantly different in the three female groups. Radial head diameter is significantly larger in female *P. t. troglodytes* than in female pygmy chimpanzees, but female *P. paniscus* and *P. t. schweinfurthii* exhibit insignificant differences in this same dimension. With respect to the width of the lumbosacral articulation, however, females of both of these subspecies of common chimpanzees are larger than female pygmy chimpanzees. Of the joint dimensions compared here, only the width of the distal articulation of the humerus exhibits sexual dimorphism in pygmy chimpanzees, but most articular variables are significantly dimorphic in the common chimpanzees.

As Zihlman and Cramer (1978) observed earlier, our results also indicate that despite significant differences in body weight, male and female pygmy chimpanzees are characterized by surprisingly little sexual dimorphism in the postcranial skeleton. The slight to nonexistent differences in many joint dimensions between the sexes are particularly noteworthy in view of the pronounced weight differences, because the male joints should increase with size in order to minimize load/area (stress) relationships (Alexander, 1980; Radin, 1980). It is possible that other parameters (e.g., cartilage thickness) of male pygmy chimpanzee joints compensate for their relatively smaller linear dimensions (Simon, 1970, 1971; Simon *et al.*, 1973). With one predictable exception, namely pubis length (Leutenegger, 1970; Steudel, 1981*b*), all linear variables of *P. t. troglodytes* are significantly sexually dimorphic. Most but not all of the same variables were also found to be sexually dimorphic in *P. t. schweinfurthii*; exceptions to this trend include hindlimb length, ilium and pubis length, and width of the lumbosacral joint.

It was noted above that most linear dimensions in male *P. t. troglodytes* (10/12) and male *P. t. schweinfurthii* (10/12) are significantly greater than in *P. paniscus* males. Similarly, most measures in female *P. t. troglodytes* (8/12) are significantly larger than in *P. paniscus* females. Substantially fewer variables distinguish female *P. t. schweinfurthii* (4/12) from *P. paniscus* females. Therefore, females of these last two taxa are quite similar in overall body size and in linear measures, whereas males differ in linear variables only. The differences observed in the nature and degree of sexual dimorphism contribute to these differences in *pattern* between *P. paniscus* and *P. t. schweinfurthii*. It should be reemphasized, however, that within the locomotor skeleton, these pattern (as opposed to allometric) differences are more strongly expressed in male-male comparisons (e.g., longer forelimbs, clavicles, ilia, and pubes as well as larger joints in male *P. t. schweinfurthii*) than in contrasts of females (e.g., shorter clavicles and ilia in female *P. paniscus*). Without detailed, sex-specific locomotor and postural data on *P. t. schweinfurthii* to compare with *P. paniscus*, it is exceedingly difficult to propose a convincing functional scenario that might explain these differences in male patterns.

3.2. Interspecific Allometry

3.2.1. Long Bones, Bony Girdles, and Vertebral Column

Relative to body weight, both forelimb length and hindlimb length scale negatively (nongeometrically, $k < 0.33$) in the African apes (Table IV, Fig. 2). This negative allometric relationship is considerably stronger in the hindlimb. The intermembral index (forelimb/hindlimb $\times$ 100) in-

Table IV. Interspecific Scaling of the Locomotor Skeleton in African Apes ($Y = \beta X^k$). Long Bones, Bony Girdles, and Vertebral Column[a]

Variable	N	Least squares k	(95% CI)	β	Major axis k	(95% CI)	β	Correlation coefficient
Forelimb length	25	0.253	(0.206–0.299)	37.56	0.255	(0.217–0.294)	36.63	0.933
Humerus length	25	0.288	(0.232–0.345)	13.47	0.292	(0.246–0.339)	12.94	0.926
Radius length	25	0.211	(0.168–0.255)	27.73	0.213	(0.178–0.249)	27.24	0.921
Hind-limb length	27	0.174	(0.131–0.216)	83.43	0.175	(0.141–0.210)	82.14	0.885
Femur length	27	0.168	(0.125–0.211)	48.36	0.169	(0.134–0.205)	47.61	0.873
Tibia length	27	0.181	(0.136–0.226)	35.31	0.183	(0.146–0.220)	34.67	0.881
Scapula breadth	25	0.367	(0.297–0.436)	2.04	0.374	(0.317–0.432)	1.89	0.931
Clavicle length	25	0.298	(0.249–0.347)	4.78	0.301	(0.261–0.341)	4.63	0.946
Ilium length	27	0.209	(0.168–0.249)	19.29	0.210	(0.177–0.243)	18.98	0.924
Pubis length	27	0.324	(0.257–0.391)	2.26	0.330	(0.275–0.386)	2.11	0.913
Ischium length	27	0.282	(0.249–0.315)	3.83	0.283	(0.256–0.310)	3.77	0.969
Lumbar length	21	0.169	(0.023–0.316)	14.71	0.183	(0.057–0.315)	12.67	0.522
Thoracic length	23	0.209	(0.139–0.279)	21.15	0.213	(0.156–0.271)	20.26	0.834

[a]X is body mass in grams; Y is length in millimeters (to nearest 0.1). Isometry (geometric similarity) = 0.333.

creases, therefore, with increasing size from *P. paniscus* and *P. t. schwein-furthii* through gorillas. Both Shea (1981) and Jungers (1984a) discerned interlimb allometric trends essentially identical to those documented here despite using somewhat inaccurate body weight values and average limb lengths rather than individual specimens of known weight. The strength of these allometric relationships (as measured by the correlation coefficient) in the present study change slightly when the subspecies of *Pan troglodytes* are treated separately. Although *P. t. schweinfurthii* is nearly identical in body weight to *P. paniscus*, the intermembral index of this subspecies is slightly greater (and closer to that of the larger sized *P. t.*

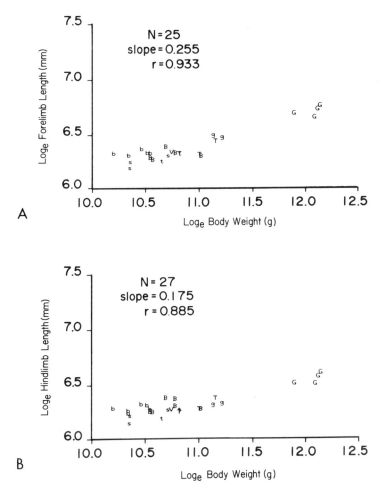

Figure 2. Bivariate interspecific allometry of (A) forelimb length and (B) hindlimb length with respect to body weight. Each point represents one adult individual. Key to symbols: B, b, male and female *P. paniscus*; V, male *P. t. verus*; s, female *P. t. schweinfurthii*; T, t, male and female *P. t. troglodytes*; G, g, male and female *Gorilla gorilla gorilla*. The major axis estimate of the exponent (slope) is indicated with the correlation coefficient.

troglodytes). Because different sexes of the same taxon have indices most similar to each other despite significant differences in overall size, one can infer a strong genetic influence on this ratio; this will probably also introduce additional noise into the size-related aspect of interlimb proportions. However, it should be stressed that the same allometric trends obtain even if the sexes are analyzed separately. Within the forelimb, radius length exhibits stronger negative allometry than does the humerus; hence the well-known lower brachial indices of gorillas. Based on our sample of 27 individual specimens, femur length is characterized by a slightly more pronounced negative allometry than is tibia length (slopes are not significantly different). Our smallest individual, a female pygmy chimpanzee of 27.0 kg, has a crural index of 83.8, whereas our largest specimen, a 187.0-kg gorilla, possesses a crural index of 87.9. In contrast, both Shea (1981) and Jungers (1984a) found that tibia length scaled more negatively than did femur length. A lower *average* crural index in gorillas lends credence to these earlier findings (Jungers, 1984a). In reference to the hindlimb, therefore, the messages are that both femur and tibia scale negatively with body mass (more so than either humerus or radius) and that the crural index does a poor job of sorting individual African apes. Relative hindlimb reduction remains the rule regardless of sample composition or statistical methods.

We believe that the scaling of interlimb proportions in African apes conforms closely to theoretical expectations based on the biomechanics of climbing in large clawless mammals. Larger bodied primates should have higher intermembral indices than smaller bodied primates if *competence* in vertical climbing is to be maintained (Cartmill, 1974; Jungers, 1978, 1979, 1984a, 1984b). In other words, the already quite high index seen in pygmy chimpanzees (>100) increases with body size in African apes to a significantly higher value in gorillas (approximately 116); a high level of competence in climbing is maintained as a consequence of this size-related change in proportions. Ecological advantages in foraging behavior accrue to this allometric pattern of changing limb proportions (MacKinnon, 1971; Kortlandt, 1975), and even the largest gorillas are frequent and adept climbers in pursuit of desirable food items (Bingham, 1932; Schaller 1963; Deschryver in Stern, 1976; Goodall, 1977). The exaggeration of this trend in orangutans (i.e., a much higher intermembral index) underscores the selective importance of *relative* forelimb elongation especially in a highly arboreal, large-bodied ape such as *Pongo* (Jungers, 1984a).

Aiello (1981) has speculated that hindlimb reduction and an increasing intermembral index in African apes might impart some biomechanical advantage in terrestrial locomotion if these proportional changes are "connected with greater weight support on the forelimbs" (Aiello, 1981, pg. 85). Although it is far from clear how this shift would necessarily imply "more efficient weight carriage and movement in terrestrial locomotion" (Aiello, 1981, p. 79), the available data on weight distribution between forelimbs and hindlimbs in primates controvert the notion that higher intermembral indices will result in more weight being borne by the forelimbs during locomotion. Forceplate data indicate that during walking the forelimb *support factors** do not increase with relatively longer forelimbs (Kimura *et al.*, 1979; Reynolds, 1981). The fraction of body weight supported by the forelimbs during walking is roughly similar in *Lemur, Macaca, Erythrocebus, Ateles,* and *Pan* (less than 50%) despite grossly different interlimb proportions. Moreover, with increasing forward velocity, both peak vertical forces on the forelimb and the forelimb support factors *decrease* in chimpanzees, presumably due to the activity of extrinsic muscles of the hindlimb (Reynolds, 1981). It can be also argued that the cranially directed glenoid fossae of chimpanzees (and gorillas) are poorly suited for a weight-bearing role in quadrupedal progression (Roberts, 1974; Reynolds, 1981; *contra* Vrba, 1979); hence the further redistribution of weight to the hindlimb with increasing speed in *Pan.*

   Shea (1981, and this volume) has also suggested that relative hindlimb reduction in large-bodied African apes might be advantageous in a terrestrial context. We concur that a more inclined rather than horizontal pronograde posture could conceivably facilitate moving into "postures of truncal erectness" by reducing the angle through which the trunk must move in order to achieve an erect position (cf. Stern, 1976). As does Shea, however, we regard this possibility as an *indirect* benefit gained through primary adaptive modifications to preserve competence in climbing. We base this conclusion on numerous parallel cases of allometric hindlimb reduction in arboreal primates (Jungers, 1978, 1979, 1980, 1984*a*, 1984*b*), both in quadrupedal (e.g., cebids, lemurids) and antipronograde (e.g., indriids, lepilemurids, and hylobatids) groups. In the most terrestrial nonhuman primates, the Cercopithecinae, the intermembral index also increases as body size increases; however, this is due to hindlimb isometry coupled with forelimb positive allometry in order to improve the crusorial adaptation of the larger members of this group (Jungers, 1984*a*). Whether increasing body size in African apes forced the largest members (gorillas) to spend more time on the ground or allowed them to exploit new terrestrial resources with less concern for predators (or both), size-related alterations in interlimb proportions have served to preserve their climbing capabilities.

   Although the interlimb differences between pygmy and common chimpanzees are small when compared to the gorilla, we agree with Shea (1981, and this volume) that the climbing model might explain the slightly

---

*Support factor is defined by Reynolds (1981) as the vertical impulse acting on that limb divided by the product of the subject's weight and the gait cycle duration. This is equal to the average fraction of the body's weight supported by a limb over the course of the gait cycle.

higher index of *P. t. troglodytes*.* Given the absence of significant body size differences, however, the proportional differences between *P. paniscus* and *P. t. schweinfurthii* cannot be explicated by the same biomechanical, size-related principles. *Pan t. schweinfurthii* is just as "long-legged" as *P. paniscus* (Shea, this volume) and the males of this species were shown to possess absolutely longer forelimbs. Only field data on positional behavior of these two can determine if these subtle morphological differences have any impact on, or relationship to, the frequency of habitual locomotor activities.

Clavical length exhibits a slightly negative ($k \simeq 0.3$) allometric relationship with body weight among the African apes (Table IV, Fig. 3). This implies that although pygmy chipmanzees have short clavicles compared to common chimpanzees, relative clavicle reduction occurs in the size series from pygmy and common chimpanzees to gorillas. This trend seems to contrast with Shea's finding that chest girth scales with strong positive allometry in the same group. Schultz (1933) also reported that relative shoulder breadth was greatest in gorillas. If clavicles of gorillas are relatively short (or even isometric) but both shoulder breadth and chest girth are relatively large, either the cranial portion of the trunk narrows considerably in gorillas or else the clavicle is less cranially elevated than in chimpanzees. Schultz's (1933) measure of "shoulder height" relative to the top of the sternum suggests that the shoulder of gorillas is in fact somewhat less cranially displaced; accordingly, a relatively shorter clavicle could conceivably span a relatively wide upper thorax. The findings of Jenkins et al. (1978) on clavicle function in spider monkeys are perhaps also relevant here: "clavicular length is directly related to the position of the scapula on the thorax and the shape of the thorax." Greater thoracic width alone does not necessarily imply a greater distance between the manubrium and acromion process of the scapula. This may also mean that the short clavicle of pygmy chimpanzees in comparison to common chimpanzees is merely a function of slightly different thoracic shape and/or shoulder height. Major differences in the locomotor function of the clavicles of African apes due to differences in relative length seem highly unlikely because all of them possess the broad thoraces and laterally facing glenoid cavities typical of hominoids (Schultz, 1930, 1933).

Scapula breath (Schultz's "morphological length" along the spinal axis) scales with positive allometry in African apes (Table IV, Fig. 3). As a consequence, the scapulae of pygmy and common chimpanzees appear

---

*The average intermembral index of six *P. t. verus* individual (sexes pooled) is approximately 106, a value similar to that of the other common chimpanzees. The lack of sex-specific data on body weights of this subspecies precludes a consideration of size-related similarities to or differences from *P. t. troglodytes* and *P. t. schweinfurthii*.

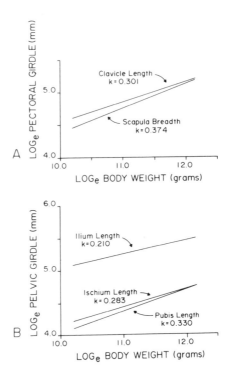

Figure 3. Bivariate interspecific allometry of the (A) pectoral and (B) pelvic girdles with respect to body weight. The major axis estimate of the exponent (slope) is indicated for each variable.

much narrower (and longer) than those of gorillas. Although members of the former group (with narrow scapulae) armswing more frequently and engage in other types of suspensory behavior to a much greater extent than do members of the second group (broad scapulae), the causal link between specific locomotor activities and scapular morphology in African apes remains to be established. We agree with Shea (this volume) that a fuller appreciation of both size and shape of the ape scapula is required before probable behavioral correlates can be accurately discerned; detailed information on *in vivo* function of the ape scapula is equally desirable.

As body size increases in African apes, the bony elements of the innominate do not scale at the same rate (Table IV, Fig. 3). Pubis length scales isometrically with body weight, whereas both ilium length and ischium length are characterized by negative allometry; the relative reduction in length of these two bones is much more pronounced in the ilium than in the ischium. In conjunction with the strong negative allometry seen in the vertebral column (see below), negative allometry of the ilium contributes to the substantially reduced relative trunk length of large-bodied apes. The mechanical impact of iliac reduction on gluteal function is not clear because the iliac blade also broadens with increasing size in pongids (Steudel, 1982). Relative ischium reduction in gorillas will not affect adversely the important extensor role of the hamstrings in climbings, because the load arm (proportional to hindlimb length) is reduced relatively even more. In other words, the *mechanical advantage* (cf. J. M. Smith and Savage, 1956; McHenry, 1975) of the hamstring muscles actually increases with body size in the African apes. The isometric scaling of the pubis suggests that the role of the adductor masculature also remains important in gorillas, presumably during climbing (cf. Howell, 1944; McArdle, 1981). Relatively longer pubes in females than in conspecific males are probably due primarily to obstetrical considerations; as such, sexual differences serve to complicate the functional significance of the observed pubic isometry in African apes.

Changes in lumbar vertebral length are poorly correlated with increases in body size in African apes (Table IV), and strong negative allometry of this region is indicated. Because the mean number of lumbar vertebrae is virtually the same for all African apes (Schultz, 1961), the observed negative allometry must be attributed to a relative shortening of individual lumbar bodies (Benton, 1976). Negative allometry also characterizes the thoracic portion of the vertebral column, although the trend is not so pronounced as in the lumbar region. Again, relative reduction of the lengths of individual thoracic vertebral bodies is most likely because there is no significant decrease in the number of vertebrae as body size increases (Schultz, 1961). Taken together, the negative interspecific allometries of both regions of the vertebral column result in significant relative shortening of the trunk in gorillas compared to pygmy or common chimpanzees. Truncal reduction serves to reduce the distance between the forelimbs and hindlimbs and thereby probably reduces the bending moments (and associated internal stresses) in the trunks of the larger animals (Slijper, 1946; Preuschoft, 1978; Preuschoft *et al.*, 1979). The especially short, well-braced lumbar column indicates that the flexible portion of the vertebral column has been sacrificed; this serves to stabilize the lumbar region and thereby reduces the chances of buckling in this region during climbing (Benton, 1976; Jungers, 1984a).

3.2.2. Articular Dimensions

Interspecific articular allometry in African apes ranges from quite negative scaling (lumbosacral joint width, $k \simeq 0.25$) to moderately positive scaling (width of the humeral trochlea, $k \simeq 0.40$). (Table V). It should be noted, however, that the 95% confidence intervals of all ten articular dimensions considered here include geometric similarity ($k = 0.33$). These

Table V. Interspecific Scaling of the Locomotor Skeleton in African Apes ($Y = \beta X^k$). Articular Dimensions[a]

Variable	N	Least squares			Major axis			Correlation coefficient
		k	(95% CI)	β	k	(95% CI)	β	
Glenoid breadth	25	0.377	(0.320–0.434)	0.415	0.382	(0.336–0.429)	0.395	0.955
Humeral head diameter	25	0.335	(0.285–0.384)	0.998	0.338	(0.298–0.379)	0.964	0.956
Articular width of distal humerus	25	0.353	(0.311–0.394)	1.002	0.355	(0.321–0.389)	0.976	0.971
Trochlear width	25	0.398	(0.328–0.469)	0.309	0.406	(0.348–0.466)	0.285	0.939
Radial head diameter	25	0.312	(0.267–0.357)	0.852	0.315	(0.278–0.352)	0.829	0.958
Acetabulum height	27	0.302	(0.248–0.356)	1.566	0.306	(0.261–0.351)	1.502	0.933
Femoral head diameter	27	0.315	(0.264–0.366)	1.124	0.318	(0.277–0.361)	1.081	0.944
Width of medial femoral condyle	27	0.341	(0.272–0.409)	0.573	0.347	(0.292–0.405)	0.532	0.918
Width of lateral femoral condyle	27	0.287	(0.213–0.360)	0.825	0.293	(0.233–0.356)	0.767	0.875
Lumbosacral articular width	26	0.246	(0.156–0.335)	2.486	0.254	(0.180–0.331)	2.264	0.792

[a]X is body mass in grams; Y is measured in millimeters (to nearest 0.1). Isometry (geometric similarity) = 0.333.

findings appear to support the theoretical and empirical model of forces in animal joints developed recently by Alexander (1980, 1981). Joint forces that are proportional to (body mass)$^{2/3}$ would imply that articular stresses are independent of body mass if animals of different size are geometrically similar, including their joint dimensions. The finding by Alexander *et al.* (1979) that limb bone lengths and diameters of animals in a size series ranging from shrew to elephant scale quite close to geometric similarity provided further support for this model. Alexander (1980) was careful to note, however, that his suggestion that maximum joint stresses are probably of the same order of magnitude in animals of all sizes "depends on the assumption that the linear dimensions of articular surfaces tend to be proportional to the overall dimensions of the bone or limb segment" (Alexander, 1980, p. 96). Such a one-to-one relationship does not hold between joint allometry and long bone (length) allometry in African apes. Most joints scale significantly faster than the long bone or bony girdle of which they are a part. The scaling of long bone diameters probably does correspond more closely to articular allometry, and this possibility demands further analysis if Alexander's model is to be fully explored and adequately evaluated. Regrettably, an investigation of this specific problem is outside the scope and primary focus of this study.

The relative scaling of different components of a given joint can also provide important insights about how articular loads are being borne by the joints of African apes as size increases. For example, consider the following three variables: articular width of the distal humerus, trochea width (the ulnar part of articular width), and radial head diameter. Articular width itself scales in a slightly positive allometric fashion; ulnar (trochlea) and radial components of the elbow scale in different directions, although their summed values correspond to the allometry for overall articular width. African apes increase the relative size of their ulnar element at the expense of their radial element (including the capitulum). The opposite trend is seen in cercopithecoids (Jungers, in preparation), with capitulum size increasing relative to trochlea size. These differences between African apes and Old World Monkeys can be interpreted functionally in the evolutionary context outlined by Jenkins (1973). As Jenkins noted, hominoids (including all pongids) have a highly derived trochlear morphology (spool-shaped) that emphasizes elbow stability in all phases of flexion, extension, and axial rotation. Our data suggest that the load-bearing role of the trochlea increases with body size in pongids; the cercopithecoids appear to be converging instead on an elbow pattern seen in other terrestrial mammalian groups wherein an increased emphasis is placed on load-bearing by the radial component with a concomitantly increased restriction of limb movements to the parasagittal plane (Jenkins, 1973). The discrepancy between the scaling trends in the medial and lateral condyles of the distal femur has important functional implications for the way the knee joint is probably loaded in African apes and how these loads must change with increasing body size. The width of the medial condyle exhibits slight positive allometry, whereas the width of the lateral condyle shows moderately negative allometry. Given the *genu varum* orientation of the African ape knee, "the medial condyle exerts greater pressure on the tibia than does the lateral and therefore its surface must be larger" (Preuschoft, 1973, p. 282). The allometric enlargement of the medial condyle in gorillas suggests that the load-bearing role of this part of the knee joint increases greatly with body size; conversely, a substantially smaller fraction of the joint load is probably being borne by the lateral condyle of the femur.

3.3. Intraspecific Allometry

3.3.1. Static Adult versus Ontogenetic Allometry in *Pan paniscus*

The "relationship" between static adult scaling and ontogenetic scaling of the four variables considered in this part of the analysis (Table VI) is virtually nonexistent (R^2 for the comparison of slopes is 0.01). The results of one type of intraspecific analysis cannot be substituted for the other, nor can one be used to predict the direction of change indicated

Table VI. Intraspecific Static and Ontogenetic Allometry in *Pan paniscus*

Variable	Allometric exponent[a]		Correlation coefficient
	LS	MA	
Static adult series (27–61 kg)			
Humerus length ($N = 10$)	− 0.009	− 0.010	− 0.057
Femur length ($N = 12$)	0.061	0.063	0.308
Clavicle length ($N = 10$)	0.079	0.082	0.393
Ilium length ($N = 12$)	0.079	0.083	0.328
Ontogenetic series (5–37 kg)			
Humeral diaphysis length ($N = 8$)	0.269	0.272	0.945
Femoral diaphysis length ($N = 9$)	0.322	0.324	0.972
Clavicle length ($N = 8$)	0.231	0.232	0.935
Ilium length ($N = 9$)	0.311	0.313	0.961

[a]LS, least squares regression; MA, major axis.

by the other. As Shea (1981, p. 192) noted in his contrast of these two types of intraspecific allometry in African apes, "reliable conclusions about ontogenetic patterns simply cannot be made with static adult data."

The flat slopes (low exponents) and nonsignificant phenotypic correlations observed for scaling within adult pygmy chimpanzees are due primarily to significant sexual dimorphism in body weight combined with the absence of sexual dimorphism in linear variables (see Tables I and III). If *genetic* correlations between these variables are comparably low in *Pan paniscus*, one could posit genetic uncoupling of size and linear dimensions; correlated responses in body size would not necessarily occur if selection favored elongation or reduction of a given linear variable. For example, using pygmy chimpanzees as a starting point, selection for the degree of forelimb length increase seen in *P. t. schweinfurthii* might occur without a correlated increase in body mass. Similarly, selection for an increase of body size into the range of *P. t. troglodytes* (again using *P. paniscus* as the initial reference) could proceed without a correlated increase in hindlimb length or ischium length. The lack of data on the genetic correlations of such parameters in African apes is only one of several possible factors that diminish the potential significance of this type of argument. Relatively low phenotypic correlations can sometimes reflect little more than the relatively constricted range of the X and Y variables in intraspecific samples (R. J. Smith, 1981b). Among closely related species such as *Pan paniscus* and *Pan troglodytes*, the possibility that some of the apparently size-related differences in metrical characters evolved instead by random genetic drift also cannot be ruled out (Lande, 1979).

3.3.2. Ontogenetic Allometries in African Apes

Intralimb ontogenetic allometries (radius on humerus, tibia on femur) tend to be close to isometry within each of the four African ape groups considered here (Table VII, Figs. 4 and 5). Similar but not identical growth patterns have been reported by Lumer (1939) and Shea (1981). With respect to the intralimb scaling of the forelimb, the 95% confidence limits of the exponent include isometry ($k = 1.0$) for pygmy chimpanzees and subspecies of the common chimpanzee but not for gorilla. Our findings suggest that forelimb shape has been preserved during growth within each group of *Pan*, but a subtle shape change (in proportions) probably occurs during gorilla growth such that the proximal element becomes relatively longer than the distal one. This growth-related change in gorillas is reinforced (in comparison to *Pan*) by a pronounced *shift* in the initial intralimb proportions (Fig. 4B); the radius is short relative to the humerus in gorillas throughout all postnatal stages of growth. The much lower interspecific

Table VII. Ontogenetic Allometries of the Limbs and Bony Girdles[a]

Variable Pair (Y on X)	Pan paniscus	Pan troglodytes schweinfurthii	Pan troglodytes troglodytes	Gorilla gorilla gorilla
Radius on humerus				
N	28	38	34	34
Exponent	0.986	0.999	1.018	0.968
95% CI	0.943–1.030	0.972–1.027	0.984–1.052	0.940–0.996
Constant	1.04	0.96	0.89	1.02
Correlation	0.994	0.997	0.996	0.997
Tibia on femur				
N	28	38	34	34
Exponent	0.972	0.955	0.969	0.980
95% CI	0.928–1.018	0.926–0.986	0.939–1.000	0.954–1.006
Constant	0.94	1.04	0.98	0.90
Correlation	0.994	0.996	0.996	0.997
Femur on humerus				
N	28	38	34	34
Exponent	1.070	1.044	1.049	0.965
95% CI	1.033–1.108	1.017–1.073	1.014–1.086	0.940–0.990
Constant	0.70	0.78	0.75	1.00
Correlation	0.996	0.997	0.995	0.997
Clavicle on scapula				
N	30	35	34	32
Exponent	0.799	0.885	0.899	0.925
95% CI	0.737–0.866	0.833–0.940	0.827–0.977	0.873–0.981
Constant	2.67	2.03	1.92	1.42
Correlation	0.979	0.986	0.974	0.988
Pubis in ilium				
N	29	36	34	30
Exponent	1.165	1.058	1.067	1.050
95% CI	1.093–1.242	0.993–1.128	1.007–1.131	1.002–1.100
Constant	0.16	0.28	0.27	0.35
Correlation	0.987	0.984	0.987	0.993

[a]Variables are those in Fig. 1. The exponent and constant are calculated by major axis fits to log-transformed data; the 95% confidence limits of the exponents are also given.

exponent for the radius compared to that for the humerus also reflects this downward transposition in the growth allometry of gorilla forelimbs. Despite these trends in gorillas, the functional analogy to graviportal species (i.e., longer proximal segments, shorter distal ones) strikes us as an overstatement (Roberts, 1974; Buschang, 1982). In comparisons of both *P. paniscus–P. t. schweinfurthii* and the *P. paniscus–P. t. troglodytes* forelimb allometries, the paired taxa appear to share the same ontogenetic

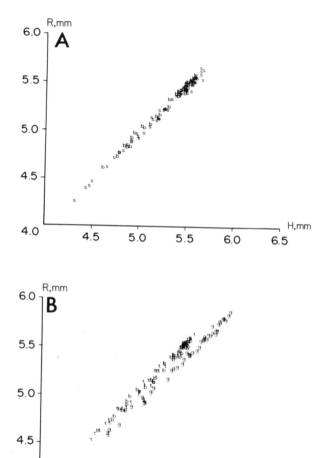

Figure 4. Bivariate ontogenetic allometry of humeral diaphysis length *H* versus radial diaphysis length *R*. (A) *Pan paniscus* (b) compared to *P. t. schweinfurthii* (s). (B) *Pan paniscus* (b), *P. t. troglodytes* (t), and *G. g. gorilla* (g) compared. See Table VII for quantitative details.

vector (Figs. 4A and 4B); however, ultimate lengths of both elements are greater in the two common chimpanzee groups.

Within the hindlimb (tibia on femur), intralimb scaling includes isometry in *P. paniscus*, *P. t. troglodytes*, and *Gorilla*, but not in *P. t. schweinfurthii* (Table VII, Fig. 5). Subtle shape alterations in the hindlimb

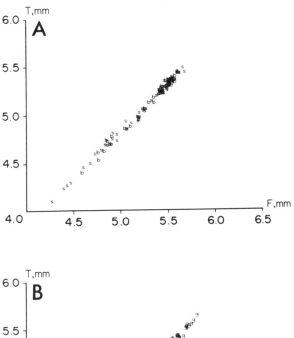

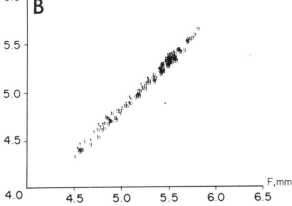

Figure 5. Bivariate ontogenetic allometry of femoral diaphysis length *F* versus tibial diaphysis length *T*. (A) *Pan paniscus* (b) compared to *P. t. schweinfurthii* (s). (B) *Pan paniscus* (b), *P. t. troglodytes* (t), and *G. g. gorilla* (g) compared. See Table VII for quantitative details.

of *P. t. schweinfurthii* are suggested by these findings; the proximal element appears to grow relatively faster than the distal one. Although this trend is more or less similar in all the African apes (i.e., the best estimate of *k* < 1.0 in all cases), it is strongest in *P. t. schweinfurthii*. The initial intralimb proportions of the hindlimb do not distinguish gorillas from the other ape taxa (i.e., no obvious transpositions) and growth introduces little change in these shape relationships, hence the very similar crural indices of all African apes (Jungers, 1984a). Once again, the phenotypic ontogenetic vectors of all three forms of *Pan* overlap extensively.

Interlimb allometries (femur on humerus) point to growth-related shape changes rather than shape conservation in all four groups considered here (Table VII, Fig. 6). The gorilla is the odd ape out in this case, with femur growth occurring at a slower rate than humerus growth ($k < 1.0$). In all three forms of *Pan* the opposite trend is observed ($k > 1.0$), with the strongest expression of this relationship occurring in pygmy chimpanzees (also cf. Table VI). The "quick test" proposed by Tsutakawa and Hewett (1977) indicates that despite considerable overlap in the ontogenetic vectors of *P. paniscus* versus *P. t. schweinfurthii* (Fig. 5A) and *P. paniscus* versus *P. t. troglodtyes* (Fig. 5B), femur length is consistently longer at a given humerus length in pygmy chimpanzee. Because adult femur length differs insignificantly among the various *Pan* taxa (cf. Table III), this persistent difference in the interlimb growth relationship is probably due to a relatively short humerus in pygmy chimpanzees. In gorillas a transpositional shift in interlimb proportions accompanies the allometric change in growth (Fig. 5B); the humerus is relatively long at any given femur length. The impact of this shift is once again reflected in significant interspecific scaling differences between humerus and femur (humerus $k >$ femur k).

Ontogenetic scaling of clavicle length on scapula breadth in African apes departs from isometry in all four taxa (Table VII, Fig. 7). The negative allometry that characterizes this growth relationship is most strongly expressed in pygmy chimpanzees and least so in gorillas. In other words, during growth scapular breadth increases much faster than does clavicle length in pygmy chimpanzees and slightly faster in gorillas. Much less overlap occurs among the allometric vectors of the pectoral girdle of pygmy and common chimpanzees in comparison to that seen for intralimb allometries. A downward shift of pygmy chimpanzees in the bivariate relationship is noticeable in comparison to both subspecies of common chimpanzees (Figs 7A and 7B); pygmy chimpanzees possess a shorter clavicle length for a given scapular breadth throughout ontogeny and on into adulthood (cf. Table III). A further downward shift separates the ontogenetic vector of gorillas from both species of *Pan* (Fig. 7B). The slightly negative interspecific allometry of the clavicle noted before (Table IV) coupled with positive interspecific allometry of scapula breadth can be explained partially by successive ontogenetic shifts in initial pectoral proportions and by differences in the slopes of the ontogenetic vectors themselves.

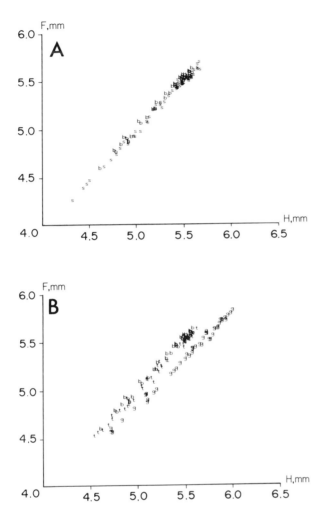

Figure 6. Bivariate ontogenetic allometry of humeral diaphysis length *H* versus femoral diaphysis length *F*. (A) *Pan paniscus* (b) compared to *P. t. schweinfurthii* (s). (B) *Pan paniscus* (b), *P. t. troglodytes* (t), and *G. g. gorilla* (g) compared. See Table VII for quantitative details.

Ontogenetic allometries of pelvic components (pubis on ilium) in African apes are all similar in sign (positive allometry), but there is at least one major difference in the degree of this departure from isometry. During growth, elongation of the pubis relative to the ilium is more pronounced in pygmy chimpanzees than in any of the other taxa (Table VII,

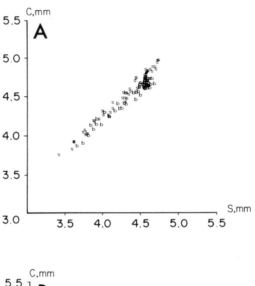

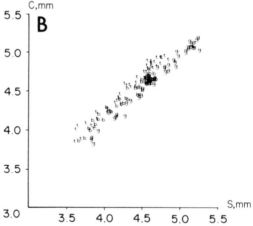

Figure 7. Bivariate ontogenetic allometry of scapula breadth *S* versus clavicle length *C*. (A) *Pan paniscus* (b) compared to *P. t. schweinfurthii* (s). (B) *Pan paniscus* (b), *P. t. troglodytes* (t), and *G. g. gorilla* (g) compared. See Table VII for quantitative details.

Fig. 8). Overlap of the ontogenetic vectors between pygmy and common chimpanzees appears quite extensive in the innominate, but the gorilla growth relationship exhibits an upward transposition (Fig. 8B). At any given iliac length, the gorilla pubis tends to be longer than that of any of the other members of *Pan*. Interspecific adult pubic isometry in conjunc-

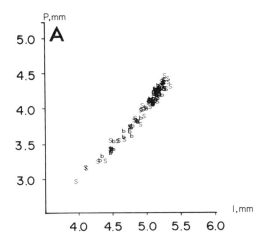

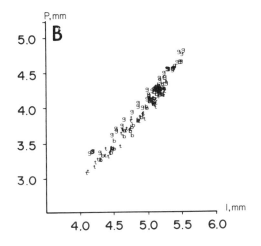

Figure 8. Bivariate ontogenetic allometry of ilium length *I* versus pubis length *P*. (A) *Pan paniscus* (b) compared to *P. t. schweinfurthii* (s). (B) *Pan paniscus* (b), *P. t. troglodytes* (t), and *G. g. gorilla* (g) compared. See Table VII for quantitative details.

tion with strongly negative allometry of ilium length is another expression of these shape differences that are evident during pelvic ontogeny and which persist into adulthood.

Overlap in phenotypic ontogenetic vectors is extensive enough in several comparisons of African ape growth patterns to warrant serious consideration of *ontogenetic extrapolation* as a possible mechanism of evolutionary change (cf. Shea, 1981, 1983c, and this volume). Truncation of the intralimb allometry of the forelimb seen in *P. t. troglodytes* would yield an ontogenetic vector similar in length and slope to that of *P. paniscus;* this might then be construed as a paedomorphic change via hypomorphosis, or "rate hypomorphosis" to be more explicit (Shea, 1983c and this volume). The ontogenetic vector overlap in the forelimbs of *P. paniscus* and *P. t. schweinfurthii* does not fit standard heterochronic categories; vector orientation is similar, but the length of the vector is greater in *P. t. schweinfurthii* despite the lack of differences in overall body size between the two taxa. Intralimb allometries of the hindlimb are similar enough in *P. paniscus, P. troglodytes,* and *Gorilla* to suggest a hypermorphic (rate) extrapolation of a common ontogenetic pattern in *Pan* to *Gorilla.* Within-pelvic allometries appear at first to recall the situation of the forelimb; truncation of the *P. t. troglodytes* ontogenetic vector might produce one similar to that seen in *P. paniscus.* The problem with *P. t. schweinfurthii* persists, however. Moreover, there is some evidence in this case that the length and slope of the ontogenetic vectors are *not* independent within *Pan* (e.g., the higher slope of the shorter pelvic vector of *P. paniscus*). Ontogenetic extrapolation seems highly unlikely for either the interlimb or pectoral girdle growth allometries; transpositional shifts in initial proportions persist throughout growth and result in different shapes in adults. As we noted in our introductory remarks, we realize that without reliable information about underlying genetic correlations, extrapolation of phenotypic allometries (ontogenetic or static) does not necessarily explain or predict the pathway of evolutionary change as body size changes within a lineage (Lande, 1979, 1984; Cheverud, 1982; Atchley, 1983). We doubt, however, that the phenotypic similarities and overlap in several groups of ontogenetic vectors across taxa (e.g., intralimb allometries) are purely coincidental; ontogenetic extrapolation remains a potential explanatory framework in these specific cases.

ACKNOWLEDGMENTS. We wish to thank the museum directors and curators of primate skeletal material for their kind cooperation and valuable assistance: Dr. F. E. Thys van den Audernaerde and D. Meirte at the Tervuren Museum; Prof. J. J. Picard (Laboratoire d'Embryologie) at Louvain-la-Neuve; D. R. Howlett and L. Barton at the Powell-Cotton Museum; Prue Napier and Paula Jenkins at the British Museum of Natural History; Dr. R. Thorington at the National Museum of Natural History; Maria Rutzmoser at the Museum of Comparative Zoology, Harvard University; Dan Russell at the American Museum of Natural History; and Dr. W. Kimbel at the Cleveland Museum of Natural History. We wish to express our sincere appreciation to Luci Betti and Leslie Jungers for the graphics and to Joan Kelly for her patience and expertise in typing this manuscript. A very special thanks is due to Dr. Brian Shea for sharing his valuable insights through numerous conversations and manuscripts in press. Many thanks also to Dr. J. T. Stern and Dr. N. Creel for computational and programming assistance, and for being such generally cool guys. This research was supported by NSF grants BNS 7924162, 8041292, 8119664, and 8217635.

References

Aiello, L. C., 1981, Locomotion in the Miocene Hominoidea, in: *Aspects of Human Evolution* (C. B. Stringer, ed.), Taylor and Francis, London, pp. 63–97.

Akeley, C. 1923, *Brightest Africa,* Garden City.

Akeley, M. L. J., 1929, *Carl Akeley's Africa,* Dodd Mead and Co., New York.

Alexander, R. McN., 1980, Forces in animal joints, *Engin. Med.* **9:**93–97.

Alexander, R. McN., 1981, Analysis of force platform data to obtain joint forces, in: *An Introduction to the Biomechanics of Joints and Joint Replacements* (D. Dowson and V. Wright, eds.), Mechanical Engin. Pub., London, pp. 30–35.

Alexander, R. McN., Jayes, A. S., Maloiy, G. M. O., and Wathuta, E. M., 1979, Allometry of limb bones of mammals from shrews *(Sorex)* to elephant *(Loxodonta), J. Zool. Lond.* **189:**305–314.

Allen, J. A., 1925, Primates collected by the American Museum Congo Expedition, *Bull. Am. Mus. Nat. Hist.* **47:**283–499.

Atchley, W. R., 1983, Some genetic aspects of morphometric variation, in: *Numerical Taxonomy* (J. Felsenstein, ed.), Springer Verlag, pp. 346–363.

Barns, T. A., 1922, *The Wonderland of the Eastern Congo,* London.

Benton, R. S., 1976, Structural patterns in the Pongidae and Cercopithecidae, *Yearb. Phys. Anthropol.* **18:**65–88.

Biegert, J., and Maurer, R., 1972, Rumpfskelettlänge, Allometrien und Körperproportionen bei catarrhinen Primaten, *Folia Primatol* **17:**142–156.

Bingham, H. C., 1932, Gorillas in a natural habitat, *Carnegie Inst. Wash. Publ.* **426:**1–66.

Bruce, E. J., and Ayala, F. J., 1979, Phylogenetic relationships between man and the apes: Electrophoretic evidence, *Evolution* **33:**1040–1056.

Buschang, P. H., 1982, The relative growth of the limb bones for *Homo sapiens*—as compared to anthropoid apes, *Primates* **23:**465–468.

Cartmill, M., 1974, Pads and claws in arboreal locomotion, in: *Primate Locomotion* (F. A. Jenkins, Jr., ed), Academic Press, New York, pp. 45–83.

Cheverud, J. M., 1982, Relationships among ontogenetic, static, and evolutionary allometry, *Am. J. Phys. Anthropol.* **59:**139–149.

Ciochon, R. L., 1983, Hominoid cladistics and the ancestry of modern apes and humans, in: *New Interpretations of Ape and Human Ancestry* (R. L. Ciochon and R. S. Corruccini, eds.), pp. 783–843. Plenum Press, New York.

Clutton-Brock, T. H., and Harvey, P. H., 1979, Home range size, population density and phylogeny in primates, in: *Primate Ecology and Human Origins* (I. S. Bernstein and E. O. Smith, eds.), Garland, New York, pp. 201–214.

Cock, A. G., 1966, Genetical aspects of metrical growth and form in animals, *Q. Rev. Biol.* **41:**131–190.

Coolidge, H. J., 1933, *Pan paniscus:* Pygmy chimpanzee from south of the Congo River, *Am. J. Phys. Anthropol.* **18**(1):1–57.

Coolidge, H. J., 1936, Zoological results of the George Vanderbilt African Expedition of 1934. Part IV. Notes on four gorillas from the Sanage River region, *Proc. Philos. Acad. Nat. Sci.* **88:**479–501.

Coolidge, H. J., and Shea, B. T., 1982, External body dimensions of *Pan paniscus* and *Pan troglodytes* chimpanzees, *Primates* **23:**245–251.

Corruccini, R. S., and McHenry, H. M., 1979, Morphological affinities of *Pan paniscus*, *Science* **204:**1341–1343.

Currey, J. D., 1977, Problems of scaling in the skeleton, in: *Scale Effects in Animal Locomotion* (T. J. Pedley, ed.), Academic Press, New York, pp. 153–167.

Cousins, D., 1972, Body measurements and weights of wild and captive gorillas, *Gorilla gorilla, Zool. Gart. N. F. Leipzig* **41:**261–277.

Cousins, D., 1978, The diminutive *Pan, Int. Zoo News* **25:**5–11.

Davis, D. D., 1962, Allometric relationships in lions vs. domestic cats, *Evolution* **16:**505–514.

Dixon, A. F., 1981, *The Natural History of the Gorilla*, Columbia University Press, New York.

Economos, A. C., 1982, On the origin of biological similarity, *J. Theor. Biol.* **94:**25–60.

Ferris, S. D., Wilson, A. C., and Brown, W. M., 1981, Evolutionary tree for apes and humans based on cleavage maps of mitochondrial DNA, *Proc. Natl. Acad. Sci. USA***78:**2432–2436.

Fink, W. L., 1982, The conceptual relationship between ontogeny and phylogeny, *Paleobiology* **8:**254–264.

Frechkop, S., 1944, *Exploration du Parc National Albert, mission S. Frechkop*, Institute des Parcs Nationale du Congo Belge, Brussels.

Galloway, A., Allbrook, D., and Wilson, A. M., 1959, The study of *Gorilla gorilla beringei* with a postmortem report, *S. Afr. J. Sci.* **55:**205–209.

Gatti, A., 1932, *Tom-toms in the Night*, London.

Giles, E., 1956, Cranial allometry in the great apes, *Hum. Biol.* **28:**43–58.

Gingerich, P. D., Smith, B. H., and Rosenberg, K., 1982, Allometric scaling in the dentition of primates and prediction of body weight from tooth size in fossils, *Am. J. Phys. Anthropol.* **58:**81–100.

Goodall, A. G., 1977, Feeding and ranging behavior of a mountain gorilla group *(Gorilla gorilla beringei)* in the Tshibinda-Kahuzi Region (Zaire), in: *Primate Ecology* (T. H. Clutton-Brock, ed.), Academic Press, London, pp. 449–479.

Goodall, A. G., and Groves, C. P., 1977, The conservation of eastern gorillas, in: *Primate Conservation* (Prince Ranier and G. H. Bourne, eds.), Academic Press, New York, pp. 599–637.

Goodman, M., 1982, Biomolecular evidence on human origins from the standpoint of Darwinian theory, *Hum. Biol.* **54:**247–264.

Gould, S. J., 1975, Allometry in primates with emphasis on scaling and the evolution of the brain, *Contrib. Primatol.* **5:**244–292.

Gregory, W. K., and Raven, H. C., 1937, *In Quest of Gorillas*, Darwin Press, New Bedford.

Groves, C. P., 1970*a*, *Gorillas*. Avco Publishing Co., New York.

Groves, C. P., 1970*b*, Population systematics of the gorilla, *J. Zool. Lond.* **161:**287–300.

Groves, C. P., and Stott, K. W., Jr. 1979 Systematic relationships of gorillas from Kahuzi, Tshiaberimu and Kayonza, *Folia Primatol.* **32:**161–179.

Grzimek, B., 1956, Masse und Gewichte von Flachland-Gorillas, *Z. Saugetierkd.* **21**:192–194.

Günther, B., 1975, Dimensional analysis and theory of biological similarity, *Physiol. Rev.* **55**:659–699.

Gyldenstolpe, N., 1928, Zoological results of the Swedish expedition to Central Africa, 1921, *Ark. Zool.* **20**:1–76.

Heglund, N. C., 1984, Comparative energetics and mechanics of locomotion: How do primates fit in?, in: *Size and Scaling in Primate Biology* (W. L. Jungers, ed.), Plenum Press, New York, pp. 319–335.

Hill, W. C. O., 1969, The nomenclature, taxonomy and distribution of chimpanzees, in: *The Chimpanzee*, Volume 1 (G. Bourne, ed.), Karger, Basel, pp. 22–49.

Hoier, R., 1955, A travers plaines et volcans au Parc National Albert, 2nd ed., Institute des Parcs Nationale du Congo Belge, Brussels.

Holloway, R. L., and Post, D. G., 1982, The relativity of relative brain measures and hominid mosaic evolution, in: *Primate Brain Evolution* (E. Armstrong and D. Falk, eds.), Plenum Press, New York, pp. 57–76.

Horn, A., 1979, A preliminary report on the ecology of the bonobo chimpanzee *(Pan paniscus,* Schwartz, 1929), *Am. J. Phys. Anthropol.* **51**:273–281.

Howell, A. B., 1944, *Speed in Animals,* Hafner, New York.

Huxley, J. S., 1924, Constant differential growth-ratios and their significance, *Nature* **114**:895.

Huxley, J. S., 1932, *Problems in Relative Growth,* Cambridge University Press, London.

Jenkins, F. A., Jr., 1973, The functional anatomy and evolution of the mammalian humero-ulnar articulation, *Am. J. Anat.* **137**:281–298.

Jenkins, F. A., Jr., Dombrowski, P. J., and Gordon, E. P., 1978, Analysis of the shoulder in brachiating spider monkeys, *Am. J. Phys. Anthropol.* **48**:65–76.

Johnson, S. C., 1981, Bonobos: Generalized hominid prototypes or specialized insular dwarfs? *Curr. Anthropol.* **22**:363–375.

Jungers, W. L., 1978, The functional significance of skeletal allometry in *Megaladapis* in comparison of living prosimians, *Am. J. Phys. Anthropol.* **19**:303–314.

Jungers, W. L., 1979, Locomotion, limb proportions and skeletal allometry in lemurs and lorises, *Folia Primatol* **32**:8–28.

Jungers, W. L., 1980, Adaptive diversity in subfossil Malagasy prosimians, *Z. Morphol. Anthropol.* **71**:177–186.

Jungers, W. L., 1984a, Scaling of the hominoid locomotor skeleton with special reference to the lesser apes, in: *The Lesser Apes: Evolutionary and Behavioural Biology* (H. Preuschoft, D. Chivers, W. Brockelman, and N. Creel, eds.), Edinburgh University Press, Edinburgh, pp. 146–169.

Jungers, W. L., 1984b, Body size and scaling of limb proportions in primates, in: *Size and Scaling in Primate Biology* (W. L. Jungers, ed.), Plenum Press, New York, pp. 345–381.

Jungers, W. L., and Fleagle, J. G., 1980, Postnatal growth allometry of the extremities in *Cebus albifrons* and *Cebus apella:* A longitudinal and comparative study, *Am. J. Phys. Anthropol.* **53**:471–478.

Jungers, W. L., and German, R. Z., 1981, Ontogenetic and interspecific skeletal allometry in nonhuman primates: Bivariate versus multivariate analysis, *Am. J. Phys. Anthropol.* **55**:195–202.

Katz, M. J., 1980, Allometry formula: A cellular model, *Growth* **44**:89–96.

Kimura, T., Okada, M., and Ishida, H., 1979, Kinesiological characteristics of primate walking: Its significance in human walking, in: *Environment, Behavior, and Morphology: Dynamic Interactions in Primates* (M. E. Morbeck, H. Preuschoft, and N. Gomberg, eds.), Gustav Fischer, New York, pp. 297–311.

Kortland, A., 1975, Ecology and paleoecology of ape locomotion, in: *Symp. 5th Congr. Int'l. Primatol. Soc.* (S. Kondo, M. Kawai, A. Ehara, and S. Kawamura, eds.), Japan Science Press, Tokyo, pp. 361–364.

Kuhry, B., and Marcus, L. F., 1977, Bivariate linear models in biometry, *Syst. Zool.* **26:**201–209.

Lambert, R., and Teissier, G., 1927, Theorie de la similitude biologique, *Ann. Physiol.* **3:**212–246.

Lande, R., 1979, Quantitative genetic analysis of multivariate evolution applied to brain:body size allometry, *Evolution* **33:**402–416.

Lande, R., 1984, Genetic and evolutionary aspects of allometry, in: *Size and Scaling in Primate Biology* (W. L. Jungers, ed.), Plenum Press, New York, pp. 21–32.

Leutenegger, W., 1970, Beziehungen zwischen der Neugeborenengrosse und dem sexual Dimorphismus an Becken bei Primaten, *Folia Primatol.* **12:**224–235.

Lindstedt, S. L., and Calder, W. A., III, 1981, Body size, physiological time, and longevity of homeothermic animals, *Q. Rev. Biol.* **56:**1–16.

Lumer, H., 1939, Relative growth of the limb bones in the anthropoid apes, *Hum. Biol.* **11:**379–392.

McArdle, J. E., 1981, Functional morphology of the hip and thigh of the Lorisiformes, *Contrib. to Primatol.* **17:**1–132.

MacKinnon, J. 1971 The orang-utan in Sabah today. Oryx *11*:141–191.

Mai, L. L., 1983, A model of chromosome evolution in primates and its bearing on cladogenesis in the Hominoidea, *In: New Interpretations of Ape and Human Ancestry* (R. L. Ciochon and R. S. Corruccini, eds.), Plenum Press, New York, pp. 87–114.

Maple, T. L., and Hoff, M. P., 1982, *Gorilla Behavior,* Van Nostrand Rheinhold, New York.

Mayr, E., 1950, Taxonomic categories in fossil hominids, *Cold Spring Harbor Symp. Quant. Biol.* **25:**109–118.

McHenry, H. M., 1975, The ischium and hip extensor mechanism in human evolution, *Am. J. Phys. Anthropol.* **43:**39–46.

McHenry, H. M., and Corruccini, R. S., 1981, *Pan paniscus* and human evolution, *Am. J. Phys. Anthropol.* **54:**355–367.

McMahon, T. A., 1973, Size and shape in biology, *Science* **197:**1201–1204.

McMahon, T. A., 1975, Using body size to understand the structural design of animals: Quadrupedal locomotion, *J. Appl. Physiol.* **39:**619–627.

McMahon, T. A., 1977, Scaling quadrupedal galloping: Frequencies, stresses, and joint angles, in: *Scale Effects in Animal Locomotion* (T. J. Pedley, ed.), Academic Press, New York, pp. 143–151.

Mills, J., 1960, Juvenile male gorilla rescued and adult male gorilla found dead on Mountain Mgahinga in the gorilla sanctuary on 23rd February, 1960, Mimeographed report by game ranger, southern range, Uganda.

Mobb, G. E., and Wood, B. A., 1977, Allometry and sexual dimorphism in the primate innominate bone, *Am. J. Anat.* **150:**531–538.

Mosimann, J. E., and James, F. C., 1979, New statistical methods for allometry with application to Florida red-winged blackbirds, *Evolution* **33:**444–459.

Napier, J. R., and Napier, P. H., 1967, *A Handbook of Living Primates,* Academic Press, London.

Oxnard, C. E., 1978, One biologist's view of morphometrics, *Annu. Rev. Ecol. Syst.* **9:**219–241.

Pedley, T. J., 1977, *Scale Effects in Animal Locomotion,* Academic Press, London.

Pilbeam, D., and Gould, S. J., 1974, Size and scaling in human evolution, *Science* **186:**892–901.

Prange, H. D., 1977, The scaling and mechanics of arthropod exoskeletons, in: *Scale Effects* in *Animal Locomotion* (T. J. Pedley, ed.), Academic Press, New York, pp. 169–183.

Preuschoft, H., 1970, Functional anatomy of the lower extremity, in: *The Chimpanzee*, Volume 3 (G. Bourne, ed.), Karger, Basel, pp. 221–294.

Preuschoft, H., 1978, Recent results concerning the biomechanics of man's acquistiion of bipedality in: *Recent Advances in Primatology. III. Evolution* (D. J. Chivers and K. A. Joysey, eds.), Academic Press, London, pp. 435–458.

Preuschoft, H., Fritz, M., and Niemitz, C., 1979, Biomechanics of the trunk in primates and problems of leaping in *Tarsius*, in: *Environment, Behavior, and Morphology: Dynamic Interactions in Primates*, (M. E. Morbeck, H. Prenschoft, and N. Gomberg, eds.) Gustav Fischer, New York, pp. 327–345.

Radin, E. L., 1980, Biomechanics of the human hip, *Clin. Orthop. Rel. Res.* **152:**28–34.

Rahm, U., 1967, Observations during chimpanzee captures in the Congo, in: *Neue Ergebnisse der Primatologie* (D. Starck, R. Schneider, and H. J. Kuhn, eds.), Gustav Fischer, Stuttgart, pp. 195–207.

Reynolds, T. R., 1981, Mechanics of Interlimb Weight Redistribution in Primates, Doctoral Dissertation, Rutgers University, New Jersey.

Roberts, D., 1974, Structure and function of the primate scapula, in: *Primate Locomotion* (F. A. Jenkins, Jr., ed.), Academic Press, New York, pp. 171–200.

Rothenfluh, E., 1976, Überprufung der Gewichtsangaben adulter Primaten im Vergleich Zwischen Gefangenschafts- und Wildfangtieren, Semester Arb., Zurich.

Sarich, V. M., 1968, The origin of the hominids: An immunological approach. in: *Perspectives on Human Evolution* (S. L. Washburn and P. Jay, eds.), Holt, Rhinehart and Winston, New York, pp. 94–121.

Schaller, G. B., 1963, *The Mountain Gorilla*, University of Chicago Press, Chicago.

Schmidt-Nielsen, K., 1975, Scaling in biology: The consequences of size, *J. Exp. Zool.* **194:**257–308.

Schmidt-Nielsen, K., 1977, Problems of scaling: Locomotion and physiological correlates, in: *Scale Effects in Animal Locomotion* (T. J. Pedley, ed.), Academic Press, New York, pp. 1–21.

Schultz, A. H., 1930, The skeleton of the trunk and limbs of higher primates, *Hum Biol.* **2:**303–438.

Schultz, A. H., 1933, Die Körperproportionen der erwachsenen catarrhinen Primaten, mit spezieller Berucksichtigung der Menschenaffen, *Anthropol. Anz.* **10:**154–185.

Schultz, A. H., 1934, Some distinguishing characters of the mountain gorilla, *J. Mammal.* **15:**51–61.

Schultz, A. H., 1961, Vertebral column and thorax. *Primatologia* **4**(5):1–66.

Schultz, A. H., 1969, The skeleton of the chimpanzee, in: *The Chimpanzee*, Volume 1 (G. Bourne, ed.), pp. 50–103, Karger, Basel.

Schwarz, E., 1929, Das Vorkommen des Schimpansen auf den linken Kongo-Ufer, *Rev. Zool. Bot. Afr.* **XVI**(4):425–426.

Shea, B. T., 1981, Relative growth of the limbs and trunk in the African apes, *Am. J. Phys. Anthropol.* **56:**179–201.

Shea, B. T., 1983a, Phyletic size change and brain/body allometry: A consideration based on the African pongids and other primates, *Int. J. Primatol.* **4:**33–62.

Shea, B. T., 1983b, Size and diet in the evolution of African ape craniodental form, *Folia Primatol.* **40:**36–68.

Shea, B. T., 1983c, Allometry and heterochromy in the African apes, *Am. J. Phys. Anthropol.* **62:**275–290.

Simon, W. H., 1970, Scale effects in animal joints. I. Articular cartilage thickness and compressive stress, *Arthritis Rheum.* **13:**244–256.

Simon, W. H., 1971, Scale effects in animal joints II. Thickness and elasticity in the deformability of articular cartilage, *Arthritis Rheumt.* **14**:493–502.

Simon, W. H., Friedenberg, S., and Richardson, S., 1973, Joint congruence. A correlation of joint congruence and thickness of articular cartilage in dogs, *J. Bone Jt. Surg.* **55A**;1614–1620.

Simpson, G. G., 1963, The meaning of taxonomic statements, in: *Classification and Human Evolution* (S. L. Washburn, ed.), Aldina, Chicago, pp. 1–31.

Slijper, E. J., 1946, Comparative biologic-anatomical investigations on the vertebral column and spinal musculature of mammals, *Verh. Kon. Ned. Akad. Wet. Sect. II.* **42**:1–128.

Smith, J. M., and Savage, R. J. G., 1956, Some locomotory adaptations in mammals, *J. Linn. Soc. (Zool.)* **42**:603–622.

Smith. R. J., 1980, Rethinking allometry, *J. Theor. Biol.* **87**:97–111.

Smith, R. J., 1981a, On the definition of variables in studies of primate dental allometry, *Am. J. Phys. Anthropol.* **55**:323–329.

Smith, R. J., 1981b, Interpretation of correlations in intraspecific and interspecific allometry, *Growth* **45**:291–297.

Stern, J. T., Jr., 1976, Before bipedality, *Yearb. Phys. Anthropol.* **19**:59–68.

Steudel, K., 1981a, Body size estimators in primate skeletal material, *Int. J. Primatol.* **2**:81–90.

Steudel, K., 1981b, Sexual dimorphism and allometry in primate ossa coxae, *Am. J. Phys. Anthropol.* **55**:209–215.

Steudel, K., 1982, Allometry and adaptation in the catarrhine postcranial skeleton, *Am. J. Phys. Anthropol.* **59**:431–441.

Susman, R. L., 1979, Comparative and functional morphology of hominoid fingers, *Am. J. Phys. Anthropol.* **50**:215–236.

Susman, R. L. and Jungers, W. L., 1981, Comment on "Bonobos: generalized hominid prototypes or specialized insular dwarfs?" *Current Anthropol.* **22**:396–370.

Susman, R. L., Badrian, N. L., and Badrian, A. J., 1980, Locomotor behavior of *Pan paniscus* in Zaire, *Am. J. Phys. Anthropol.* **53**:69–80.

Szalay, F. S., and Delson, E., 1979, *Evolutionary History of the Primates*, Academic Press, New York.

Taylor, C. R., 1977, The energetics of terrestrial locomotion and body size in vertebrates, in: *Scale Effects in Animal Locomotion* (T. J. Pedley, ed.), Academic Press, New York, pp. 127–141.

Templeton, A. R., 1983, Phylogenetic inference from restriction endonuclease cleavage site maps with particular reference to the evolution of humans and the apes, *Evolution* **37**:221–244.

Tsutakawa, R. K., and Hewett, J. E., 1977, Quick test for comparing two populations with bivariate data, *Biometrics* **33**:215–219.

Tuttle, R. H., 1967, Knuckle-walking and the evolution of hominoid hands, *Am. J. Phys. Anthropol.* **26**: 171–206.

Tuttle, R. H., 1969, Knuckle-walking and the problem of human origins, *Science* **166**:953–961.

Tuttle, R. H., 1975, Parallelism, brachiation, and hominoid phylogeny, in: *Phylogeny of the Primates: A Multidisciplinary Approach* (W. P. Luckett and F. S. Szalay, eds.), Plenum Press, New York, pp. 447–480.

Vrba, E. S., 1979, A new study of the scapula of *Australopithecus africanus* from Sterkfontein, *Am. J. Phys. Anthropol.* **51**:117–130.

Willoughby, D. P., 1950, The gorilla—Largest living Primate, *Sci. Monthly* 1950(January):48–57.

Wood, B. A., 1979, Relationship between body size and long bone lengths in *Pan* and *Gorilla*, *Am. J. Phys. Anthropol.* **50**:23–25.

Wrangham, R. W., and Smuts, B. B., 1980, Sex differences in the behavioural ecology of chimpanzees in the Gombe National Park, Tanzania, *J. Reprod. Fertil. Suppl.* **28**:13–31.

Zihlman, A. L., 1979, Pygmy chimpanzee morphology and the interpretation of early hominids, *S. Afr. S. Sci* **75**:165–168.

Zihlman, A. L., and Cramer, D. L., 1978, Skeletal differences between pygmy *(Pan paniscus)* and common chimpanzees *(Pan troglodytes), Folia Primatol.* **29**:86–94.

Zihlman, A. L., Cronin, J. E., Cramer, D. L., and Sarich, V. M., 1978, Pygmy chimpanzee as a possible prototype for the common ancestor of humans, chimpanzees, and gorillas, *Nature* **275**:744–746.

CHAPTER 8

Body Build and Tissue Composition in Pan paniscus and Pan troglodytes, with Comparisons to Other Hominoids

ADRIENNE L. ZIHLMAN

1. Introduction

Pygmy chimpanzees (*Pan paniscus*) have received a great deal of attention as a possible prototype for the African hominoid ancestor (Zihlman *et al.*, 1978; Johnson, 1981; Latimer *et al.*, 1981), but they are interesting for additional reasons. First, comparing body build and tissue composition between *Pan paniscus* and *Pan troglodytes* documents in detail the extent of morphological difference between the two species. Further comparison with Asian apes provides insights into the basic pongid adaptation and contributes to analysis of the correlation between body build and locomotor behavior. Finally, comparison with fossil and living hominids offers clues on the changing body build correlated with the origin of bipedal locomotion.

In his early studies of pygmy chimpanzees, Coolidge (1933) raised three central issues: (1) he argued for the specific status of *Pan paniscus;* (2) he concluded that the two chimpanzee species differed in morphology and that *P. paniscus* was more generalized than the common species; and (3) he believed that *P. paniscus* "may approach more closely to the common ancestor of chimpanzees and man than does any living chimpanzee" (Coolidge, 1933, p. 56). Other researchers supported Coolidge's conclu-

ADRIENNE L. ZIHLMAN ● Department of Anthropology, University of California at Santa Cruz, Santa Cruz, California 95064.

sions about the existence of two species (Frechkop, 1935; Tratz and Heck, 1954; Vandebroek, 1959).

Recent and more elaborate laboratory and field research documents distinct behavioral patterns in communication and social organization (Savage *et al.*, 1977; Savage-Rumbaugh and Wilkerson, 1978; Kuroda, 1979, 1980; Kano, 1980; see also subsequent chapters in this volume). For example, in *P. paniscus* affiliative behavior is highest between male and female adults, and next highest among adult females, contrasted with strong male–male bonds and female "unsociability" among *P. troglodytes* (Kano, 1980; Wrangham, 1979). Kuroda (1979) views pygmy chimpanzees as different from common chimpanzees in their social behavior. Although both species have a flexible social organization, Kuroda considers the pygmy chimpanzee's social structure as the least specialized of all African apes.

That two distinct species exist has been subject to disagreement. On the basis of a small sample of limb bone dimensions, which overlap in the two species, Schultz (1969) correctly questioned the term "pygmies" as applied to *P. paniscus*. Osman Hill (1969) accepted the two-species designation, whereas Horn (1979) and Groves (1982) argue for subspecific status for all chimpanzees, based on overlapping skeletal dimensions. Horn argues that even though statistical differences in the metric analysis of some skeletal elements exists, these are not sufficient to support the hypothesis that biological differences separate the two groups. There is no doubt that many measurements do overlap, and that their full interpretation cannot be made until the various racial or subspecific chimpanzee groups have been compared (Groves, 1981). However, these rejections of *P. paniscus* as a distinct species have not taken all the evidence into account.

The analysis of morphotypes within *Pan* has relied most often upon measurements of the cranium and mandible (Coolidge, 1933; Fenart and Deblock, 1973; Cramer, 1977). On the one hand, the average difference in cranial capacity between *P. paniscus* and *P. troglodytes* is 350 cm³ versus 390 cm³, respectively, but overlap is considerable (Cramer, 1977). Most cranial and facial measurements are smaller in *P. paniscus*, but here, too, there is overlap. On the other hand, the two species can be perfectly discriminated on mandibular breadth (Cramer, 1977). The molecular data—based upon two samples of captive pygmy chimpanzee groups and several samples of common chimpanzees—indicate that the two groups should be considered two species (Cronin, 1983; see also Sarich, this volume).

In contrast, the postcranial data on the two species are less clearcut. The sample of skeletons used in Coolidge's study was small, but even in later studies with larger sample sizes, the linear dimensions and body weight* overlapped in the two species. The relative dimensions of a number of skeletal parts suggest that the two species differ in body proportions, for example, in clavicle length (and therefore girth of the upper trunk), in overall pelvic length and breadth (and therefore in girth of the lower trunk), and in relative limb lengths (Zihlman and Cramer, 1978). However, metric studies do not definitively distinguish the species in the postcranial skeleton and it is desirable to seek other methodologies that can quantify additional anatomical information and thereby tease out more clearcut morphological differences.

In this chapter I present another set of data that help to assess both the morphological differences between the two chimpanzee species and their possible locomotor behavioral correlates. The methodology is an elaboration of that developed by Grand (1977a,b, 1978a,b) and consists in determining (1) the weight of tissue types relative to total body weight (TBW) and (2) the distribution of tissue types within body segments: the upper limbs, lower limbs, and head/trunk. The distribution of body weight to the body segments gives a profile of an animal's body build, and Grand has demonstrated that these parameters correlate with locomotor type.

Grand's methodology also quantifies the distribution of tissue types over the various body segments. Tissue composition can vary. For example, the skin of sloths, howler monkeys, and macaques contributes more than 12% of body weight, whereas in greyhound dogs the skin is only 5% of body weight (Grand, 1977a). In sloths and howler monkeys, both arboreal folivores, musculature comprises 25% of body weight; in macaques it is about 40%; whereas greyhound dogs and agoutis have over 50% muscle (Grand, 1977a, 1978a). Distribution of body weight over segments also varies. Pottos, which are slow climbers, and galagos, specialized leapers, may be similar in total body weight, but have very different limb weights: in pottos the forelimbs and hindlimbs are almost equal (12.4% versus 14%), whereas in galagos the hindlimb is 22.6% of body weight and the forelimbs only 9% (Grand, 1977a). This type of body build

*Present information on body weights of *Pan paniscus* and *Pan troglodytes* shows a great deal of overlap between them. Based on weights I compiled, I reported an average for *P. paniscus* of about 35 kg, with females about 31 kg and males about 38 kg (Zihlman and Cramer, 1978). These figures are consistent with four additional *P. paniscus* body weights. However, if we include the body weight of 61 kg for an adult male *P. paniscus* reported by Coolidge and Shea (1982), this average estimate, particularly for males, may be too low. Male *P. paniscus* may achieve greater body weight than previously thought. The average weights for *P. paniscus* are nearly identical to the weights from Gombe chimpanzees (*P. troglodytes*) reported in Zihlman and Cramer (1978), with the female average about 31 kg and the male 38 kg. From other data on male *P. troglodytes*, they frequently exceed 60 kg in adult weight.

analysis shows that convergences occur in segmental distribution patterns. Sloths converge with pottos and lorises, marmosets with tree shrews, owl monkeys with galagos, and cebus monkeys with macaques (Grand, 1977*a*).

The analysis makes comparisons more illuminating: betweem mammalian orders (e.g., primates, carnivores, edentates, rodents); within orders (e.g., Primates); within genera (e.g., *Macaca*); between animals of similar body weight; or between animals occupying a similar niche. Species within the genus *Macaca* show an overall similarity is tissue composition, but species differences in segment distribution can be discerned among *M. fuscata, M. mulatta,* and *M. fasicularis* (Grand, 1977*a*). These data suggest the possibility that differences in body build between the two chimpanzee species may be detected even though many of their linear skeletal dimensions are so similar. Further, because tissue and segmental weights and distribution correlate with locomotor pattern, such information may direct attention to possible species differences in the motor behavior of chimpanzees.

2. *Pan paniscus* and *Pan troglodytes:* A Comparison of Body Builds

Coolidge's designation of two species was based on external characteristics, such as skin and hair, and skeletal features, especially cranial. He initially compared one adult female *P. paniscus* skeleton with a *P. troglodytes* sample from Schultz (1930), and an infant of each species. In clavicular, scapular, and pelvic dimensions *P. paniscus* was smaller and also had a lower intermembral index. In nine skull measurements and in cranial capacity *P. paniscus* was also the smaller.

A larger sample of adult *P. paniscus* skeletons confirmed and elaborated these early findings (Fenart and Deblock, 1973; Cramer, 1977; Cramer and Zihlman, 1978; Zihlman and Cramer, 1978). In addition, there was shown to be less sexual dimorphism in *P. paniscus* than in *P. troglodytes* (Cramer and Zihlman, 1978). In comparison with gorillas and common chimpanzees, pygmy chimpanzees have longer upper and lower limbs relative to trunk length (Shea, 1981). This may have been the primitive African ape condition. Pygmy chimpanzees also have smaller chest girth in relation to height (Coolidge and Shea, 1982). Thus, there is considerable evidence that the body builds of the two chimpanzee species represent two morphotypes that are not merely allometric variants as argued by Corruccini and McHenry (1979) and McHenry and Corruccini (1981). (See also, Jungers and Susman, this volume; Shea, this volume).

Body proportions are most often deduced from lengths of skeletal elements (Schultz, 1937, 1973), which define distances between moving parts and hence the lever systems related to locomotor adaptation. To extend metric studies I collected data on tissue composition and segment weights of *P. paniscus* and *P. troglodytes*. Linear dimensions correlate with body size, so that it is critical to compare animals of the same sex and similar ages and body weights. When these variables are held constant, the similarities and differences in body composition provide a check, or refinement of, skeletal dimensions. If differences are demonstrated, they suggest species differences in morphology. The study looks at the body as a whole and at individual muscular–skeletal groups, and compares individuals of similar body weight in closely related species.

3. Methods

Two female chimpanzees were dissected according to the methods developed and described by Grand (1977*a*). The female *P. paniscus* Anne Marie* was about 24 years old and weighed 29.5 kg. The female *P. troglodytes* Prude† was adult and weighed 31.5 kg at autopsy. The cause of death for both animals was accidental rather than from disease. Their bodies were frozen and dissected fresh.

Measurements of segment lengths and circumferences were taken externally before dissection. In both animals, the left side of the body was dissected segmentally; cuts were made at the joints, and wet weights were recorded of the skin, muscle, and bone of each segment. Tissue composition of the whole body, as well as each segment of the lower limb (thigh, leg, and foot) and upper limb (arm, forearm, hand) could be calculated. The right sides of the two animals were dissected side by side; corresponding muscles were taken from origin and insertion at the same

---

*The female *Pan paniscus* Anne Marie was wild-born and arrived in captivity at an estimated age of 4 years. Initially she was housed at the Antwerp Zoo and subsequently at the Fort Wayne, Indiana, Zoo. I obtained her from Dr. Sue Savage-Rumbaugh for study. Anne Marie was about 24 years at death and in excellent health. Body weight prior to death was 29.5 kg. Anne Marie has been pictured in a number of publications (e.g., Hill, 1969) and her history is outlined in Gijzen (1974).

†The female *Pan troglodytes* Prude was a captive animal obtained for study from the Laboratory for Experimental Medicine and Surgery in Primates (LEMSIP), New York University Medical Center. She was adult, and weighed 31.5 kg at autopsy, although a few months before death she weighed 34 kg. She was in reasonably good health at the time of death. Compared to the female *P. paniscus*, she had a higher percentage of body fat, about 10% compared to 3%.

time and weighed wet. After cleaning the bones, cranial capacities were taken, bones measured, and photographs taken.

4. Results

There are striking differences in the two chimpanzee females in the overall size of the skull, degree of craniofacial robusticity, mandibular length, and cranial capacity. Prominent brow ridges and facial bones are evident in *P. troglodytes* in Fig. 1. Table I gives cranial capacities, dimensions, and mandibular lengths for the two females.

For the postcranial skeleton, three kinds of information are provided: linear dimensions (Table II), photographs of the bones for visual comparison with the linear and mass data (Figs. 2–5), and the results of tissue and segment weights (Figs. 6–10).

In linear dimensions, the clavicular, scapular, and pelvic dimensions (but for sacral breadth) are smaller in *P. paniscus,* as noted in Table II and as can be seen in Figs. 2 and 3. Similarly, lengths of the upper limb

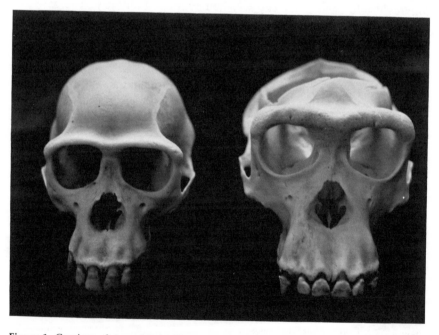

Figure 1. Cranium of Anne Marie (*Pan paniscus*) on left and Prude (*Pan troglodytes*) on right (skull cap missing). See Table I for measurements.

Table I.   Comparison of Craniofacial Measurements of *Pan paniscus* and *Pan troglodytes*

| Variable | *Pan paniscus* | | *Pan troglodytes* | |
	Anne Marie	Average[a]	Prude	Average[a]
Vault length, mm	119.	122.8	128.	135.3
Vault breadth, mm	104.	105.7	117.	121.4
Vault height, mm	87.	85.4	97.[b]	89.3
Interorbital breadth, mm	12.4	14.5	20.5	20.0
Postorbital breadth, mm	63.7	65.6	69.0	69.7
Bizygomatic breadth, mm	105.5	110.9	120.0	124.9
Palatine breadth, mm	49.1[c]	51.5	64.1	59.7
Palatine length, mm	60.3	58.1	69.0	76.3
Mandibular breadth, mm	93.5	93.4	95.0	105.2
Mandibular length, mm	111.	102.4	136.5	130.
Mandibular height, mm	48.	55.4	63.	66.9
Cranial capacity, cm³	358	350	415	390

[a]Population averages for adults of both species from Cramer (1977).
[b]Estimate only; skull cap missing.
[c]M¹, M², M³ abscessed or missing.

bones are smaller in Anne Marie (*P. paniscus*) than in Prude (*P. troglodytes*) (Fig. 4), although when population averages are included the two species appear closer than do these two individuals. The two female chimpanzees are very similar in lengths of the femur, tibia, and fibula (Table II and Fig. 5).

The weight-of-tissue and weight-of-segment data add another dimension and give a more detailed picture of body composition. The distribution of tissue types—muscle, bone, skin—shows that *P. paniscus* has a greater percentage of muscle (Fig. 6) than *P. troglodytes*. Although this might reflect individual variation, data on two other *P. paniscus* do confirm a large amount of muscle tissue relative to total body weight (A. Zihlman, unpublished data).

How the total body weight is distributed among the segments is seen in Figs. 7 and 8. The weight of the upper limb segments (Fig. 7) relative to total body weight is the same, at 15.8% in both females. Within the upper limb, one can see that the arm, forearm, and hand segments are similar.

The lower limbs, however, are quite different in the two chimpanzee females. In *P. troglodytes* each lower limb is 9.2% total body weight,

Table II. Comparison of Skeletal Measurements in *Pan paniscus* (Both Sexes) and *Pan troglodytes* (Females Only)

| Variable, mm | Pan paniscus | | | Pan troglodytes | | |
	Anne Marie	Average[a]	SD	Prude	Average[a]	SD
Clavicle length	102	105	4.3	116	122	8.
Scapula length	132	138	11.0	145	140	8.8
Scapula breadth	67	72	5.4	74.5	76	4.8
Innominate length	242	253	15.	264	259	13.
Iliac breadth	81.5	97	8.7	111	116	12.
Sacral breadth	67.5	63	4.8	61.3	67.5	6.
Humerus length	270	285	12.	284	286	18.
Radius length	249	262	14.	260	264	14.
Ulna length	256	274	13.	260	278	16.
Femur length	280	293	10.	282	281	13.
Tibia length	228	242	10.	228	233	11.5
Fibula length	201	218	7.5	205	218	12.

[a]Population averages and standard deviations from Zihlman and Cramer (1978).

18.4% for both limbs, whereas in *P. paniscus,* each limb is 12.1%, 24.2% for both. As shown in Fig. 8, the difference lies, not in the foot segment, but in the thigh and calf segments.

The upper and lower limbs can be compared in another way, by looking at the distribution of total muscle and of total bone to the various body segments. Figure 9, reporting bone distribution, shows that *P. paniscus* has slightly more bone in the lower than in the upper limb. In *P. troglodytes* the reverse is true and more bone occurs in the upper than in the lower limb. Similarly, muscle distribution (Fig. 10) in *P. troglodytes* shows greater muscle weight in the upper than lower limb, whereas in *P. paniscus* there is relatively more muscle in the lower than in the upper limb.

Thus, even though these two female chimpanzees weigh the same, they differ considerably in how their weight is distributed. The relative weights of their upper and lower limbs differ more than would have been predicted from the lengths of the limb bones alone. The limbs may have more or less muscle for the same amount of bone, and *P. paniscus* apparently has more muscle in general than *P. troglodytes,* especially in the lower limbs. These findings are significant, but of course a larger sample is needed to quantitatively verify the variation within each species as well as between species.

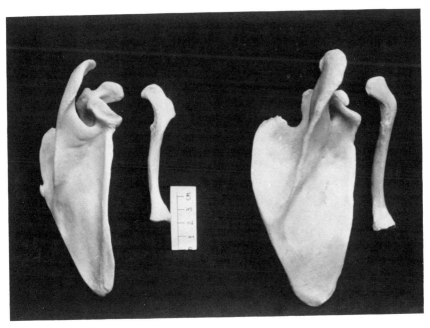

Figure 2. Clavicle and scapula of Anne Marie (*Pan paniscus*) on left and Prude (*Pan troglodytes*) on right. See Table II for measurements.

The difference in weight of segments between *P. paniscus* and *P. troglodytes* raises the issue how morphology correlates with locomotor behavior. It also raises the contested issue of whether *P. paniscus* and *P. troglodytes* do differ in locomotor behavior, and if so, how. Before addressing these issues, it is instructive to look at examples from other hominoid species where the correlation between body build and locomotor behavior is more straightforward.

5. Body Build and Locomotor Pattern: Case Studies

The correlation between tissue and segment weights, motor behavior, and locomotor pattern has been demonstrated for a variety of primates and other mammals (Grand, 1977a, 1978a). Monkeys and apes, compared to other mammals, characteristically have more weight in the forearm and hand, a feature reflecting the use of the hand in grasping, feeding,

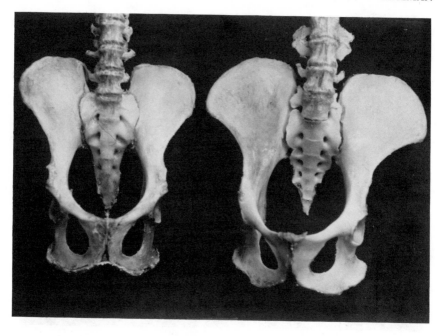

Figure 3. Pelvis of Anne Marie (*Pan paniscus*) on left and Prude (*Pan troglodytes*) on right. See Table II for measurements.

and manipulation. A heavy foot is also characteristic of prosimians, monkeys, and apes and reflects its grasping functions.

The relative weight of upper versus lower limbs correlates with substrate and locomotor type. For example, the hindlimbs of the macaque, comprising 25% of total body weight, are well adapted for propulsion. This pattern is similar to other ground-dwelling cursorial animals, such as the dog (20.2%), cat (20.8%), and agouti (20%). The forelimbs, however, may be as little as 4.6% in the agouti, 8.6% in the cat, and 9.4% in the dog, compared with 13% in the macaque (Grand, 1977a, 1978a,b).

This kind of comparison helps define the hominoid adaptation in general and delineate possible variations within the group.

5.1. *Symphalangus*

Long and heavy upper limbs (20% total body weight) and well-developed lower limbs (18%) characterize an adult female siamang.* Of total

*This adult female siamang weighed 7.5 kg at death and was obtained by Ted Grand from the Washington Park Zoo in Portland. The animal was dissected by Lynda Brunker at the Oregon Regional Primate Center.

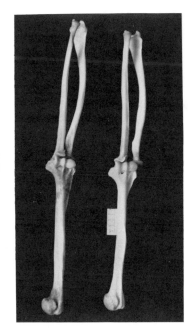

Figure 4. Humerus, radius, and ulna of Anne Marie (*Pan paniscus*) on left and Prude (*Pan troglodytes*) on right.

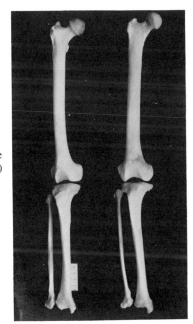

Figure 5. Femur, tibia, and fibula of Anne Marie (*Pan paniscus*) on left and Prude (*Pan troglodytes*) on right. See Table II for measurements.

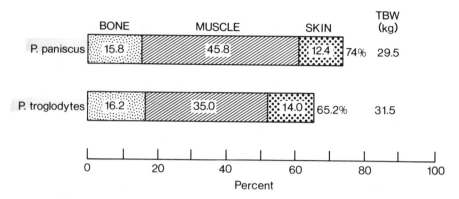

Figure 6. Distribution of tissue types (bone, muscle, skin) in a female *Pan paniscus* and a female *Pan troglodytes* relative to total body weight (TBW).

body musculature, almost half (47%) lies in the upper limbs (including scapular muscles) and 33.9% in the lower limbs. Of total bone, the upper limbs have 24% compared with 22% in the lower limbs. Siamangs rely heavily upon upper limb motions—brachiation, climbing, and hanging (Fleagle, 1976). Brachiation accounts for 51% of locomotor activity and 67% of distance traveled. Their well-developed upper limbs correlate with such locomotor behavior. Bipedalism and leaping each accounted for 6% of locomotor activity, and the well-developed lower limbs (18%) correlate with these bipedal and leaping modes.

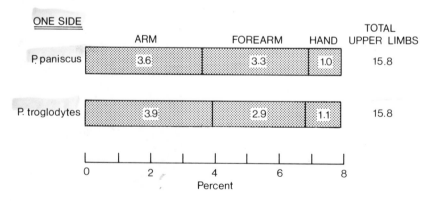

Figure 7. Upper limb segments (arm, forearm, hand) relative to total body weight in a female *Pan paniscus* and a female *Pan troglodytes*.

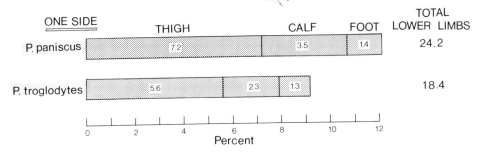

Figure 8. Lower limb segments (thigh, calf, foot) relative to total body weight in a female *Pan paniscus* and a female *Pan troglodytes*.

5.2. *Pongo*

The upper and lower limbs of a young adult female orangutan* each represent 18% of body weight. Muscle and bone tissue, however, are not equally distributed. There is more muscle in the upper than in the lower limb (47% versus 40%) and more bone in the hindlimb (23.5% versus 20.5%). Orangutans have extremely mobile hip, knee, and ankle joints with a great deal of rotation, and the femoral head is round and lacks a ligamentum teres. Their locomotor activity is that of a versatile, highly arboreal climber, although the large males occasionally move on the ground (MacKinnon, 1974; Galdikas, 1979). Orangutans often move below the branches using their large and prehensile hands and feet, and thereby they achieve somewhat of an equal balance of upper and lower limb activity during locomotion. The pattern of relatively equal upper and lower limbs is characteristic of other arboreal climbers, such as pottos, lorises, and sloths (Grand, 1977a, 1978a).

5.3. *Pan troglodytes* and *Pan paniscus*

The length and relative weight of the upper limbs (15.8%) in both species reflect their importance in climbing and hanging during arboreal locomotion, and the lower limbs reflect propulsive functions. Both species knuckle walk, as do gorillas. *Pan troglodytes* feeds, sleeps, climbs, and hangs in trees but moves on the ground to get from tree to tree. They

---

*The female orangutan Bunga was born and raised at Yerkes Regional Primate Center. She weighed 27.8 kg at autopsy and was 9 years old. Ted Grand acquired the carcass for study and the animal was dissected by Lynda Brunker at the Oregon Regional Primate Center. The 27.8 kg body weight lies within the adult female orangutan range reported by Schultz (1941), although the average is 37 kg.

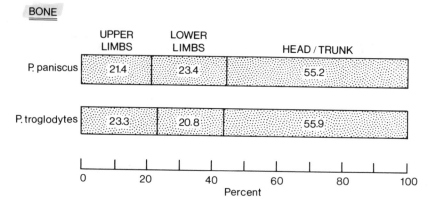

Figure 9. Distribution of bone to body segments relative to total bone weight in a female *Pan paniscus* and a female *Pan troglodytes*.

sometimes move bipedally in the trees and on the ground, and if injured are capable of sustained bipedal locomotion (Goodall, 1968; Bauer, 1977).

Pan paniscus has a similar range of locomotor behaviors—climbing and hanging in trees, quadrupedal locomotion on the ground and in the trees, leaping, and bipedal behaviors (Susman *et al.*, 1980; Susman, this volume). The relatively longer and heavier lower limbs in *P. paniscus* suggests a greater reliance upon motor behaviors involving the lower limb than does *P. troglodytes*, although comparable locomotor data are not available on the two species.

The anatomical data do raise the question once again of whether one chimpanzee species is more or less arboreal than the other, and therefore

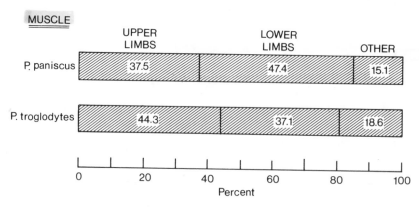

Figure 10. Distribution of muscle to body segments relative to total muscle weight in a female *Pan paniscus* and a female *Pan troglodytes*.

whether one species is more or less specialized for forest living. On the one hand, *P. paniscus* is more agile in movements in the trees; they move and sleep higher in the canopy and may flee through the forests using the high canopy (MacKinnon, 1978; Kano, 1979; Horn, 1980). On the other hand, *P. paniscus* often move through the forests on the ground and make use of the ground in feeding, digging, fishing, and constructing ground nests, leading Kano (1979) to conclude that *P. paniscus* is primarily terrestrial. On the issue of forest-living, Reynolds (1967) argues that pygmy chimpanzees have probably lived longer and more completely inside the forest than common chimpanzees, whereas MacKinnon (1978) argues that *P. paniscus* has reverted to the forests because it would be less ground-dependent and more like the arboreal Asian apes if it had enjoyed a long and uninterrupted evolution in the rain forest. No resolution of these issues is possible at this time.

5.4. *Homo sapiens*

Our own species has a unique body build among the hominoids. The lower limb has become longer and heavier, whereas the upper limb, although similar in absolute length to that of chimpanzees (Schultz, 1968), has become much lighter relative to body weight. In a human female* the upper limbs represent 8% of total body weight, the lower limbs 30.4%. There is three times more bone and over 3.5 times more muscle in the lower than in the upper limbs. During locomotion the lower limbs perform all functions (propulsion, balance). The upper limbs do not bear weight, though they do assist in balance.

6. *Pan paniscus:* Its Place among the Hominoids

The data on body build of hominoids presented here are summarized in Fig. 11. It is possible to conclude that long and heavy upper limbs are a hominoid, though not a hominid trait. The *P. paniscus* and *P. troglodytes* patterns are similar, although the former has more weight in the lower limb and less in the head/trunk. The Asian hominoids are a *departure* from the chimpanzee pattern in morphology and locomotor behavior. In contrast, the hominids, in developing lighter upper limbs, are a departure from the African as well as the Asian hominoids. Although hominids have

---

*Previously published data on segment weights (Zihlman and Brunker, 1979) were calculated from Dempster and Gaughran (1967) using adult males only (upper limbs 9.4% total body weight, lower limbs 32%). The data presented here are calculated from an adult human female weighing 69.5 kg, which I dissected.

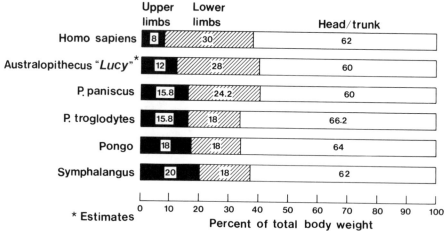

Figure 11. Distribution of body weight to body segments in five female hominoids. Included is an estimate of body weight distribution in Lucy.

retained absolute length, the relative weight of the upper limbs has decreased as part of a reorganization of the body build for bipedalism. The heavy lower limbs of modern humans greatly surpass those of the apes.

Because bipedalism is the definitive hominid characteristic, the means by which it arose from a quadrupedal ape ancestor remains a central question. The transition to bipedalism may have occurred in 1.5 million years or less, given a divergence time of 5 million years ago (MYA) according to the molecular data (Zihlman and Brunker, 1979). Although modern hominids and living pygmy chimpanzees seem widely separated in body build (Fig. 12), the appropriate comparison with the earliest hominids of 3–3.5 MYA is with an ape. But which ape? Tuttle has maintained that we cannot choose among alternative models for "the proximate ancestors of emergent Hominidae" (Tuttle, 1977, p. 292).

To help decide among alternatives, the early hominid Lucy is compared with a female orangutan, siamang, and pygmy chimpanzee (Figs. 13–15). Because *P. paniscus* and Lucy are so similar in overall body size, cranial capacity, and lower limb length, the comparison makes it possible to estimate possible relative masses of upper and lower limbs (Zihlman, 1979; Zihlman and Lowenstein, 1982). I estimate Lucy's upper limbs as about 12%, a considerable reduction from the apes, whereas her lower limb may have approached or even surpassed 28% (see Fig. 11). *Pan paniscus* provides a better morphological link between the early hominids and ape ancestor than does the siamang or orangutan. *Pan troglodytes* is also a better model than either Asian ape for the common ancestor and was used extensively prior to more detailed knowledge about *P. paniscus*. Given the direction in which *P. paniscus* varies with respect to other apes—heavier lower limbs—this species refines the ancestral model and clarifies what might have happened in the morphological transformation. The possible existence of an ape ancestor like *P. paniscus* suggests no great morphological leap from the quadrupedal ape ancestor to hominids, and perhaps less of a behavioral leap than previously thought.

For some time, the ape like—in particular, chimpanzeelike—features of the australopithecine pelvic and limb bones have been noted (Le Gros Clark, 1947, 1967; Zuckerman, 1966). More recently studies on Hadar and Olduvai hominid fossils indicate they not only had chimpanzee-like features, but that these were developed to such an extent that the animals were capable of climbing (Stern and Susman, 1983; Susman and Stern, 1982). Although the hominid morphology reflects possible climbing abilities, the decrease in length and relative weight of the upper limbs, as indicated by Lucy, would certainly have made climbing more difficult. Given the morphology and behavioral abilities of pygmy chimpanzees, it is no longer necessary to hypothesize an unusual gait in order to characterize the locomotor pattern of early hominids (*contra* Charteris *et al.*, 1981). The early hominids at the very least must have been biomechanically different from modern hominids in their locomotor pattern, which

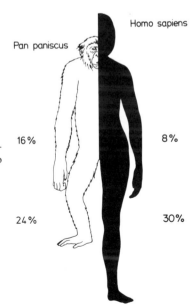

Figure 12. Comparison of body build of *Pan paniscus* and *Homo sapiens* drawn approximately to scale.

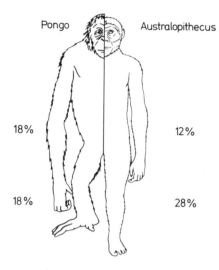

Pongo   Australopithecus

18%   12%

18%   28%

Figure 13. Comparison of body build of *Pongo* and that estimated for *Australopithecus* (Lucy). Drawn approximately to scale.

may or may not have affected gait per se, and the difference in body build between the early and modern hominids provides further anatomical support for this conclusion (Jungers, 1982; Zihlman, 1978).

Whether or not dentition or sexual dimorphism of pygmy chimpanzees is suitable for comparison with early hominids (e.g., Latimer *et al.*, 1981), and regardless of whether the common chimpanzee is just as suitable a model (McHenry and Corruccini, 1981), the morphological evidence of body build and behavioral correlates make it more difficult to deny the usefulness of pygmy chimpanzees in contructing the transition from ape

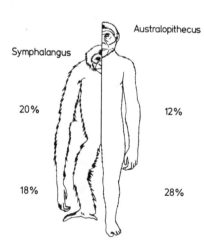

Australopithecus

Symphalangus

20%   12%

18%   28%

Figure 14. Comparison of body build of *Symphalangus* and that estimated for *Australopithecus* (Lucy). Drawn approximately to scale.

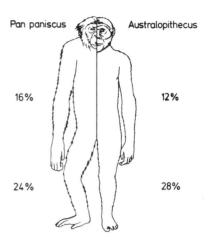

Pan paniscus   Australopithecus

16%   12%

24%   28%

Figure 15. Comparison of body build of *Pan paniscus* and that estimated for *Australopithecus* (Lucy). Drawn approximately to scale.

to human. The siamang and orangutan body build show a trend in the opposite direction from hominids—toward heavier, not lighter, upper limbs. Such conclusions cannot be inferred from bones alone; the methodology of segment and tissue weight analysis makes it possible to go beyond the limitations of the metric and skeletal data. The results of body build studies argue even more convincingly than do the bones that pygmy chimpanzees may well represent in locomotor adaptation the precursor to a small bipedal hominid of 3.5 MYA.

ACKNOWLEDGMENTS. This study has spanned a number of years and is part of a larger project involving the cooperation of a number of people. I am especially grateful to Lynda Brunker for her dissection skills, her contribution to the development of the methodology, and her comments and support during the past several years. Ted Grand, whose methodology forms the basis of this study, has provided assistance and comments during the project and generously made the siamang and orangutan available for study. Sue Savage-Rumbaugh, by offering the pygmy chimpanzee for study, was responsible for the genesis of this project, and I thank her for her cooperation and interest over the years. I also thank LEMSIP for providing the common chimpanzee for study; the Washington Park Zoo in Portland, Oregon, for the siamang; and Yerkes Regional Primate Research Center for the orangutan.

Catherine Borchert and Ted Grand made helpful comments on the manuscript, and Carla Simmons drews Figs. 12–15. Grants from the Faculty Research Committee, University of California, Santa Cruz, are acknowledged with appreciation.

References

Bauer, H. R. 1977. Chimpanzee bipedal locomotion in the Gombe National Park, East Africa, *Primates* **18**:913–921.

Charteris, J., Wall, J. C., and Nottrodt, J. W., 1981, Functional reconstruction of gait from the Pliocene hominid footprints at Laetoli, northern Tanzania, *Nature* **290**:496–498.

Coolidge, H. J., 1933, *Pan paniscus:* Pygmy chimpanzee from south of the Congo River, *Am. J. Phys. Anthropol.* **18**(1):1–57.

Coolidge, H., and Shea, B., 1982, External body dimensions of *Pan paniscus* and *Pan troglodytes* chimpanzees, *Primates* **23**(2):245–251.

Corruccini, R. S., and McHenry, H. M., 1979, Morphological affinities of *Pan paniscus*, *Science* **204**:1341–1343.

Cramer, D. L., 1977, Craniofacial morphology of *Pan paniscus:* A morphometric and evolutionary appraisal, *Contrib. Primatol.* **10**:1–64.

Cramer, D. L., and Zihlman, A. L., 1978, Sexual dimorphism in the pygmy chimpanzee, *Pan paniscus,* in: *Recent Advances in Primatology,* Volume 3, *Evolution* (D. J. Chivers and K. A. Joysey, eds.), Academic Press, London, pp. 487–490.

Cronin, J. E., 1983, Apes, humans and molecular clocks: A reappraisal, in: *New Interpretations of Ape and Human Ancestry* (R. Ciochon and R. Corruccini, eds.), Plenum, New York pp. 115–151.

Dempster, W. T., and Gaughran, G. R. L., 1967, Properties of body segments based on size and weight, *Am. J. Anat.* **120**:33–54.

Fenart, R., and Deblock, R., 1973, *Pan paniscus* et *Pan troglodytes* craniométrie: Étude comparative et ontogénique selon les méthodes classiques et vestibulaires. Tome I, *Mus. R. Afr. Centr. Tervuren, Belg. Ann. Ser. Octavo Sci. Zool.* **204**:1–473.

Fleagle, J. G., 1976, Locomotion and posture of the Malayan siamang and implications for hominoid evolution, *Folia Primatol.* **26**:245–269.

Frechkop, S., 1935, Notes sur les mammifères. XVII. A propos du Chimpanzé de la rive gauche du Congo, *Mus. R. Hist. Nat. Belg. Bull.* **11**:1–43.

Galdikas, B. M. F., 1979, Orangutan adaptation at Tanjung Puting Reserve: Mating and ecology, in: *The Great Apes* (D. Hamburg and E. S. McCown, eds.), Benjamin/Cummings, Menlo Park, California, pp. 194–233.

Gijzen, A., 1974, Studbook of *Pan paniscus* Schwarz, 1929, *Acta Zool. Pathol. Antverp.* **61**:119–164.

Goodall, J., 1968, The behavior of free living chimpanzees in the Gombe Stream Reserve, *Anim. Behav. Monogr.* **1**:161–311.

Grand, T. I., 1977a, Body weight: Its relation to tissue composition, segment distribution, and motor function. I. Interspecific comparisons, *Am. J. Phys. Anthropol.* **47**:211–239.

Grand, T. I., 1977b, Body weight: Its relation to tissue composition, segment distribution, and motor function. II. Development of *Macaca mulatta, Am. J. Phys. Anthropol.* **47**:241–248.

Grand, T. I., 1978a, Adaptations of tissues and limb segments to facilitate moving and feeding in arboreal folivores, in: *The Ecology of Arboreal Folivores* (G. G. Montgomery, ed.), Smithsonian Institution Press, Washington, D. C., pp. 231–241.

Grand, T. I., 1978b, The anatomical basis of locomotion, *Am. Biol. Teacher* **38**(3):150–156.

Groves, C., 1981, Reply to Johnson, S. 1981: "Bonobos: Hominid prototypes or specialized insular dwarfs?," *Curr. Anthropol.* **22**(4):366.

Groves, C. P., 1982, The geography of apes in Africa, *Int. J. Primatol.* **3**(3):249 (Abstract).

Hill, W. C. O., 1969, The nomenclature, taxonomy and distribution of chimpanzees, in: *The Chimpanzee*, Volume 1, Karger, Basel, pp. 22–49.

Horn, A. D., 1979, The taxonomic status of the bonobo chimpanzee, *Am. J. Phys. Anthropol.* **51**:273–282.

Horn, A. D., 1980, Some observations on the ecology of the bonobo chimpanzee (*Pan paniscus* Schwarz, 1929) near Lake Tumba, Zaire, *Folia Primatol.* **34**:145–169.

Johnson, S. C., 1981, Bonobos: Generalized hominid prototypes or specialized insular dwarfs?, *Curr. Anthropol.* **22**:363–375.

Jungers, W. L., 1982, Lucy's limbs: Skeletal allometry and locomotion in *Australopithecus afarensis*, *Nature* **297**:676–678.

Kano, T., 1979, A pilot study on the ecology of pygmy chimpanzees, *Pan paniscus*, in: *The Great Apes* (D. A. Hamburg and E. R. McCown, eds.), Benjamin/Cummings, Menlo Park, California, pp. 122–135.

Kano, T., 1980, Social behavior of wild pygmy chimpanzees (*Pan paniscus*) of Wamba: A preliminary report, *J. Hum. Evol.* **9**:243–260.

Kuroda, S., 1979, Grouping of the pygmy chimpanzees. *Primates* **20**(2):161–183.

Kuroda, S., 1980, Social behavior of the pygmy chimpanzees, *Primates* **21**(2):181–197.

Latimer, B. M., White, T. D., Kimbel, W. H., and Johanson, D. C., 1981, The pygmy chimpanzee is not a living missing link in human evolution, *J. Hum. Evol.* **10**:475–488.

Le Gros Clark, W. E., 1947, Observations on the anatomy of the fossil australopithecinae, *J. Anat.* **81**:300–333.

Le Gros Clark, W. E., 1967, *Man-apes or Ape-men?*, Holt, Rinehart and Winston, New York.

MacKinnon, J., 1974, The behavior and ecology of wild orangutans (*Pongo pygmaeus*), *Anim. Behav.* **22**:3–74.

MacKinnon, J., 1978, *The Ape within Us*, Holt, Rinehart and Winston, New York.

McHenry, H. M., and Corruccini, R. S., 1981, *Pan paniscus* and human evolution, *Am. J. Phys. Anthropol.* **54**:355–367.

Reynolds, V., 1967, *The Apes*, Dutton, New York.

Savage, E. S., Wilkerson, B. J., and Bakeman, R., 1977, Spontaneous gestural communication among conspecifics in the pygmy chimpanzee (*Pan paniscus*), in: *Progress in Ape Research* (O. H. Bourne, Ed.), Academic Press, New York, pp. 97–116.

Savage-Rumbaugh, E. S., and Wilkerson, B. J., 1978, Socio-sexual behavior in *Pan paniscus* and *Pan troglodytes:* A comparative study, *J. Hum. Evol.* **7**:327–344.

Schultz, A. H., 1930, The skeleton of the trunk and limbs of higher primates, *Hum. Biol.* **9**(3):303–438.

Schultz, A. H., 1937, Proportions, variability and asymmetries of the long bones of the limbs and the clavicles in man and apes, *Hum. Biol.* **9**:281–328.

Schultz, A. H., 1941, Growth and development of the orang-utan, *Contrib. Embryol.* **29**(Carnegie Inst. Wash. Publ., No. 525):57–110.

Schultz, A. H., 1968, The recent hominoid primates, in: *Perspectives on Human Evolution*, Volume I (S. L. Washburn and P. C. Jay, eds.), Holt, Rinehart and Winston, New York, pp. 122–195.

Schultz, A. H., 1969, The skeleton of the chimpanzee, in: *The Chimpanzee*, Volume 1, Karger, Basel, pp. 50–103.

Schultz, A. H., 1973, Age changes, variability and genetic differences in body proportions of recent hominoids, *Folia Primatol.* **19**:338–359.

Shea, B. T., 1981, Relative growth of the limbs and trunk in the African apes, *Am. J. Phys. Anthropol.* **56**:179–201.

Stern, J. T., and Susman, R. L., 1983, The locomotor anatomy of *Australopithecus afarensis*, *Am. J. Phys. Anthropol.* **60**:279–318.

Susman, R. L., and Stern, J. T., 1982, Functional morphology of *Homo habilis*, *Science* **217**:931–934.

Susman, R. L., Badrian, N. L., and Badrian, A. J., 1980, Locomotor behavior of *Pan paniscus* in Zaire, *Am. J. Phys. Anthropol.* **53**:69–80.

Tratz, E., and Heck, H., 1954, Der afrikanische anthropoide "Bonobo," eine neue Menschenaffengattung, *Säugetierkd. Mitt.* **2**:97–101.

Tuttle, R. H., 1977, Naturalistic positional behavior of apes and models of hominid evolution, 1929–1976, in: *Progress in Ape Research* (G. H. Bourne, ed.), Academic Press, New York, pp. 277–296.

Vandebroek, G., 1959, Notes ecologiques sur les anthropoldes africains, *Ann. Soc. R. Zool. Belg.* **LXXXIX**(1):203–211.

Wrangham, R. W., 1979, Sex differences in chimpanzee dispersion, in: *The Great Apes* (D. A. Hamburg and E. R. McCown, eds.), Benjamin/Cummings, Menlo Park, California, pp. 481–489.

Zihlman, A. L., 1978, Interpretations of early hominid locomotion, in: *Early Hominids of Africa* (C. J. Jolly, ed.), Duckworth, London, pp. 361–377.

Zihlman, A. L., 1979, Pygmy chimpanzees and early hominids, *S. Afr. J. Sci.* **75**(4):165–168.

Zihlman, A., and Brunker, L., 1979, Hominid bipedalism: Then and now, *Yearb. Phys. Anthropol.* **22**:132–162.

Zihlman, A. L., and Cramer, D. L., 1978, Skeletal differences between pygmy (*Pan paniscus*) and common chimpanzees (*Pan troglodytes*), *Folia Primatol.* **29**:86–94.

Zihlman, A. L., and Lowenstein, J., 1983, *Ramapithecus* and *Pan paniscus*: Significance for human origins, in: *New Interpretations of Ape and Human Ancestry* (R. Ciochon and R. Corruccini, eds.), Plenum Press, New York, pp. 677–694.

Zihlman, A. L., Cronin, J. E., Cramer, D. L., and Sarich, V. M., 1978, Pygmy chimpanzee as a possible prototype for the common ancestor of humans, chimpanzees and gorillas, *Nature* **275**:744–746.

Zuckerman, S., 1966, Myths and methods in anatomy, *J. R. Coll. Surg. Edinburgh* **11**:87–114.

CHAPTER 9

The Common Ancestor

A Study of the Postcranium of Pan paniscus, Australopithecus, and Other Hominoids

HENRY M. McHENRY

1. Introduction

Pan paniscus is a fascinating animal in its own right, but what makes it especially dear to some of us is the possibility suggested in 1933 by Harold Coolidge and debated ever since that, "it may approach more closely to the common ancestor of chimpanzees and man than does any living chimpanzee hitherto discovered and described" (Coolidge, 1933, p. 56). If the pygmy chimpanzee resembles our common ancestor, then one might expect it to show a special resemblance to the earliest known hominids grouped in the genus *Australopithecus*. Fifty years ago Schwarz (1932) pointed out that it might be important in the interpretation of *Australopithecus*, and in 1934 he wrote, "the importance of this race is evident, as it presents analogies to *Australopithecus*, which appears to be only a dwarfed gorilla" (Schwarz, 1934, p. 583). All that was known of *Australopithecus* at that time was the juvenile skull and dentition of the Taung baby (Dart, 1925). Subsequently, the sample of these early hominids has increased to the point that we have several specimens of most joints of the postcranial skeleton and hundreds of individuals represented by teeth and crania.

What makes *Pan paniscus* and *Australopithecus* especially important to studies of human and ape evolution is the apparent fact that very

---

HENRY M. McHENRY ● Department of Anthropology, University of California at Davis, Davis, California 95616.

recently in the past there was a trifurcation in the evolutionary lineage leading to the African hominoids (human, chimp, and gorilla). This apparent trifurcation is demonstrated by microcomplement fixation of several serum proteins and by DNA hybridization (Sarich and Cronin, 1977). Other methods of genetic comparison show either the *Gorilla* lineage splitting off slightly before the division of *Pan* and *Homo* (Yunis and Prakash, 1982) or the division of the *Homo* lineage before the division of the *Pan* and *Gorilla* lineages (Ferris *et al.,* 1981; Goodman, 1971). Whatever the exact order of branching is, the fact appears to be well documented that the African hominoids including humans share a common ancestor that is unlikely to have predated the middle Miocene [*ca.* 16 MYA (Pilbeam, 1978, 1982)] and probably is much more recent [5–8 MYA (Cronin *et al.,* 1981; Greenfield, 1980; McHenry and Corruccini, 1980; Sarich and Cronin, 1977)]. This is long after the separation of the Asian great ape, *Pongo pygmaeus* (Andrews, 1982). *Australopithecus* is especially important in the problem of African Hominoidea because it is by far the best known fossil African hominoid and is closest in time to the common ancestor. In some ways *Pan paniscus* resembles *Australopithecus* and may be the least derived of all of the African hominoids in many body parts.

The purpose of this study is to determine how similar *Australopithecus* is to *Pan paniscus* in postcranial morphology relative to other pongids and hominids and what this similarity or lack of similarity implies for hominoid evolution.

In two previous studies (Corruccini and McHenry, 1979; McHenry and Corruccini, 1981) the differences between *H. sapiens, P. paniscus,* and *P. troglodytes* were assessed in comparison to some of the early hominid fossils. What follows is based on entirely new analyses of all large-bodied hominoid species and new pre-2 million-year-old hominid fossils, and a new size-standardization method.

2. Materials and Methods

The results reported here are based on over 20,000 measurements taken on the postcranium of the great apes, humans, and fossil hominids. The original purpose of taking so many measurements was to describe the known Plio-Pleistocene hominid postcranial anatomy. Most of these measurements are of joints and aim to describe morphological adaptations to joint motions. The dimensions are sometimes idiosyncratic because they were devised for fragmentary fossils.

Measurements are emphasized over traits, which are better described in other ways (e.g., verbally, pictorially), to reduce subjectivity, take into account variability within species, and identify effects of overall size on morphology. The latter is an especially difficult problem. The distal humerus, for example, is shaped somewhat differently in *Gorilla* and *Pan* even though it functions in very much the same way in the two genera. Presumably the differences are due to different biomechanical demands of different body sizes, but these effects are hard to understand.

The sample sizes vary from joint to joint due to the fact that the data were collected over a dozen years for many different studies of early hominid postcranial anatomy. I report here a six-way comparison (with the exception of the capitate): *Homo sapiens, Pan paniscus, Pan troglodytes, Gorilla gorilla, Pongo pygmaeus,* and pre-2-million-year Plio-Pleistocene hominids. *Pongo* is included as an out-group to the African Hominoidea. Such an out-group is useful in determining the polarity of the morphocline in many instances: if some African hominoids share a trait with *Pongo* but others do not, then that trait *may* be a symplesiomorphy (shared primitive). I must emphasize the word "may" because a great deal of common sense is usually necessary in finding the level of certainty in reconstructing evolutionary sequences.

Trait-by-trait comparisons are supplemented by multivariate analyses. The method used is canonical variate analysis and the resulting Mahalanobis D distances between groups (Blackith and Reyment, 1971). Canonical variates are derived from the extant hominoid data and the fossils are evaluated in the resulting multidimensional space defined by the contrasts between modern hominoids. The method has advantages and disadvantages and it is important to apply the results to appropriate questions. The method is most effectively used in attempting to falsify hypotheses. Canonical variates describe essential contrasts between known groups in interpretable ways. The notion that one can assess the overall functional affinities of joints by this method is optimistic. When an unknown (a fossil in this case) is evaluated among those known groups it is assumed that it belongs to one of the groups. This assumption limits interpretation. If the fossil falls outside of the 95% range of variation of *H. sapiens*, for example, one can say that in those features that best distinguish the *H. sapiens* joint from all other hominoid joints, the fossil is unlike *H. sapiens*. But if the fossil falls within the range of variation of a group, one is on less secure grounds in stating that the specimen is similar to the members of that group, as Bronowski and Long warned long ago, "for the measurements might, for example, be matched on an object which turns out to be a pebble or a fake" (Bronowski and Long, 1975[1], p. 794). Overall size of specimens often dominates the multivariate discrimination by making small fossil hominids, for example, resemble similar sized apes due to similarity in overall size and masking subtler contrasts in shape that may be important in biomechanical function. Recently, size correction procedures have proliferated and have become very sophisticated (e.g., Albrecht, 1980; Corruccini, 1978a; Susman and Creel, 1979). Each method has drawbacks and advantages. The best procedure might be to use several different size-correcting procedures and choose the results that make the most sense based on other criteria. In this study I have chosen a very simple procedure for partially adjusting for overall size contrasts. The average of all standardized measurements of each individual bone is simply subtracted from all measurements on that bone and the results are standardized across the entire sample. This method only removes the gross effects of size, of course. Traits will vary due to allometric changes in very subtle ways. But the simpler procedure has the advantage of interpretability: results are not obscured by so much manipulation of the data that hidden problems are not revealed. This simple method of size adjustment is checked by computing canonical variates using raw data and using regression-adjusted data in a more elaborate size-correcting procedure relying on size-related changes within one species (in this case *Gorilla gorilla*) to adjust variables in all other species.

The samples of *H. sapiens* are primarily derived from an Amerindian population excavated from Mexico and stored at the Peabody Museum, Harvard University. I also include seven San skeletons from the British Museum (Natural History). The *P. troglodytes* and *G. gorilla* samples are mostly from the Powell–Cotton Museum. *Pongo pygmaeus* is from the Smithsonian Institution and from the British Museum. The *P. paniscus* sample is from the collection at the Koninklijk Museum voor Midden Afrika, Tervuren, and the Museum of Comparative Zoology, Harvard University. All specimens are adult as judged by fusion of epiphyses. The fossil hominids were measured by the author at the Transvaal Museum, Department of Anatomy of the University of Witwatersrand, National Museum of Kenya, National Museum of Ethiopia, and the Cleveland Museum of Natural History. Details about the samples and measurements are given elsewhere (McHenry, 1972, 1973, 1974, 1975, 1976, 1978; McHenry *et al.*, 1976, 1980; McHenry and Corruccini, 1975a,b, 1976a,b, 1978).

3. Results

3.1. Shoulder*

The shoulder data set consists of nine measurements designed to compare the two best preserved early hominid shoulders: Sts 7 from

---

*Variables used in the analysis of the shoulder: glenoid width, coracoid neck thickness, glenoid height, spinoglenoid notch height, humeral head (*m–l* and *a–p*), intertubercular groove width, extent of articular surface of the humerus, projection of greater tuberosity.

Sterkfontein, representing *Australopithecus africanus* (Ciochon and Corruccini, 1976; Robinson, 1972; Vrba, 1979), and AL 288-11 ("Lucy") from
Hadar, representing *A. afarensis* (Johanson *et al.*, 1982). The shoulder of
the pygmy chimpanzee is clearly smaller than that of the common chimpanzee, but when the standard size variable is subtracted, most of the
differences disappear. Only the depth of the spinoglenoid notch is significantly different at a level of probability below 0.01. The two fossils are
distinctive in the narrowness of their glenoid fossae. The specimens differ
in several respects, including the shape of the glenoid fossa (Sts 7 is
exceptionally elongated, whereas AL 288-11 is very short in its infero-
superior dimension). The anteroposterior diameter of the humeral head
is relatively small in AL 288-11 but average for hominoids in Sts 7.

When the size-adjusted measurements are used to construct the principal contrasts between the various extant hominoid groups (human, pygmy
and common chimpanzees, gorilla, and orangutan) via canonical variates
analysis, it is the contrast between the terrestrial quadrupedal group
(chimpanzee and gorilla) and the arboreal orangutan that explains over
three-quarters of the variation. Humans fall midway between these two
extremes. The trait with highest correlation with this axis is the projection
of the greater tubercle over the humeral head (low in orangutans). Curi-
ously, the AL 288-11 shoulder falls well away from the African great apes
on this axis. In contrast, Sts 7 is indistinguishable from the African apes.
Subsequent canonical variates have very low eigenvalues (less than one)
and account for much less of the total dispersion.

The average Mahalanobis D distances derived from this canonical
variates analysis are presented in Table I. Of all the extant hominoids,
the two chimpanzee species are the most similar to one another. With
this data set the human shoulder is closer to *Pongo* than *Pan*, a result
similar to that of Oxnard (1977, and references therein). The Sts 7 specimen is closest to the pygmy chimpanzee mean, and the AL 288-11 shoul-

Table I.   Shoulder: Mahalanobis D Distances between the Means Based on
Nine Size-Standardized Measurements of the Shoulder

	Human	Pygmy chimp	Common chimp	Gorilla	Orang	Sts 7	AL 288-11
Human							
Pygmy chimp	3.5						
Common chimp	3.8	2.3					
Gorilla	2.7	2.5	2.6				
Orang	2.3	4.9	5.5	4.3			
Sts 7	4.3	1.8	3.5	2.8	5.1		
AL 288-11	2.0	4.1	4.6	4.0	1.4	4.6	

der is closest to the orangutan mean. This does not necessarily imply that
Sts 7 is pygmy chimpanzee-like and AL 288-11 is orang-like, but some
possibilities can be tentatively eliminated. In my sample there are no
humans or orangutans that match the combination of features seen in Sts
7. That combination includes a relatively narrow and elongated glenoid
fossa, a shallow spinoglenoid notch, and large humeral head. On individual
traits, the ranges of humans and orangs overlap values seen in the fossil,
but no single individual has the combination of traits like *A. africanus*.
My test is unable to reject the hypothesis that Sts 7 is outside the range
of variation seen in extant African apes, however. Several gorillas (12.5%)
are further from the gorilla centroid than is Sts 7 from that centroid. Within
the *P. troglodytes* sample one specimen is further away from the common
chimpanzee centroid than is Sts 7 from that centroid. *Although in overall
Mahalanobis D distance Sts 7 is closer to the pygmy chimpanzee than
any other group, it is much further from the pygmy chimpanzee mean
than is any member of that species.* The fossil's distance from the pygmy
chimpanzee centroid is due largely to the fourth canonical variate, which
has an insignificant eigenvalue of 0.00003, and accounts for less than
0.0001% of the total dispersion. But it may be important to take this
apparently trivial axis into account. It is quite obvious from direct comparisons that Sts 7 is *not* shaped like a pygmy chimpanzee. It is unique.
It is closer to the pygmy chimpanzee centroid than is any other in my
sample, but it is not within the range of that species.

Unlike Sts 7, the Lucy shoulder joint (AL 288-11) is unlike the African
great apes. Its Mahalanobis *D* distance from the pygmy chimpanzee mean
is *much* greater than any *P. paniscus* shoulder in my sample. The same
is true for *P. troglodytes*. Two male gorillas are slightly further from the
gorilla centroid than is AL 288-11, however. Nor is this fossil particularly
like any *H. sapiens* in my sample: its Mahalanobis distance from the
human mean is great and only one human shoulder is as far away from
that mean (a San). Of extant hominoids sampled, *Pongo* is the most
similar. The fossil projects relatively close to the *Pongo* mean and 42%
of the orang sample is further away from that mean than is AL 288-11.
It is a complex combination of traits that aligns AL 288-11 with *Pongo*
in this analysis of nine variables, but one important feature is the relatively
narrow humeral head. The anteroposterior diameter of the humeral head
has a *Z* score of − 1.7, which is well below the observed range of variation
in all samples except *Pongo*.

In a previous study (McHenry and Corruccini, 1981) the results were
similar, although the measurements included clavicular length, scapular
lengths and breadths, and other traits described in Corruccini and Ciochon
(1976). Clavicular length is significantly shorter in *P. paniscus* even when

Table II.   Humerus: Mahalanobis *D* Distances between the Means Based on
16 Size-Standardized Variables on the Humerus

	Human	Pygmy chimp	Common chimp	Gorilla	Orang	AL 288-1m	AL 137-48A	Kanapoi
Human								
Pygmy chimp	4.0							
Common chimp	4.0	3.0						
Gorilla	5.2	4.1	4.7					
Orang	4.8	3.4	5.1	4.6				
AL 288-1m	2.4	2.5	3.9	2.2	4.2			
AL 137-48A	1.9	3.5	3.7	3.8	5.4	2.3		
Kanapoi	1.1	3.7	3.0	2.7	4.8	2.6	2.2	

the data are size-adjusted. A short clavicle appears to be diagnostic of pygmy chimpanzees (Coolidge, 1933; Tratz and Heck, 1954; Schultz, 1969; Zihlman and Cramer, 1978; Zihlman *et al.*, 1978; Zihlman, 1979).

3.2. Distal Humerus*

Although the pygmy chimpanzee humerus is significantly smaller than that of the common chimpanzee in most dimensions, the shape differences are minor. With size-adjusted data the relative width of the capitulum and the shaft diameter are significantly smaller and the lateral epicondyle is placed in a more proximal position in *P. paniscus* than in *P. troglodytes*. In overall shape as measured by Mahalanobis *D* distances, the distal humeri of the two chimpanzees are closer to one another than to any extant hominoid (Table II).

The three oldest known hominid distal humeri include one from Kanapoi dated at 5–5.5 MY (B. Patterson and Howells, 1967) and two from Hadar (AL 288-1m and AL 137-48A) dated at 3.5 MY and classified as *A. afarensis* (Johanson *et al.*, 1982; Lovejoy *et al.*, 1982*a*). The Kanapoi humerus is extraordinarily close to the modern *H. sapiens* centroid in this

---

*Variables used to describe the shape of the humerus: head width (*a–p*), trochlea width (*m–l, a–p*), lateral trochlea ridge width (*a–p*), capitulum width (*m–l*), capitulum height, articular surface width, biepicondyle width, trochlea-medial epicondyle diameter, trochlea-suponcondyle width, capitulum-lateral epicondyle, olecranon width (*m–d*), olecranon depth, medial olecranon width (*m–l*), lateral olecranon width (*m–d*), shaft width (*a–p*), medial epicondyle width (*a–p*), total length (see McHenry and Corruccini, 1975*b*).

and all other multivariate tests (McHenry, 1972, 1976; McHenry and Corruccini, 1975b). One might suspect the specimen is misdated, but its placement in the stratigraphy is not seriously doubted by those who were
involved in the discovery and study of its geological context (Behrensmeyer, 1976). The placement of the two A. afarensis fossils helps to clarify
the issue. They are also quite close to the human mean projection in the
four-dimensional space defined by the canonical variates derived from
size-adjusted data. This implies that the human pattern of distal humerus
construction emerged early. It also implies that the elbow is not particularly diagnostic, especially considering the fact that AL 288-1m, for
example, is equally close to the gorilla, pygmy chimpanzee, and human
centroids. Later fossil hominids vary considerably in their overall affinities
among Hominoidea, with TM 1517 closer to common chimpanzees than
anything else, KNM-ER 739 closer to orangutans, and KNM-ER 1504
closest to pygmy chimpanzees (McHenry and Corruccini, 1975b). One
diagnostic trait of the African apes is that steepness of the lateral border
of the olecranon fossa, which is high and straight in the African apes and
low and rounded in modern humans. All of the fossils show the human
configuration in this trait.

Results here support those of Senut (1978, 1979, 1980, 1981a,b), who
found some early hominids to be more similar to Homo (i.e., Kanapoi,
AL 137-48A) and others to be more like the pongids (AL 288-1m). My
results differ from those reported by Feldesman (1982). In his multivariate
study of the distal humerus the Homo sample was closest to Pongo, and
the Homo–Pongo cluster was closest to P. paniscus. All of the extant
large-bodied Hominoidea were closer to one another than they were to
the fossil hominids in his dendrogram of Mahalanobis D values. My results
agree that AL 137-48A and the Kanapoi humerus are similar to one another, but Feldesman's analysis results in separating Homo further from
these two fossils than from Hylobates.

Although Zihlman and Cramer (1978) emphasize that the fact that
the proximal humeral head is significantly smaller in P. paniscus than it
is in P. troglodytes, my results show that this is not true relative to other
skeletal dimensions. The probability that the actual head diameters of the
two chimpanzees derive from the same population is indeed small (less
than 1×10^{-9}), but when data are size-adjusted the difference is insignificant. In this case the size variable used to adjust the data is the average
of 47 measurements of the humerus, radius, ulna, femur, tibia, and talus.
Most of these measurements are of joints, so that the size-adjusted humeral head diameter represents the size of this feature relative to other
joint sizes in the body.

Table III. Ulna: Mahalanobis D Distances between Means Based on 12 Size-Standardized Variables of the Ulna

	Human	Pygmy chimp	Common chimp	Gorilla	Orang	Omo L40-19	O.H. 36
Human							
Pygmy chimp	4.6						
Common chimp	5.0	2.5					
Gorilla	4.0	4.8	4.8				
Orang	4.6	6.0	5.1	5.0			
Omo L40-19	4.5	3.4	2.0	3.1	5.0		
O.H. 36	5.6	3.9	3.4	5.9	7.7	3.6	

3.3. Ulna*

The ulna is represented by several specimens, but the two most complete are those from Olduvai [O.H. 36 (Oakley *et al.*, 1977)] and from the Omo [L40-19 (Howell and Wood, 1974; McHenry *et al.*, 1976)]. Twelve measurements were taken on these specimens. The two species of chimpanzee are very similar when data are size-standardized. Three traits differ at a level of probability less than 0.01: length of the trochlea, position of the bicipital tuberosity, and the mediolateral width of the shaft at the proximal end. In its proximal ulnar morphology the pygmy chimpanzee is the most derived of the hominoids. The fact that the fossil hominids are more similar in these features to the common than to the pygmy chimpanzee is further evidence that *Pan paniscus* displays the derived condition.

The Mahalanobis D distance between common and pygmy chimpanzees (2.5, Table III) is less than that between any other pair of extant hominoids. A plot of the first two canonical variates reveals *extensive* overlap.

Both of the fossil ulnae are unusual for hominoids. In Mahalanobis D distance both resemble *P. troglodytes* more closely than any other hominoid. Neither is particularly close to the human centroid. Both resemble the common chimpanzee in having strong shaft curvature, narrow trochleae, low projections of their coronoid processes, and anteroposterior thickening of their shafts at the center of their trochlea. But the two fossils differ from one another in several respects, especially in the rel-

---

*Variables used to describe the ulna: depth curvature: midshaft width (m–l), midshaft width (a–p), trochlea width (m–l), trochlea width (a–p), coronoid height, olecranon width (a–p), trochlea length, tuberosity position, olecranon length, proximal shaft width (a–p), proximal shaft width (m–l) (see McHenry *et al.*, 1976).

Table IV.   Capitate: Mahalanobis D Distances between Means Based on 11
Size-Standardized Variables on the Capitate

	Human	Pygmy chimp	Common chimp	Gorilla	TM 1526	AL 288-1w	AL 333-40
Human							
Pygmy chimp	7.5						
Common chimp	7.1	1.3					
Gorilla	9.4	5.9	4.6				
TM 1526	1.8	5.9	5.7	8.7			
AL 288-1w	2.3	5.5	5.4	8.7	.5		
AL 333-40	2.2	7.1	7.2	10.4	1.7	1.8	

atively proximodistal diameter of the trochlea in O.H. 36, the close proximity of the bicipital tuberosity to the center of the trochlea in O.H. 36
(a trait characteristic of *P. paniscus*), the anteroposterior flattening of the
proximal shaft in Omo L40-19, and the mediolateral widening of the shaft
in the latter.

The placement of the fossils confirms earlier morphometric work on
the ulna (McHenry *et al.*, 1976; Feldesman, 1979) and it underscores the
uniqueness of both the Omo ulna and the even more unusual Olduvai
specimen, neither of which fossil resembles the other.

3.4. Capitate*

The capitates of the two species of chimpanzee are essentially identical except for a slight size difference and relatively deeper cupping of
the distal articular surface in *P. troglodytes*. Three fossil hominid capitates
are known from the Plio-Pleistocene record and these contrast sharply
with the chimpanzee specimens: TM 1526 (Robinson, 1972), AL 288-1w
(Johanson *et al.*, 1982), and AL 333-40 (Bush *et al.*, 1982). Their closest
affinity is clearly with *Homo sapiens* in the 12 measurements used in the
analysis (Table IV). They are also remarkably similar to one another. The
fossils do have some resemblances to pongids, however, that distinguish
them from modern hominids. Most conspicuous is the posterior position
of the trapezoid facet, which is present in most apes but only rarely in
Homo sapiens. Two other features that are pongid-like in the fossils are
the lack of an indentation for the styloid process of the second metacarpal

---

*Variables used to describe the capitate: height, maximum length to MV facet, minimum
length to MV facet, styloid projection, distal cupping, head width, neck width, metacarpal
II facet width ($a–p$), metacarpal II facet height, distal width (see McHenry, 1983).

Table V. Pelvis: Mahalanobis D Distances between Means Based on 17 Size-Standardized Variables of the Pelvic Bone

	Human	Pygmy chimp	Common chimp	Gorilla	Orang	Sts 14	AL 288-lao
Human							
Pygmy chimp	15.1						
Common chimp	14.4	1.1					
Gorilla	11.2	7.6	6.4				
Orang	11.9	6.2	5.1	6.2			
Sts 14	4.3	11.4	10.6	7.4	7.6		
AL 288-lao	5.7	10.2	9.4	6.4	6.3	1.3	

and the deep excavation on the distomedial side for the capitometacarpal ligament. Both species of chimpanzee have these characteristics. Other aspects of the early hominid hands display pongid-like features as well (Broom, 1942; Broom and Robinson, 1949; Day and Scheuer, 1973; O. J. Lewis, 1973, 1977; Marzke, 1971; Napier, 1962; Rightmire, 1972; Susman, 1978, 1979; Susman and Creel, 1979; Susman and Stern, 1979; Tuttle, 1981; McHenry, 1983).

3.5. Pelvis*

Early hominid pelvic bones are reorganized from the generalized hominoid precursor to the extent that in all major respects they are comparable only to *H. sapiens*. But the fossil hominids are *not* identical to *H. sapiens* by any means (McHenry, 1982, and references therein; McHenry and Temerin, 1979).

Of all of the living hominoids other than *H. sapiens*, the pelvic bones of *Pongo* and *Gorilla* resemble the fossil hominids most closely (Table V). The os coxae of the pygmy chimpanzee is essentially indistinguishable from that of the common chimpanzee except for a thinner iliac blade and a deeper acetabulum. The gorilla pelvic bone, however, is somewhat peculiar, which probably relates to the demands of its much larger size rather than any behavioral difference.

The two fossil hominid pelvic bones [AL 288-lao described by Johanson *et al.* (1982) and Sts 14 described by Robinson (1972)] are almost identical (Suzman, 1982). They share the same fundamental characteris-

*Variables used in the analysis of the pelvis: acetabulum height, acetabulum, width of ant. acetabulum, ilium width, minimum ilium width, ilium thickness, sciatic thickness, acetabulum (posterosuperior), acet-iliop, antsup.-iliop., antsup.-postinf., postinf.-iliop., acetabulum depth, length of pubis, length of ischium (see McHenry, 1975).

Table VI. Proximal Femur: Mahalanobis D Distances between Means Based on Ten Size-Standardized Variables of the Proximal Femur

	Human	Pygmy chimp	Common chimp	Gorilla	Orang	AL 288-lap	AL 333-3
Human							
Pygmy chimp	3.8						
Common chimp	3.8	0.3					
Gorilla	5.0	2.9	2.9				
Orang	4.4	4.2	4.0	3.8			
AL 288-lap	2.6	2.9	2.7	5.0	3.8		
AL 333-3	3.4	2.8	2.5	4.6	3.1	1.1	

tics of *H. sapiens,* but they also share unique features: they both have relatively small acetabula, reduced width of bone behind the acetabulum, large breadth of iliac blade, greater distance between posterosuperior iliac spine to posteroinferior spine, reduced width of the sacral surface, and long superior ramus of pubis.

3.6. Proximal Femur*

Ten measurements were used to describe the proximal femur. Although the two species of chimpanzee are significantly different in every trait when raw data are used, the distinctions completely fall away when the data are size-standardized. Their similarity to one another is conspicuous in the Mahalanobis D distance that separates them, which is a tiny 0.3 (Table VI).

The fossil hominid femora [AL 288-lap described by Johanson *et al.* (1982) and AL 333-3 described by Lovejoy *et al.* (1982*b*)] are very similar to one another (Table VI), but not particularly close to any extant hominid.

The overall reorganization of the os coxae in fossil hominids is not so apparent in this portion of the hip. There is clearly a major reorganizational stage between the common African hominoid and the modern hominid hip. Both AL 288-lap and AL 333-3 share a suite of unique characteristics, such as an exceptionally small femoral head (about two standard deviations below the human and hominoid average), thick shaft (particularly in the mediolateral direction), exceptionally long neck, and a lesser trochanter, which is more proximally located than any other

---

*Variables used in the analysis of the femur: head diameter, neck vert. diameter, neck *a–p* diameter, transverse diameter of the shaft, diameter, shaft *a–p* diameter, prox. width, neck length, lesser trochanter–head, lesser trochanter–neck, lesser trochanter–greater trochanter, length, midshaft width (*m–l*), midshaft width (*a–p*), medial condyle width (*m–l*) (see McHenry and Corruccini, 1978).

hominid. In overall resemblance they are nearly identical (when overall size is subtracted) and nearly equally distant from humans and pygmy and common chimpanzees. This is not to say that their morphology is an intermediate step between *Pan* and *Homo*. The two fossils are quite unique. If *A. afarensis* is to be considered a common ancestor to later species of hominids, then it is clear that the evolution of the human hip went through some complicated reorganizational stages. This is particularly obvious when later fossil hominids are considered. The Swartkrans femora (SK 87 and SK 92) and those from Lake Turkana (KNM-ER 1481, 1472, 1504) show their own peculiarities that are difficult to interpret (Jenkins, 1972; McHenry and Corruccini, 1978).

3.7. Distal Femur

The distal femur is difficult to describe by simple measurements. Using seven traits that describe the widths and breadths of the shaft and condyles, one finds that the two best preserved *A. afarensis* distal femora [AL 129-1a and AL 333-4 described in Lovejoy *et al.* (1982*b*)] are very similar to one another when compensation is made for their size difference. By these measurements these two specimens are most similar to the chimpanzee species and *Pongo* and not to *H. sapiens* despite the demonstration that these specimens are adapted functionally to bipedalism (Day, 1973; Heiple and Lovejoy, 1971; Johanson *et al.*, 1976) (Table VII). The Sterkfontein specimen TM 1513 (Robinson, 1972) is in between humans and the two chimpanzee groups.

These results support most of the findings of Tardieu (1981), who found that TM 1513 was more like modern humans than AL 129-1a, which she relegated to a more primitive (pongidlike) group.

Table VII. Distal Femur: Mahalanobis *D* Distances between Means Based on Seven Size-Standardized Variables of the Distal Femur

	Human	Pygmy chimp	Common chimp	Gorilla	Orang	TM 1513	AL 129-1a	AL 333-4
Human								
Pygmy chimp	5.5							
Common chimp	5.6	0.8						
Gorilla	7.2	3.2	2.4					
Orang	5.2	0.4	0.6	3.0				
TM 1513	3.2	3.4	3.9	6.2	4.5			
AL 129-1a	4.8	0.9	0.9	3.0	0.5	3.2		
AL 333-4	3.1	2.6	2.6	4.2	2.3	3.2	1.7	

3.8. Foot*

Forty-six variables were used to describe the tarsals and first meta-tarsal in *P. paniscus, P. troglodytes,* and the Olduvai Hominid 8 foot. The selection of measurements was guided by what was preserved on O.H. 8 (Day and Napier, 1964; O. J. Lewis, 1980, 1981; Susman and Stern, 1982). As with the other parts of the postcranium, the common chimpanzee is significantly larger than the pygmy chimpanzee. When size is standardized, however, only seven measurements are significantly different at probabilities of less than 0.01. In all but two of these the O.H. 8 approaches the pygmy more closely than it does the common chimpanzee.

A tonguelike projection of the cuboid fits into a depression of the calcaneus in human feet but not in apes, except for *P. paniscus,* which has a slight depression on the medial side of the cuboid facet of the calcaneus (but not nearly as deep as *H. sapiens* or O.H. 8). The width of the lateral cuneiform facet on the navicular is smaller in *P. paniscus* than it is in *P. troglodytes* and the adjacent cuboid facet is much wider. Both of these traits align O.H. 8 with *P. paniscus.* The two chimps differ in two other measurements of their naviculars: the distance from the point at which lateral and cuboid facets meet dorsally to the most distant point on the talar facet and to the closest point on the talar facet are smaller in *P. paniscus.* Both the proximal and distal mediolateral breadths of the cuboid are smaller in *P. paniscus,* traits that align them with *Homo* and especially O.H. 8.

Multivariate analysis of these 46 variables results in the Mahalanobis *D* distances shown in Table VIII. Of all the extant hominids, the two species of chimpanzee approach each other most closely. *Homo sapiens* is nearest to *P. paniscus.* The fossil does not resemble any extant hominoid very closely. It is slightly closer to the pygmy chimpanzee than to modern humans. This result is surprising, since the Olduvai specimen clearly has a larger number of shared derived traits with *Homo sapiens* (Susman and Stern, 1982). Apparently the foot still retains many features that recall its heritage despite its reorganization for adaptation to bipedalism (Oxnard and Lisowski, 1980).

The principal contrasts between the O.H. 8 foot and that of *H. sapiens* is the fossil's relatively short anterior half of the calcaneus (like *P. paniscus*), its exceptionally narrow talar facet on its navicular bone (exceptional), and its relatively high navicular facet on its lateral cuneiform (exceptional). Several other traits distinguish the fossil from *H. sapiens* (15 out of 46 differ by at least one standard deviation).

*Variables used in the analysis of the foot include those on the talus, calcaneus, navicular, cuboid, lateral, and medial cuneiforms, and first metatarsal.

Table VIII. Foot: Mahalanobis D Distances between Means Based on 46 Size-Standardized Variables of the Tarsals and First Metatarsals

	Human	Pygmy chimp	Common chimp	Gorilla	Orang	O.H. 8
Human						
Pygmy chimp	12.9					
Common chimp	13.9	6.5				
Gorilla	16.0	9.7	7.2			
Orang	14.8	11.7	13.1	12.8		
O.H. 8	8.2	7.2	11.8	14.1	13.3	

Judging by the mean deviations from the average values for all hominoid feet tested here, the two chimpanzees have the least derived tarsals and first metatarsals of all the great apes and humans. The architecture of the human, gorilla, and orangutan foot appears to be very specialized for particular adaptations (bipedalism, heavy-bodied terrestrialism, and high-mobility arborealism, respectively). The close proximity of the Olduvai fossil to *P. paniscus* in the canonical variate space might imply that the *P. paniscus* foot is the least derived of all of the large-bodied Hominoidea.

There are numerous *A. afarensis* foot bones that I have not yet analyzed by the same methods as I have for O.H. 8 (Latimer *et al.*, 1982; Stern and Susman, 1983). Some of these specimens are remarkably primitive (Stern and Susman, 1983; Tuttle, 1981). *Pan paniscus* will be an invaluable comparative sample in their full analysis. The 3.7-MY old Laetoli footprints of hominids look perfectly modern to most observers (Charteris *et al.*, 1982; Clarke, 1979; Day and Wickens, 1980; Leakey, 1978, 1981; White, 1980), but chimpanzee-like to some (Stern and Susman, 1983).

3.9. Proportions

Forty-seven variables describing the shape of the humerus, ulna, radius, femur, and talus are combined into a single study to describe the postcranium of AL 288-1. The primary contrast among hominoid species in this comparison is between modern hominids and the apes (Table IX). The eigenvalue of this first canonical variate is 89.4, accounting for 80% of the total dispersion. Not unexpectedly, the traits most heavily weighted in this variate are femur and humerus length. Relative robusticity of the forelimb and hindlimb are also important in this primary axis of discrimination. The AL 288-1 partial skeleton has a relatively small femur and large humerus for a hominid (Jungers, 1982), which explains the fact that

Table IX. Proportions: Mahalanobis D Distances between Means Based on 47 Size-Standardized Variables of the Humerus, Radius, Ulna, Femur, Tibia, and Talus

	Human	Pygmy chimp	Common chimp	Gorilla	Orang	AL 288-1
Human						
Pygmy chimp	15.6					
Common chimp	17.6	6.2				
Gorilla	21.3	11.0	10.5			
Orang	25.9	14.1	13.2	10.6		
AL 288-1	10.3	7.6	8.5	11.7	16.1	

on this first axis it projects intermediate between apes and human. On subsequent axes the fossil is not outside the human range of variation. Its overall affinities are closest to the pygmy chimpanzee and slightly further from the common chimpanzee and still further from modern humans. The relatively larger femur of *P. paniscus* appears to be the main cause of its close relationship to Lucy (AL 288-1). In this feature, the pygmy chimpanzee seems to be less derived than other hominoids.

As in previous analyses, the two chimpanzees are closer to one another than are any other pair of hominoid species, and the pygmy chimpanzee is closer to humans than is any other hominoid.

Shea (1981) presents an explanation for the relatively longer hindlimbs and shorter forelimbs of *P. paniscus* that rules out ontogenetic scaling. He argues that biomechanical scaling may explain the forelimb and hindlimb proportions in the African Pongidae: to maintain functional equivalence, larger bodied forms have relatively longer forelimbs and shorter hindlimbs required for climbing for a large-bodied clawless animal.

4. Discussion

There is a growing body of evidence that Coolidge (1933) was correct when he separated the two kinds of chimpanzee at the species level. The two chimpanzees are distinct genetically (Borgaonkar *et al.,* 1971; Bogart and Benirschke, 1977; Chiarelli, 1963, 1970; Cronin, 1977; K. Gottlieb, personal communication; Hamerton *et al.,* 1963; Goodman *et al.,* 1970; Khudr *et al.,* 1973; Klinger *et al.,* 1963; Moor-Jankowski *et al.,* 1972, 1975; Socha, this volume; Wiener *et al.,* 1973, 1975), cranially (Coolidge, 1933; Cramer, 1977; Deblock, 1973; Deblock and Fenart, 1973, 1977; Delattre *et al.,* 1968; Fenart and Deblock, 1973; Laitman and Heimbuch, this volume; Rode, 1937, 1941; Vandebroek, 1959, 1969), dentally (Almquist, 1974;

Corruccini and McHenry, 1979; Johanson, 1974; Kinzey, this volume; McHenry and Corruccini, 1981), in soft anatomy (Brehme, 1975; Izor *et al.*, 1981; Miller, 1952; Schouteden, 1930, 1948), and behaviorally (Badrian and Badrian, 1977, and this volume; Badrian *et al.*, 1981; Bourne, 1976; Fiedler, 1956; Handler *et al.*, this volume; Horn, 1975, 1977, 1979, 1980; Hubsch, 1970; Itani, 1980; Kano, 1979, 1980; Kuroda, 1979; MacKinnon, 1976; Neugebauer, 1980; Nishida, 1972; T. L. Patterson, 1979; Rempe, 1961; Savage-Rumbaugh and Wilkerson, 1978; Susman, 1980; Susman *et al.*, 1980; van Bree, 1963). There are many reports of postcranial contrasts as well (Coolidge, 1933; Coolidge and Shea, 1982; Frechkop, 1935; Roberts, 1974; Rode, 1941; Urbain and Rode, 1940; Schultz, 1969; Susman, 1979; Tratz and Heck, 1954; Zihlman, 1979; Zihlman and Cramer, 1978; Zihlman *et al.*, 1978).

When pygmy and common chimpanzees are compared to the other Hominoidea, the differences in postcranial anatomy of the two species of *Pan* are not very great, especially when overall size is subtracted. Trait by trait, the two species are more similar to one another than are any other two species within Hominoidea. Their similar postcranial anatomy is demonstrated by the close affinity displayed in the multivariate analyses reported above. Among the large-bodied Hominoidea, the closest pair of species is invariably the two species of chimpanzee in all of the nine tests except for the distal femur, in which *Pongo* is about equally close to the two chimpanzees.

The close affinity of the postcranial morphology of the two *Pan* species should be of no surprise to those who have studied their locomotor behavior: both arboreally (when support size is large enough) and terrestrially, *P. troglodytes* and *P. paniscus* are predominantly knuckle walkers (Susman *et al.*, 1980; Susman, this volume). Both species are capable of armswinging and quadrumanous climbing, but the pygmy chimpanzee appears to be more agile at this (Badrian and Badrian, 1977; Horn, 1975, 1977, 1980; MacKinnon, 1976; Neugebauer, 1980; Nishida, 1972; T. L. Patterson, 1979; Susman, 1980; Susman *et al.*, 1980, 1981). The greater arboreal acrobatics of the pygmy chimpanzee might be due in part to its lighter and more linear build.

There *are* differences between the postcranial anatomy of the two species of chimpanzee, however, and these contrasts make an interesting study. Over two dozen traits survived my efforts to standardize for overall size and remained significantly different at probabilities of less than 0.01. Some of these differences, such as the narrower chests and relatively long hindlimbs of the pygmy chimpanzee, are obvious to any perceptive zoo visitor. Some of these, such as the position of the medial epicondyle of the humerus, the reduced cupping of the distal articular surface of the capitate, or the more deeply excavated cuboid facet on the calcaneus in the pygmy chimpanzee, are subtle. The meaning of these differences is beyond the scope of this study.

Are these postcranial contrasts related to phylogeny in any interpretable way? One might expect the common ancestor of the African Hominoidea to have a morphology that requires the least number of parallel, convergent, or reversed evolutionary events. A key might be the affinities of extant African hominoids to the earliest known fossil that postdates the divergence of the extant lineages. In Africa, this fossil group is *Australopithecus,* since the *Ramapithecus* fossils have a dubious connection with the African Hominoidea (Andrews, 1982; Pilbeam, 1982).

In my study no single extant hominoid has an exclusive claim on affinities to *Australopithecus. Pan paniscus* shoulders, feet, and proportions resemble some of the australopithecine fossils more closely than do any other extant hominoid, but in some parts *P. troglodytes* is closer (ulna), in others the orang (some distal femora), in one the gorilla (AL 288 humerus), and in most *H. sapiens* (other distal femora, pelvis, capitates, and most early humeri). The obvious conclusion from this is that the common ancestor of the African Hominoidea was not precisely like any modern hominoid and its reconstruction must derive its form from clues provided by all extinct and extant Hominoidea.

Darwin (1871, p. 520) warned that "we must not fall into the error of supposing that the early progenitor of the whole Simian Stock, including man, was identical with, or even closely resembled, any existing ape or monkey." This consideration may be behind much of the negative reaction to the claim that *P. paniscus* is the "best prototype of the prehominid ancestor" (Zihlman *et al.,* 1978, p. 744) [for criticism see especially Johnson (1981) and comments therein, and Latimer *et al.* (1981)]. The critics often overlook the fact that Zihlman (1979), Zihlman and Cramer (1978), and Zihlman *et al.* (1978) explicitly state that the pygmy chimpanzee is the best prototype *among living Hominoidea and extinct Hominidae.* In some ways, particularly in forelimb and hindlimb proportions and in the tarsus, the pygmy chimpanzee might approach the form of the common ancestor. In other traits, such as its diminutive masticatory system or the pattern of sexual dimorphism (McHenry and Corruccini, 1981; Latimer *et al.,* 1981; Susman and Jungers, 1981), it appears to be derived.

So what did the common ancestor of the African Hominoidea look like? This is an old and well-worn subject.

Darwin's "hairy, bearded, tailed creatures that could move their pointed ears freely" (Tuttle, 1974, p. 389) have received few supporters. Keith's (1923) large-bodied orthograde brachiating forms have taken up knuckle walking in the opinion of others (Sarich, 1973; Washburn, 1967).

Morton's (1927) hylobation model with small-bodied generalized climbers has received support from Tuttle (1974) in his excellent review of the issues. There is fresh evidence, however, that needs to be interpreted: among many other things, *P. paniscus* is much better known now than a few years ago, the pre-2 MY-old fossil hominid record has expanded exponentially in the last decade, behavioral, functional–morphological, and developmental aspects of living hominoids are much better known, and confirmation and acceptance of the genetic comparisons between living species is becoming nearly universal.

With these new lines of evidence in mind, Figs. 1–3 have been constructed to illustrate three hypotheses about the common ancestor. The first (Fig. 1) can be called the knuckle-walking hypothesis, in which the common ancestor was a knuckle-walker before the trifurcation into lineages leading to *Gorilla, Pan,* and *Homo.* The second (Fig. 2) is titled the parallelism hypothesis, which would have the common ancestor looking like a generalized hominoid arboreal climber without any terrestrial specializations. In this model, knuckle walking arose in the lineages leading to *Pan* and *Gorilla* independently. Finally (Fig. 3) the bifurcation hypothesis is the scheme in which the common ancestor was a generalized hominoid climber without knuckle-walking adaptations. This ancestor was ancestral to a bifurcation of the lineages leading to African apes and another to *Homo.* The African apes shared a common ancestor after this bifurcation and it was during this time that knuckle walking developed.

Each of these alternatives has strengths and weaknesses. The knuckle-walking hypothesis appeals because of its simplicity. All African apes are knuckle walkers and by most measures of genetic distance humans are as closely related to chimps and gorillas as chimps and gorillas are to one another. The origin of bipedalism involved giving up a quadrupedal terrestrial gait in favor of free hands and striding. At least two disadvantages to this theory are apparent, however. First, there are several special adaptations to knuckle walking that can be seen in the forelimb of the African apes and *none* of these are retained in the earliest hominid fossils. The 3.5-MY old Hadar metacarpals, for example, might be expected to have retained the dorsal ridges on their distal ends, but they have not (Tuttle, 1969, 1981). The capitate might be expected to show some of the extension-limiting features that are characteristic of knuckle walkers (Jenkins, 1981; Jenkins and Fleagle, 1975; Corruccini, 1978b), but the two specimens from Hadar and the one from Sterkfontein do not. The expectation of primitive retentions is not vacuous: the hindlimbs of these early hominids are reorganized for bipedalism, but retain many primitive features found in the hominoids but not in modern hominids (Stern and Susman, 1983; Tuttle, 1981). Another disadvantage to the knuckle walking

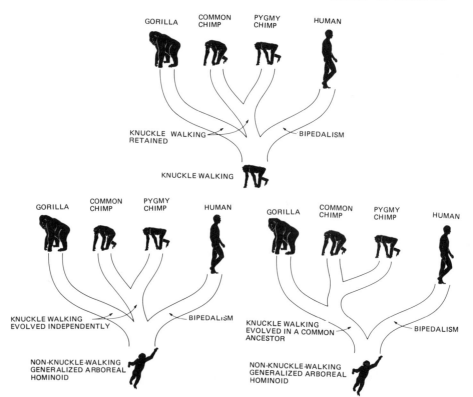

Figure 1. Knuckle-walking trifurcation: an evolutionary tree assuming the lineages leading to *Gorilla*, *Pan*, and *Homo* deverged at approximately the same time and from a common ancestor that was adapted to knuckle walking.

Figure 2. Parallelism trifurcation: a phylogeny assuming that the common ancestor of the African apes and humans was not a knuckle walker and that the lineages leading to *Gorilla*, *Pan*, and *Homo* diverged at approximately the same time, implying that knuckle walking evolved twice independently.

Figure 3. Bifurcation: an evolutionary tree assuming that the lineage leading to *Homo* diverged before the separation of the African ape lineages leading to *Pan* and *Gorilla*.

hypothesis is the step to bipedalism: once a generalized arboreal hominoid with a habitually orthograde trunk adapted to climbing develops a specialized pronograde terrestrial quadrupedalism like knuckle walking, the origin of bipedalism requires an evolutionary reversal of trunk posture and the loss of one mode of terrestrial locomotion in favor of another (Fleagle *et al.*, 1981; Rodman and McHenry, 1980; Stern and Susman, 1981).

The theory that knuckle walking evolved independently in the two

African ape lineages after the evolutionary trifurcation of *Homo, Pan,* and *Gorilla* (the parallelism hypothesis, Fig. 2) has the obvious weakness of supposing that such a peculiar mode of locomotion could have evolved twice. Knuckle-walking is energetically inefficient (Taylor and Rowntree, 1973) and appears to be a compromise solution for being quadrupedal and keeping long digits for climbing. But evolutionary alternatives are not limitless (Gould and Lewontin, 1979). Perhaps the flat-chested, mobile-forelimbed, arboreally adapted common ancestor of the African Hominoidea could only adapt to terrestrialism in a limited number of ways. Given its long manual digits, the forelimb could only become part of a quadrupedal prop if the distal two phalanges of each digit were curled under.

Despite most molecular evidence to the contrary, the appeal of the bifurcation hypothesis (Fig. 3) is understandable (Ferris *et al.,* 1981). The peculiarities of knuckle walking imply that it was derived from a common ancestor that was also a knuckle walker, yet the idea of a stage of knuckle walking in human evolution is problematic (Fleagle *et al.,* 1981; Tuttle, 1969, 1974, 1981). If *Pan* and *Gorilla* shared a common ancestor that was a knuckle-walker after the divergence of the hominid evolutionary lineage, then the evolution of both bipedalism and knuckle walking is more easily explained. The common ancestor of all African Hominoidea was a generalized arboreal climber without any terrestrial specializations (Tuttle, 1974, 1981). Large-bodied variants, perhaps, tended toward quadrupedalism when on the ground. Smaller bodied variants could balance on their hindlimbs during terrestrial locomotion. The eventual evolutionary outcome was one lineage of terrestrial quadrupeds that soon bifurcated into the lineages leading to *Pan* and *Gorilla,* and one lineage of small-bodied hominoids that adapted more and more to bipedalism when on the ground. Bipedalism is not an unusual alternative for a small-bodied, arboreally adapted, short-backed, long- and flexible-forelimbed hominoid. All lesser apes are habitual bipeds when terrestrial and even young chimpanzees are energetically just as efficient while walking bipedally as quadrupedally (Taylor and Rowntree, 1973). The disadvantage of this theory is that most molecular evidence indicates a trifurcation. The range of error could incorporate several alternatives, of course. And at least one method (cleavage mapping of mitochondrial DNA) shows that African apes could well have shared a common ancestor after the divergence of the hominid lineage (Ferris *et al.,* 1981).

Whatever model is chosen for the common ancestor of the African Hominoidea, the study of *P. paniscus* will yield invaluable clues. It represents an alternative chimpanzee with some more derived and some more primitive characteristics than the common chimpanzee. If it is true that the reduced masticatory system and paedomorphic face and skull of the pygmy chimpanzee are derived from a more robust ancestor, then its pattern of evolution parallels that of the sequence from *Australopithecus* to *Homo sapiens*.

5. Conclusions

The purpose of this study has been to compare the postcranial skeleton of *P. paniscus* with the other large-bodied hominoids, including the pre-2-MY-old hominids. This comparison shows that the pygmy chimpanzee is much more similar to the common chimpanzee than it is to any other hominoid. When correction is made for the fact that the pygmy chimpanzee is slightly smaller in skeletal size than the common chimpanzee, very few traits are significantly different between the two species of chimpanzee. When traits are compared together in multivariate analyses, the two species of chimpanzee are also more similar to one another than are any other pair of hominoid species.

The postcranium of the pre-2-MY-old fossil hominids most closely resembles modern hominids in most respects, but there are many individual traits that are more similar to other hominoids. Some of these ape-like traits are most like *P. paniscus*, but certainly not all. The multivariate affinities of the fossils are usually closer to modern hominids, although there are numerous exceptions. With some of these fossils the overall shape resembles *P. paniscus* most closely, whereas with others, *P. troglodytes, G. gorilla,* or *Pongo pygmaeus* is a closer match.

These results imply that the reconstruction of the common ancestor of apes and humans must draw on evidence from all living and extinct Hominoidea and that no living species can be considered as being the least derived in all of its anatomy. In several aspects of its postcranium, the pygmy chimpanzee may indeed be the least changed from the common ancestor of Hominoidea, particularly in its overall limb proportions and in its feet. Its reduced masticatory apparatus and paedomorphic skull are probably not primitive, however, but instead are derived from a more robust ancestor.

ACKNOWLEDGMENTS. I thank C. K. Brain, E. Vrba, and the staff of the Transvaal Museum, Pretoria, P. V. Tobias, A. Hughes, and the staff of the Department of Anatomy, University of Witwatersrand, Johannesburg; R. E. F. Leakey, M. D. Leakey, the late L. S. B. Leakey, L. Jacobs, and the staff of the National Museums of Kenya, Nairobi; F. C. Howell, D. C. Johanson, and the staff of the Cleveland Museum of Natural History;

and Tadessa Terfa, Mammo Tessena, and the staff of the National Museum of Ethiopia, for permission to study the original fossil material in their charge. I thank L. Barton, K. Nicklin, and C. Powell-Cotton of the Powell-Cotton Museum, Birchington; B. Lawrence, C. Mack, and M. Rutzmoser of the Museum of Comparative Zoology, Harvard; R. Thorington and the staff of the Division of Mammalogy, Smithsonian Institution; W. W. Howells, M. Poll, D. F. E. T. van den Audenaerde, M. Lovette, and the staff of the Musée Royale de l'Afrique Centrale, Tervuren; and J. Biegert and the staff of the Anthropologisches Institute, Zurich, for permission to study the comparative material in their charge; N. McLaughlin for typing; and L. J. McHenry for advice and assistance. Partial funding was provided by the Committee on Research, University of California, Davis.

References

Albrecht, G. H., 1980, Multivariate analysis and the study of form, with special reference to canonical variate analysis, *Am. Zool.* **20**(4):679–693.

Almquist, A. J., 1974, Sexual differences in the anterior dentition in African primates, *Am. J. Phys. Anthropol.* **40**:359–368.

Andrews, P., 1982, Hominoid evolution, *Nature* **295**:185–186.

Badrian, A., and Badrian, N., 1977, Pygmy chimpanzees, *Oryx* **13**(5):463–468.

Badrian, N., Badrian, A., and Susman, R. L., 1981, Preliminary observations on the feeding behavior of *Pan paniscus* in the Lomako Forest of Central Zaire, *Primates* **22**(2):173–181.

Behrensmeyer, A. K., 1976, Lothagam, Kanapoi, and Ekora: A general summary of stratigraphy and fauna, in: *Earliest man and environments in the Lake Rudolf Basin* (Y. Coppens, F. C. Howell, G. L. Isaac, and R. E. F. Leakey, eds.), University of Chicago Press, Chicago, pp. 163–172.

Blackith, R. E., and Reyment, R. A., 1971, *Multivariate Morphometrics,* Academic Press, New York.

Bogart, M. H., and Benirschke, K., 1977, Chromosomal analysis of the pygmy chimpanzee (*Pan paniscus*) with a comparison to man, *Folia Primatol.* **27**(1):60–67.

Borgaonkar, D. S., Sadasivan, G., and Ninan, T. A., 1971, *Pan paniscus* Y chromosome does not fluoresce, *J. Hered.* **62**(1):245–246.

Bourne, G. H., 1976, Pygmy chimpanzees at the Yerkes Center, *Yerkes Newsl.* **13**(1):2–7.

Brehme, H., 1975, Epidermal patterns of the hands and feet of the pygmy chimpanzee (*Pan paniscus*), *Am. J. Phys. Anthropol.* **42**:255–262.

Broom, R., 1942, The hand of the ape-man, *Paranthropus robustus, Nature* **149**:513.

Broom, R., and Robinson, J. T., 1949, Thumb of the Swartkrans ape-man, *Nature* **164**:841–842.

Bronowski, J., and Long, W. M., 1951, Statistical methods in anthropology, *Nature* **168**:794.

Bush, M. E., Lovejoy, C. O., Johanson, D. C., and Coppens, Y., 1982, Hominid carpal, metacarpal, and phalangeal bones recovered from the Hadar Formation: 1974–1977 collections, *Am. J. Phys. Anthropol.* **57**(4):651–678.

Charteris, J., Wall, J. C., Nottrodt, J. W., 1982, Pliocene hominid gait: New interpretations based on available footprint data from Laetoli, *Am. J. Phys. Anthropol.* **58**(2):133–144.

Chiarelli, B., 1963, Sensitivity to P.T.C. in primates, *Folia Primatol.* **1**(2):88–94.

Chiarelli, B., 1970, The chromosomes of the chimpanzee, in: *The Chimpanzee,* Volume 2 (G. Bourne, ed.), Karger, Basel, pp. 254–264.

Ciochon, R. L., and Corruccini, R. S., 1976, Shoulder joint of Sterkfontein *Australopithecus, S. Afr. J. Sci.* **72:**80–82.

Clarke, R. J., 1979, Early hominid footprints from Tanzania, *S. Afr. J. Sci.* **75:**148–149.

Coolidge, H. J., 1933, *Pan paniscus:* Pigmy chimpanzee from south of the Congo River, *Am. J. Phys. Anthropol.* **17**(1):1–57.

Coolidge, H. J., and Shea, B. T., 1982, External body dimensions of *Pan paniscus* and *Pan troglodytes* chimpanzees, *Primates* **23**(2):245–251.

Corruccini, R. S., 1978a, Morphometric analysis: Uses and abuses, *Yearb. Phys. Anthropol.* **21:**134–150.

Corruccini, R. S., 1978b, Comparative osteometrics of the hominoid wrist joint, with special reference to knuckle-walking, *J. Hum. Evol.* **7:**307–321.

Corruccini, R. S., and Ciochon, R. L., 1976, Morphometric affinities of the human shoulder, *Am. J. Phy. Anthropol.* **45**(1):19–38.

Corruccini, R. S., and McHenry, H. M., 1979, Morphological affinities of *Pan paniscus, Science* **204:**1341–1343.

Cramer, D. L., 1977, Craniofacial morphology of *Pan paniscus, Contrib. Primatol.* **10:**1–64.

Cronin, J. E., 1977, Pygmy chimpanzee (*Pan paniscus*) systematics, *Am. J. Phys. Anthropol.* **47:**125.

Cronin, J. E., Boaz, N. T., Stringer, C. B., and Rak, Y., 1981, Tempo and mode in hominid evolution, *Nature* **292:**113–122.

Dart, R. A., 1925, *Australopithecus africanus,* the man-ape of South Africa, *Nature* **115:**195–199.

Darwin, C., 1871, *The Descent of Man and Selection in Relation to Sex,* Murray, London.

Day, M. H., 1973, Locomotor features of the lower limb in hominids, *Symp. Zool. Soc. Lond.* **33:**29–51.

Day, M. H., and Napier, J. R., 1964, Hominid fossils from Bed I, Olduvai Gorge, Tanganyika: Fossil foot bones, *Nature* **201:**969–970.

Day, M. H., and Scheuer, J. L., 1973, A new hominid metacarpal from Swartkrans, *J. Hum. Evol.* **2**(6):429–438.

Day, M. H., and Wickens, E. H., 1980, Laetoli Pliocene hominid footprints and bipedalism, *Nature* **286:**385–387.

Deblock, R., 1973, Craniométrie comparée *Pan paniscus* et *Pan troglodytes,* Thèse de doctorat en sciences naturelles, Lille.

Deblock, R., and Fenart, R., 1973, Differences sexuelles sur crânes adultes chez *Pan paniscus, Bull. Assoc. Anat. (Nancy)* **57**(157):299–306.

Deblock, R., and Fenart, R., 1977, Les angles de la base du crâne chez les chimpanzées, *Bull. Assoc. Anat. (Nancy)* **61**(173):183–188.

Delattre, A., Fenart, R., and Deblock, R., 1968, Os wormiens et anomalies suturales dans une collection de crânes de *Pan paniscus, Assoc. Anat. Compt. Rend.* **53:**1794–1796.

Feldesman, M. R., 1979, Further morphometric studies of the ulna from the Omo Basin, Ethiopia, *Am. J. Phys. Anthropol.* **51**(3):409–416.

Feldesman, M. R., 1982, Morphometric analysis of the distal humerus of some Cenozoic catarrhines: The late divergence hypothesis revisisted, *Am. J. Phys. Anthropol.* **59:**73–95.

Fenart, R., and Deblock, R., 1973, *Pan paniscus* et *Pan troglodytes* craniométrie: Étude comparative et ontogénique selon les méthodes classiques et vestibulaire. Tome 1, *Mus. R. Afr. Cent. Tervuren Belg. Ann. Octavo Sci. Zool.* **204:**1–473.

Ferris, S. D., Wilson, A. C., and Brown, W. M., 1981, Evolutionary tree for apes and humans based on cleavage maps of mitochondrial DNA, *Proc. Natl. Acad. Sci. USA* **78**(4):2432–2436.

Fiedler, W., 1956, Ubersicht über das system der Primates, in: *Primatologia,* Volume 1 (H. Hofer, A. H. Schultz, and D. Starck, eds.), Karger, Basel, pp. 1–266.

Fleagle, J. G., Stern, J. T., Jr., Jungers, W. L., Susman, R., Vangor, A. K., and Wells, J. P., 1981, Climbing: A biomechemical link with brachiation and with bipedalism, *Symp. Zool. Soc. Lond.* **48**:359–375.

Frechkop, S., 1935, Notes sur les mammifères. XVII. A propos du chimpanzé de la rive gauche du Congo, *Mus. R. Hist. Nat. Belg. Bull.* **11**(2):1–43.

Goodman, M., 1971, Immunodiffusion systematics of the primates, I. The catarrhini, *Syst. Zool.* **20**:19–62.

Goodman, M., Moore, G. W., Farris, W., and Poulik, E., 1970, The evidence from genetically informative macromolecules on the phylogenetic relationships of the chimpanzee, in: *The Chimpanzee* Volume 2 (G. Bourne, ed.), Karger, Basel, pp. 318–360.

Gould, S. J., and Lewontin, R. C., 1979, The spandrels of the San Marco and the Panglossian paradigm: A critique of the adaptationist programme, *Proc. R. Soc. Lond. B* **205**:581–598.

Greenfield, L. O., 1980, A late divergence hypothesis, *Am. J. Phys. Anthropol.* **52**(3):351–366.

Hamerton, J. L., Klinger, H. P., Mutton, D. E., and Lang, E. M., 1963, The somatic chromosomes of the Hominoidea, *Cytogenetics* **2**(4–5):240–263.

Heiple, K. G., and Lovejoy, C. O., 1971, The distal femoral anatomy of *Australopithecus, Am. J. Phys. Anthropol.* **35**:75–84.

Horn, A., 1975, Adaptations of the pygmy chimpanzee (*Pan paniscus*) to the forests of the Zaire Basin, *Am. J. Phys. Anthropol.* **42**(2):307.

Horn, A. D., 1977, A preliminary report on the ecology and behavior of the bonobo chimpanzee (*Pan paniscus,* Schwarz 1929) and a reconsideration of the evolution of the chimpanzee, Ph.D. Dissertation, Yale University, New Haven, Connecticut.

Horn, A. D., 1979, The taxonomic status of the bonobo chimpanzee, *Am. J. Phys. Anthropol.* **51**(2):273–282.

Horn, A. D., 1980, Some observations on the ecology of the bonobo chimpanzee (*Pan paniscus,* Schwarz 1929) near Lake Tumba, Zaire, *Folia Primatol.* **34**:145–169.

Howell, F. C., and Wood, B. A., 1974, Early hominid ulna from the Omo basin, Ethiopia, *Nature* **249**:174–176.

Hubsch, I., 1970, Einiges zum Verhalten der Zwergschimpansen (*Pan paniscus*) und der Schimpansen (*Pan troglodytes*) im Frankfurter Zoo, *Zool. Gart.* **38**(3/4):107–132.

Itani, J., 1980, Social structures of African great apes, *J. Reprod. Fertil. Suppl.* **28**:33–41.

Izor, R. J., Walchuk, S. L., and Wilkins, L., 1981, Anatomy and systematic significance of the penis of the pygmy chimpanzee, *Pan paniscus, Folia Primatol.* **35**(2–3):218–224.

Jenkins, F. A., Jr., 1972, Chimpanzee bipedalism, *Science* **178**:877–879.

Jenkins, F. A., Jr., 1981, Wrist rotation in primates: A critical adaptation for brachiators, *Symp. Zool. Soc. Lond.* **48**:429–451.

Jenkins, F. A., Jr., and Fleagle, J. G., 1975, Knuckle-walking and the functional anatomy of the wrists in living apes, in: *Primate Functional Morphology and Evolution* (Russell H. Tuttle, ed.), Mouton, The Hague, pp. 213–227.

Johanson, D. C., 1974, Some metric aspects of the permanent and deciduous dentition of the pygmy chimpanzee (*Pan paniscus*), *Am. J. Phys. Anthropol.* **41**:39–48.

Johanson, D. C., Lovejoy, C. O., Burstein, A. H., and Heiple, K. G., 1976, Functional implications of the Afar knee joint, *Am. J. Phys. Anthropol.* **44**(1):188 (Abstract).

Johanson, D. C., Lovejoy, C. O., Kimbel, W. H., White, T. D., Ward, S. C., Bush, M. E., Latimer, B. M., and Coppens, Y., 1982, Morphology of the Pliocene partial hominid skeleton (A.L. 288-1) from the Hadar Formation, Ethiopia, *Am. J. Phys. Anthropol.* **57**(4):403–452.

Johnson, S. C., 1981, Bonobos: Generalized hominid prototypes or specialized insular dwarfs?, *Curr. Anthropol.* **22**(4):363–375.

Jungers, W. L., 1982, Lucy's limbs: Skeletal allometry and locomotion in *Australopithecus afarensis*, *Nature* **297**:676–678.

Kano, T., 1979, A pilot study on the ecology of pygmy chimpanzees, *Pan paniscus*, in: *The Great Apes* (D. A. Hamburg and E. McCown, eds.), Benjamin/Cummings, Menlo Park, California, pp. 123–135.

Kano, T., 1980, Social behavior of wild pygmy chimpanzees (*Pan paniscus*) of Wamba: A preliminary report, *J. Hum. Evol.* **9**:243–260.

Keith, A., 1923, Man's posture: Its evolution and disorders, *Br. Med. J.* **1**:451–54, 499–502, 545–48, 387–90, 624–26, 669–72.

Khudr, G., Benirschke, K., and Sedgwick, C. J., 1973, Man and *Pan paniscus:* A karyologic comparison, *J. Hum. Evol.* **4**:323–331.

Klinger, H. P., Hamerton, J. L., Mutton, D., and Lang, E. M., 1963, The chromosomes of the Hominoidea, in: *Classification and Human Evolution* (S. L. Washburn, ed.), Aldine, Chicago, pp. 235–242.

Kuroda, S., 1979, Grouping of the pygmy chimpanzee, *Primates* **20**:161–183.

Latimer, B. M., White, T. D., Kimbel, W. H., Johanson, D. C., and Lovejoy, C. O., 1981, The pgymy chimpanzee is not a living missing link in human evolution., *J. Hum. Evol.* **10**:475–488.

Latimer, B. M., Lovejoy, C. O., Johanson, D. C., and Coppens, Y., 1982, Hominid tarsal, metatarsal, and phalangeal bones recovered from the Hadar Formation: 1974–1977 collections, *Am. J. Phys. Anthropol.* **57**(4):701–720.

Leakey, M., 1978, Pliocene footprints in Laetolil, Northern Tanzania, *Antiquity* **52**(205):133.

Leakey, M. D., 1981, Tracks and tools, *Philos. Trans. R. Soc. Lond. B* **292**:95–102.

Lewis, O. J., 1973, The hominid os capitatum, with special reference to the fossil bones from Sterkfontein and Olduvai Gorge, *J. Hum. Evol.* **2**:1–12.

Lewis, O. J., 1980, The joints of the evolving foot, Part I, The ankle joint, *J. Anat.* **130**:527–543; 2, The intrinsic joints, *J. Anat.* **130**(4):833–850; 3, The fossil evidence, *J. Anat.* **131**(1/2):275–298.

Lewis, O. J., 1977, Joint remodelling and the evolution of the human hand, *J. Anat.* **123**(1):157–201.

Lewis, O. J., 1981, Functional morphology of the joints of the evolving foot, *Symp. Zool. Soc. Lond.* **46**:169–188.

Lovejoy, C. O., Johanson, D. C., and Coppens, Y., 1982a, Hominid upper limb bones recovered from the Hadar Formation: 1974–1977 collections, *Am. J. Phys. Anthropol.* **57**(4):537–650.

Lovejoy, C. O., Johanson, D. C., and Coppens, Y., 1982b, Hominid lower limb bones recovered from the Hadar Formation: 1974–1977 collections, *Am. J. Phys. Anthropol.* **57**(4):679–700.

MacKinnon, J., 1976, Mountain gorillas and bonobos, *Oryx* **13**:372–382.

Marzke, M. W., 1971, Origin of the human hand, *Am. J. Phys. Anthropol.* **34**:61–84.

Marzke, M. W., 1983, Joint functions and grips of the *Australopithecus afarensis* hand with special reference to the region of the capitate, *J. Hum. Evol.* **12**(2):197–211.

McHenry, H. M., 1972, The Postcranial Anatomy of Early Pleistocene Hominids, Ph.D. Dissertation, Harvard University, Cambridge, Massachusetts.

McHenry, H. M., 1973, Early hominid humerus from East Rudolf, Kenya, *Science* **180**:739–741.

McHenry, H. M., 1974, How large were the australopithecines?, *Am. J. Phys. Anthropol.* **40**:329–340.

McHenry, H. M., 1975, A new pelvic fragment from Swartkrans and the relationship between the robust and gracile australopithecines, *Am. J. Phys. Anthropol.* **53**:245–262.

McHenry, H. M., 1976, Multivariate analysis of early hominid humeri, in: *The Measures of Man* (E. Giles and J. S. Friedlander, eds.), Peabody Museum Press, Harvard University, Cambridge, Massachusetts, pp. 338–371.

McHenry, H. M., 1978, Fore- and hindlimb proportions in Plio-Pleistocene hominids, *Am. J. Phys. Anthropol.* **49**(1):15–22.

McHenry, H. M., 1982, The pattern of human evolution: Studies on bipedalism, mastication, and encephalization, *Annu. Rev. Anthropol.* **11**:151–173.

McHenry, H. M., 1983, The capitate of *Australopithecus afarensis* and *A. africanus*, *Am. J. Phys. Anthropol.* **62**:187–198.

McHenry, H. M., and Corruccini, R. S., 1975*a*, Multivariate analysis of early hominid pelvic bones, *Am. J. Phys. Anthropol.* **43**(2):263–270.

McHenry, H. M., and Corruccini, R. S., 1975*b*, Distal humerus in hominoid evolution, *Folia Primatol.* **23**:227–244.

McHenry, H. M., and R. S. Corruccini, 1976*a*, Affinities of Tertiary hominoid femora, *Folia Primatol.* **26**(2):139–150.

McHenry, H. M., and Corruccini, R. S., 1976*b*, Fossil hominid femora and the evolution of walking, *Nature* **259**:657–658.

McHenry, H. M., and Corruccini, R. S., 1978, The femur in early human evolution, *Am. J. Phys. Anthropol.* **49**(4):473–488.

McHenry, H. M., and Corruccini, R. S., 1980, Late Tertiary hominoids and human origins, *Nature* **285**:397–398.

McHenry, H. M., and Corruccini, R. S., 1981, *Pan paniscus* and human evolution, *Am. J. Phys. Anthropol.* **54**:355–367.

McHenry, H. M., and Temerin, L. A., 1979, The evolution of hominid bipedalism: Evidence from the fossil record, *Yearb. Phys. Anthropol.* **22**:105–131.

McHenry, H. M., Corruccini, R. S., and Howell, F. C., 1976, Analysis of an early hominid ulna from the Omo Basin, Ethiopia, *Am. J. Phys. Anthropol.* **44**(2):295–304.

McHenry, H. M., Andrews, P., and Corruccini, R. S., 1980, Miocene hominoid palatofacial morphology, *Folia Primatol.* **33**:241–252.

Miller, R. A., 1952, The musculature of *Pan paniscus*, *Am. J. Anat.* **91**:183–232.

Moor-Jankowski, J., Wiener, A. A., Socha, W. W., Gordon, E. B., and Mortelmans, J., 1972, Blood groups of the dwarf chimpanzee (*Pan paniscus*), *J. Med. Primatol.* **1**(1):90–101.

Moor-Jankowski, J., Weiner, A. S., Socha, W. W., Gordon, E. B., Mortelmans, J., and Sedgwick, C. J., 1975, Blood groups of pygmy chimpanzee (*Pan paniscus*): Human-type and simian-type, *J. Med. Primatol.* **4**:262–267.

Morton, J. J., 1927, Human origin, correlation of previous studies on primate feet and posture with other morphological evidence, *Am. J. Phys. Anthropol.* **10**:173–203.

Napier, J. R., 1962, Fossil hand bones from Olduvai Gorge, *Nature* **196**:409–411.

Neugebauer, W., 1980, The status and management of the pygmy chimpanzee *Pan paniscus* in European zoos, *Int. Zoo Yearb.* **20**:64–70.

Nishida, T., 1972, Preliminary information of the pygmy chimpanzees (*Pan paniscus*) of the Congo Basin, *Primates* **13**(4):415–425.

Oakley, K. P., Campbell, B. G., and Molleson, T. I., 1977, *Catalogue of Fossil Hominids, Part I: African*, 2nd ed., British Museum (Natural History), London.

Oxnard, C. E., 1977, Morphometric affinities of the human shoulder, *Am. J. Phys. Anthropol.* **46**(2):367–374.

Oxnard, C. E., and Lisowski, F. P., 1980, Functional articulation of some hominoid food bones: Implications for the Olduvai (Hominid 8) foot, *Am. J. Phys. Anthropol.* **52**(1):107–118.

Patterson, B., and Howells, W. W., 1967, Hominid humeral fragment from early Pleistocene of northwestern Kenya, *Science* **156**:64–66.

Patterson, T. L., 1979, The behavior of a group of captive pygmy chimpanzees (*Pan paniscus*), *Primates* **20**(3):341–354.

Pilbeam, D., 1978, Rethinking human origins, *Discovery* **13**(1):2–9.

Pilbeam, D., 1982, New hominoid skull material from the Miocene of Pakistan, *Nature* **295**:232–234.

Rempe, Udo, 1961, Einige beobachtungen an bonobos, *Pan paniscus* Schwarz, 1929, *Z. Wiss. Zool.* **165**(1–2):81–87.

Rightmire, G. P., 1972, Multivariate analysis of an early hominid metacarpal from Swartkrans, *Science* **176**:159–161.

Roberts, D., 1974, Structure and function of the primate scapula, in: *Primate Locomotion* (F. Jenkins, Jr., ed.), Academic Press, New York, pp. 171–200.

Robinson, J. T., 1972, *Early Hominid Posture and Locomotion*, University of Chicago Press, Chicago.

Rode, P., 1937, Les races géographiques du chimpanzé, *Mammalia (Paris)* **1**:165–177.

Rode, P., 1941, Étude d'un chimpanzé pygmée adolescent (*Pan satyrus paniscus* Schwarz), *Mammalia (Paris)* **5**:50–68.

Rodman, P. S., and McHenry, H. M., 1980, Bioenergetics and the origin of hominid bipedalism, *Am. J. Phys. Anthropol.* **52**(1):103–106.

Sarich, V. M., 1973, Just how old is the hominid line?, *Yearb. Phys. Anthropol.* **17**:98–112.

Sarich, V. M., and Cronin, J. E., 1977, Molecular systematics of the primates, in: *Molecular Anthropology* (M. Goodman and R. E. Tashian, eds.), Plenum Press, New York, pp. 141–170.

Savage-Rumbaugh, E. S. and Wilkerson, B. J., 1978, Sociosexual behavior in *Pan paniscus* and *Pan troglodytes:* A comparative study, *J. Hum. Evol.* **7**:327–344.

Schouteden, H., 1930, Le chimpanzé de la rive gauche du Congo, *Bull. Cerc. Zool. Congl.* **VII**(4):114–119.

Schouteden, H., 1948, Faune du Congo Belge et du Ruanda-Urundi, I. Mammifères, *Mus. R. Congo Belge, Tervuren Belg. Ann. Sci. Zool.* **1**:1–313.

Schultz, A. H., 1969, The skeleton of the chimpanzee, in: *The Chimpanzee*, Volume 1 (G. H. Bourne, ed.), University Park Press, Maryland, pp. 50–103.

Schwarz, E., 1932, Discussion of *Australopithecus*, *Z. Saugetierk.* **7**:10.

Schwarz, E., 1934, On the local races of the chimpanzee, *Ann. Mag. Nat. Hist. (London)* **13**(Ser. 10):576–583.

Senut, B., 1978, Révision de quelques piéces humérales Plio-Pléistocénes sud-africaines, *Bull. Mem. Soc. Anthropol. Paris* **5**:223–229.

Senut, B., 1979, Comparaison des hominidés de Gomboré 1B et de Kanapoi: Deux piéces de genre *Homo?*, *Bull. Mem. Soc. Anthropol. Paris* **6**(1 Ser. 13):111–117.

Senut, B., 1980, New data on the humerus and its joints in Plio-Pleistocene hominids, *Coll. Anthropol.* **4**(1):87–93.

Senut, B., 1981*a*, Humeral outlines in some hominoid primates and in Plio-Pleistocene hominids, *Am. J. Phys. Anthropol.* **56**(3):275–284.

Senut, B., 1981*b*, Outlines of the distal humerus in hominoid primates: Application to some Plio-Pleistocene hominids, in: *Primate Evolutionary Biology* (A. B. Chiarelli and R. S. Corruccini, eds.), Springer-Verlag, Berlin, pp. 81–92.

Shea, B. T., 1981, Relative growth of the limbs and trunk in the African apes, *Am. J. Phys. Anthropol.* **56**(2):179–202.

Stern, J. T., and Susman, R. L., 1981, Electromyography of the gluteal muscles in *Hylobates*,

Pongo, and *Pan:* Implications for the evolution of hominid bipedalism, *Am. J. Phys. Anthropol.* **55:**153–166.

Stern, J. T., Jr., and Susman, R. L., 1983, The locomotor anatomy of *Australopithecus afarensis, Am. J. Phys. Anthropol.* **60:**279–317.

Susman, R. L., 1978, Functional morphology of homnoid metacarpals, in: *Recent Advances in Primatology,* Volume 3, *Evolution* (D. J. Chivers, and K. A. Joysey, eds.), Academic Press, London, pp. 77–80.

Susman, R. L., 1979, Comparative and functional morphology of hominoid fingers, *Am. J. Phys. Anthropol.* **50**(2):215–236.

Susman, R. L., 1980, Acrobatic pygmy chimpanzees, *Nat. Hist.* **89**(9):32–39.

Susman, R. L., and Creel, N., 1979, Functional and morphological affinities of the subadult hand (O.H. 7) from Olduvai Gorge, *Am. J. Phys. Anthropol.* **51**(3):311–332.

Susman, R. L., and Jungers, W. L., 1981, Comment, *Curr. Anthropol.* **22**(4):369–370.

Susman, R. L., and Stern, J. T., Jr., 1979, Telemetered electromyography of flexor digitorum profundus and flexor digitorum superficialis in *Pan paniscus* and implications for interpretations of the O.H. 7 hand, *Am. J. Phys. Anthropol.* **50**(4):565–574.

Susman, R. L., and Stern, J. T., 1982, Functional morphology of *Homo habilis, Science* **217:**931–934.

Susman, R. L., Badrian, N. L., and Badrian, A. J., 1980, Locomotor behavior of *Pan paniscus* in Zaire, *Am. J. Phys. Anthropol.* **53**(1):69–80.

Susman, R. L., Badrian, N., Badrian, A., and Handler, N. T., 1981, Pygmy chimpanzee in peril, *Oryx* **16:**180–183.

Suzman, I. M., 1982, A comparative study of the Hadar and Sterkfontein australopithecine innominates, *Am. J. Phys. Anthropol.* **57**(2):235.

Tardieu, C., 1981, Morpho-functional analysis of the articular surfaces of the knee-joint in primates, in: *Primate Evolutionary Biology* (A. B. Chiarelli and R. S. Corruccini, eds.), Springer-Verlag, Berlin, pp. 68–80.

Taylor, C. R., and Rowntree, V. J., 1973, Running on two or four legs: which consumes more energy?, *Science* **179:**186–187.

Tratz, E., and Heck, H., 1954, Der afrikanische anthropoide "bonobo," ein neue menschenaffen-gartung, *Saugetierkd. Mitt.* **2:**97–101.

Tuttle, R. H., 1969, Quantitative and functional studies on the hands of Anthropoidea I. The Hominoidea, *J. Morphol.* **128:**309–63.

Tuttle, R. H., 1974, Darwin's apes, dental apes, and the descent of man: Normal science in evolutionary anthropology, *Curr. Anthropol.* **15**(4):389–398.

Tuttle, R. H., 1981, Evolution of hominid bipedalism and prehensile capabilities, *Philos. Trans. R. Soc. Lond. B* **292:**89–94.

Urbain, A., and Rode, P., 1940, Un chimpanzé pygmée au Parc Zoologique du Bois de Vincennes, *Mammalia (Paris)* **4:**12–14.

Van Bree, P. H. H., 1963, On a specimen of *Pan paniscus* Schwarz, 1929, which lived in the Amsterdam Zoo from 1911 till 1916, *Zool. Gart. (NF)* **27:**292–295.

Vandebroek, G., 1959, Notes écologiques sur les anthropoïdes Africains, *Ann. Soc. R. Zool. Belg.* **89:**203–211.

Vandebroek, G., 1969, Évolution des Vertébrés *de leur origine à l'homme,* Masson, Paris.

Vrba, E. S., 1979, A new study of the scapula of *Australopithecus africanus* from Sterkfontein, *Am. J. Phys. Anthropol.* **51**(1):117–129.

Washburn, S. L., 1967, Behavior and the origin of man: The Huxley memorial lecture, 1967, *Proc. R. Anthropol. Inst.* **1967:**21–27.

White, T. D., 1980, Evolutionary implications of Pliocene hominid footprints, *Science* **208:**175–176.

Wiener, A. S., Socha, W. W., and Sedgwick, C. J., 1973, Blood groups of the pygmy chimpanzee (*Pan paniscus*): Further observations, *Lab. Prim. Newsl.* **12**(2):6–8.

Wiener, A. S., Socha, W. W., Moor-Jankowski, J., and Gordon, E. B., 1975, Family studies on the simian-type blood groups of chimpanzees, *J. Med. Primatol.* **4**(1):45–50.

Yunis, J. J., and Prakash, O., 1982, The origin of man: A chromosomal pictorial legacy, *Science* **215**:1525–1530.

Zihlman, A. L., 1979, Pygmy chimpanzee morphology and the interpretation of early hominids, *S. Afr. J. Sci.* **75**:165–168.

Zihlman, A. L., and Cramer, D. L., 1978, Skeletal differences between pygmy (*Pan paniscus*) and common chimpanzees (*Pan troglodytes*), *Folia Primatol.* **29**:86–94.

Zihlman, A. L., Cronin, J. E., Cramer, D. L., and Sarich, V. M., 1978, Pygmy chimpanzee as a possible prototype for the common ancestor of humans, chimpanzees and gorillas. *Nature* **275**:744–746.

PART II

Behavior of Pan paniscus

CHAPTER 10

Feeding Ecology of the Pygmy Chimpanzees (Pan paniscus) of Wamba

TAKAYOSHI KANO AND MBANGI MULAVWA

1. Introduction

In recent years, intensive field studies on pygmy chimpanzees (*Pan paniscus*), have been carried out in at least four localities within their range in the Zaire (Congo) forest: Wamba, Lomako, Lake Tumba, and Yalosidi (Fig. 1). Of these four localities, reports from study sites other than Wamba have centered around the ecology of pygmy chimpanzees [e.g., Badrian and Badrian (1977) and Badrian *et al.* (1981) from Lomako; Horn (1980) from Tumba; Kano (1983) from Yalosidi]. At Wamba, field studies have been in progress since 1974, focusing on behavior (Kuroda, 1980, and this volume; Kano, 1980, 1982*a, b,* and 1984) and social organization (Kuroda, 1979; Kano, 1982*a;* Kitamura, 1983), with ecology a secondary concern.

Artificial provisioning was begun at Wamba in 1974. One of the unit groups inhabiting the Wamba region, the E group, has been continuously provisioned since 1976. The P group has been provisioned with sugar cane since 1980, and the K group since 1981. The purpose of the present report is to present data on pygmy chimpanzee ecology collected at Wamba from 1974 to the present time. In spite of the fragmentary nature of these data, this report should provide reference for future ecological research at Wamba, as well as providing comparatative data for studies comparing work at Wamba with ecological research results from other study sites.

TAKAYOSHI KANO • Department of Human Ecology, College of Medicine, University of the Ryukyus, 207, Uehara, Nishihara Okinawa, Japan. *MBANGI MU-LAVWA* • Institut de Recherche Scientifique, Centre de Mabali, B.P. 36 Bikoro, Equateur, Zaire.

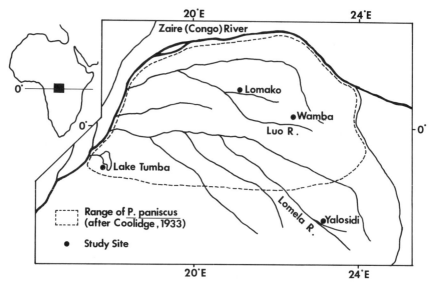

Figure 1. Distribution of *Pan paniscus* and study sites. Yalosidi is outside of Coolidge's (1933) range, but according to a survey by T. Kano (unpublished), pygmy chimpanzees occur as far south as the left bank of the Lomela.

1.1. Subjects and Methods

Five pygmy chimpanzee unit groups at Wamba have part or all of their home ranges within the study area (Fig. 2). One of them, the E group, has become almost fully habituated. The P, B, and K groups are partly habituated. The last group, the S group, is as yet unhabituated. The sizes of the E, P, B, and K groups are estimated at 65, 50–60, 80–100, and 90–120, respectively, but as young nulliparous females transfer from group to group with comparative freedom (Kano, 1982a), the membership of the groups is somewhat unstable. The size of the unhabituated S group is not clear. With regard only to identified individuals, the adult male membership of the unit groups is extremely stable.

Several subunits can be seen in the E group (Kano, 1982a). Some of these subunits have as their core a mother and her male offspring, while others appear to center around a partnership of males or of females (Kano, 1982a). Within the E group the units can be roughly divided into two subgroups (Kitamura, 1983). The first are mainly nomadic in the southern half of the E group range (the central and southern regions; Fig. 2) and the other nomadic subgroup is mainly in the northern half (the eastern and northern regions) of the range. The analysis of the subgroup structures of the unit groups will be left to a future study. Examination of the E

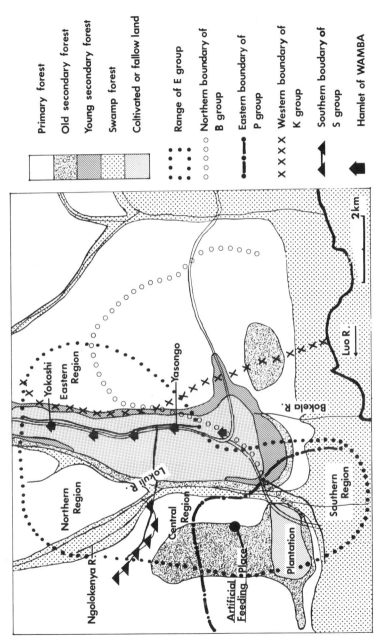

Legend:
- Primary forest
- Old secondary forest
- Young secondary forest
- Swamp forest
- Cultivated or fallow land
- Range of E group
- Northern boundary of B group
- Eastern boundary of P group
- Western boundary of K group
- Southern boudary of S group
- Hamlet of WAMBA

Figure 2. Distribution of vegetation types in the study area.

group subunits is as yet insufficient, and the existence of such subunits in the other unit groups has not yet been verified.

The ranges of the unit groups vary annually. The total range recorded for the E group covers 58 km^2, 66% (38 km^2) of which overlaps with the ranges of other groups (30 km^2 with one and 8^2 km with two groups). The area for the E group of 38 km^2 results in a density of 1.7 pygmy chimpanzees per km^2.

Field studies were carried out during five occasions: October 31, 1975–February 9, 1976; September 17, 1976–January 18, 1977; October 23, 1977–March 5, 1978; January 1, 1979–February 24, 1979; and October 20, 1981–March 2, 1982. (One of the authors—MM—participated only in the last study period.) Data on food selection were collected during all periods, but much of the material presented in this report, in particular the data on feeding techniques and ranging behavior, was collected only during the final study. The data collected were seasonally biased: investigation was carried out repeatedly in the period from October or November to February or March, but the period from April to September is not represented.

The principal observation method used was the "random sampling method" (Altmann, 1974). There were two observation areas, the natural habitat and the artificial feeding site. Almost all results obtained from the artificial feeding site have been eliminated from this report. A total 1470 hr of observation was made by TK, about 962 hr in the natural habitat, 431 hr in the artifical feeding places, and 77 hr under the mobile provisioning state.*

1.2. Study Area

1.2.1. Vegetation

The study area comprised about 100 km^2 of forest surrounding the five hamlets that constitute the village of Wamba (Fig. 2). After 1977, however, when the E group became the main object of study, the investigation was concentrated in the area occupied by that group, namely the banks of the Lokuli River and the region extending for 3 km to the east of Yasongo–Yokoshi hamlets (Fig. 2).

The vegetation of the study area was roughly divided into the following five categories.

a. Primary Forest. The primary forest that dominates the study area is a dry and closed forest standing on a firm foundation. The highest trees

---

*Artificial food items are provisioned to pygmy chimpanzees wherever they are encountered (Nishida, 1979).

in this forest reach a height of 50 m. There is a high density of low (5–15 m), medium (15–25 m), and tall (over 25 m) trees, but in general the height of the trees is continuous and a clear stratification is not in evidence. As the sunlight is largely cut off before it reaches the forest floor, the shrub and herb layers are relatively sparse. The tree layers are rich in species, and in many cases there is no definitely dominant species. In the area close to the swamp forest, however, *Gilbertiodendron dewevrei* tends to become numerous in the tall tree layer.

   b. Swamp Forest. Swamp forest is seen to a greater or lesser extent along practically all the river courses. Along the Luo River there is an extensive swamp forest, attaining a width of several kilometers. The swamp forest belt on the Lokuli River within the home range of the E group has a width of 300–500 m. The foundation of the swamp forest is weak, the density of tall trees low, and, because a greater amount of sunlight reaches the forest floor than in the primary forest, the undergrowth is more luxuriant.

   c. Aged Secondary Forest. The density of trees, especially medium and tall trees, is lower than that of the primary forest. In places where direct sunlight reaches the forest floor, there is a thick growth of the giant Marantaceae spp., e.g., *Sarcophrynium macrostachyum* and *Haumania liebrechtsiana,* in the herb layer. Much of this type of forest consists of aged secondary forest regenerating after disturbance by humans. There is, however, also some open, secondary dry forest that is not the result of human activity. In this chapter, however, these two types are treated together as aged secondary forest. Many of the tree species found here are common to the primary forest as well.

   d. Young Secondary Forest. This is forest growing on fallow land where at least 5 years has elapsed since clearance by humans. Species characteristic of this type of forest are *Musanga smithii, Albizia gummifera, Croton haumanianus, Macaranga* spp., and so on, which dominate the medium and tall tree layers. It is bright, open forest, with a luxuriant undergrowth composed of Marantaceae spp., *Aframomum* spp., etc.

   e. Secondary Shrub. This is vegetation on land that has been fallow for less than 5 years. Its physiognomy is characterized by a thick growth of herbs and shrubs, which are mainly *Aframomum* spp.

1.2.2. Topography

   The topography of the study area is extremely simple. The region comprising the village, the road, and the area to the north of these is a flat tableland, while between this region the swamp forest of the Luo also stands on extremely flat land.

1.2.3. Climate

Though the meteorological observations were insufficient in this study, we may assume that the climatic conditions at Wamba are practically the same as at Djolu, 80 km to the north. According to the data obtained at the weather station at Djolu between 1953 and 1963, the absolute maximum monthly temperature was between 32.6 and 36°C, the absolute minimum monthly temperature was between 12.7 and 17.1°C, and the temperatures in the shade reached maxima of 29.3–31.4°C and minima of 18.2–19.2°C (Vuanza and Crabbe, 1975). The average annual rainfall in Djolu from 1936 to 1959 was 2005 mm (range, 1368–2310 mm, $N = 24$). There is no month of the year without rainfall. Rainfall is greatest from September to November (about 200 mm per month), while from December to February there is often a comparatively small amount of rainfall (less than 100 mm per month). There is considerable variation from year to year (Vuanza and Crabbe, 1975).

2. Results

2.1. Food Repertoire

The pygmy chimpanzees of Wamba obtain their food from four sources: wild plants, domesticated plants, animals, and inorganic matter. Among these, wild plants are the richest in variety and constitute the principal food source.

The number of wild plant species consumed by the pygmy chimpanzees was recorded at 114, including 13 unidentified species (Table I). Among these, 93 species were verified by direct observation of the pygmy chimpanzees, while the remaining species were verified through the fresh food remains found along trails and by fecal analysis. A total of 133 food types were recorded. These include pulp (pericarp) of 67 species, immature and mature leaves of 24 species, 12 species of seeds, eight species of shoots, five species of petioles, four species of stems, the piths of four species, three mushrooms, two types of pods, one species of bark, and one species of flower. Two types of plants were consumed while in a dead and decaying state. With regard to the life form of food plants, tall trees (44 species) and woody vines (23) were the most numerous, followed by low trees (12), medium trees (ten), herbs (nine), herbaceous vines (three), fungi (three) and shrubs (two).

Crops cultivated by the villagers of Wamba include manioc, maize, rice, coffee, sugar cane, sweet potato, oil-palm, yam, banana, pineapple, avocado, papaya, cacao, grapefruit, and other fruits and vegetables. Among

Table I. Wild Food Plants of the Wamba Pygmy Chimpanzee

Plant	Part eaten	Life form	Main habitat[a]
Anacardiaceae			
Antrocaryon micraster A. CHEV. & CUILLAUM.	Pulp	Tall tree	PF
Annonaceae			
Anonidium mannii (OLIV.) ENGL. et DIELS	Pulp	Medium tree	PF
Polyalthia suaveolens ENGL. et DIELS	Pulp	Medium tree	PF
Apocynaceae			
Alstonia boonei DE WILD.	Pulp	Low tree	PF, ASF
Ancylobotrys pyriforms PIERRE	Pulp	Woody vine	PF
Baissea thollonii HUA.	Pulp	Woody vine	PF, ASF
Dictyophleba ochracea (K.SCHUM. ex HALLIER f.) PICHON	Pulp	Woody vine	PF
Landolphia angolensis (STAPF) A.RICH.	Pulp	Woody vine	ASF, PF
Landolphia congolensis (STAPF) PICHON	Pulp	Woody vine	ASF
Landolphia ligustrifolia (STAPF) M. PICHON	Pulp	Woody vine	PF
Landolphia owariensis P. BEAUV.	Pulp	Woody vine	PF, ASF
Landolphia violacea (K.SCHUM. ex HALLIER f.) PICHON	Pulp	Woody vine	PF
Burseraceae			
Canarium schweinfurthii ENGL.	Pulp	Tall tree	PF
Dacryodes edulis (G. DON) H. J. LAM	Pulp	Tall tree	PF
Santiria trimera (OLIV.) AUBRÉV.	Pulp	Tall tree	PF
Caesalpiniaceae			
Afzelia bipindensis HARMS	Leaf, pod, seed	Tall tree	PF
Brachystegia laurentii (DE WILD.) LOUIS	Seed	Tall tree	PF
Crudia harmsiana DE WILD.	Seed	Tall tree	PF
Cynometra alexandri C. H. WRIGHT	Pod, seed	Tall tree	PF
Cynometra hankei HARMS	Leaf	Tall tree	PF
Dialium corbisieri STANER	Pulp, leaf	Tall tree	PF
Dialium excelsum LOUIS ex STEYAERT	Pulp, young leaf	Tall tree	PF
Dialium pachyphyllum HARMS	Pulp	Tall tree	PF
Dialium zenkeri HARMS	Pulp	Tall tree	PF

(continued)

Table I. (*Continued*)

Plant	Part eaten	Life form	Main habitat[a]
Erythrophleum suaveoleus (GUILL. & PERR.) J. BRENAN	Seed	Tall tree	PF
Gilbertiodendron dewevrei (DE WILD.) J. LEONARD	Seed	Tall tree	PF
Leonardoxa romii (DE WILD.) AUBREV.	Young leaf, stem	Medium tree	PF
Macrolobium coeruleoides DE WILD.	Seed	Low tree	PF
Macrolobium fragrans (BAK. f.)	Seed	Tall tree	PF
Scorodophloeus zenkeri HARMS	Seed, leaf, flower	Tall tree	PF, ASF
Capparidaceae			
Pentadiplandra brazzeana BAILL.	Pulp	Vine	YSF
Commelinaceae			
Palisota ambigua (P. BEAUV.) C.B.CL.	Herbaceous stem	Herb	PF, ASF, SWF
Palisota brachythyrsa MILDBR.	Herbaceous stem	Herb	PF, ASF, YSF
Palisota schweinfurthii C. B. CL.	Stem	Herb	PF, ASF, SWF
Ebenaceae			
Diospyros alboflavescens F. WHITE	Pulp	Low tree	PF, ASF
Euphorbiaceae			
Croton haumanianus J. LEONARD	Pulp	Tall tree	YSF
Dichostemma glaucescens PIERRE	Pulp	Low tree	SWF
Drypetes angustifolia PAX et K. HOFFM.	Pulp	Tall tree	PF
Drypetes louisii J. LEONARD	Pulp	Tall tree	PF
Macaranga spinosa MÜLL. ARG.	Leaf	Tall tree	YSF
Manniophyton fulvum MÜLL. ARG.	Leaf	Woody vine	PF, ASF
Uapaca guineensis MÜLL. ARG.	Pulp, petiole	Tall tree	ASF, SWF
Flacourtiaceae			
Caloncoba welwitschii (OLIV.) GILG	Pulp	Low tree	YSF, SB
Guttiferae			
Mammea africana SABINE	Pulp	Medium tree	PF
Symphonia globulifera L. f.	Pulp	Tall tree	SWF
Hippocrateaceae			
Cuervea macrophylla (VAHL) R. WILCZEK ex HALLÉ	Seed, leaf	Woody vine	PF
Salacia alata DE WILD.	Pulp	Vine or herb	PF

Species	Food part	Growth form	Forest type
Irvingiaceae			
Irvingia gabonensis (AUBRY LECOMTE ex O'RORKE) BAILL.	Pulp	Tall tree	PF
Ixonanthaceae			
Ochthocosmum africanus HOOK F.	Leaf	Tall tree	PF, ASF
Lauraceae			
Beilschmiedia corbisieri (ROBYNS) ROBYNS et WILCZEK	Pulp	Tall tree	PF
Beilschmiedia variabilis ROBYNS et WILCZEK	Pulp	Tall tree	PF
Malvaceae			
Sida rhombifolia L.	Pulp	Shrub	YSF
Marantaceae			
Ataenidia conferta (BENTH.) MILNE-REDH.	Shoot, pulp	Herb	PF, SWF
Haumania liebrechtsiana (DE WILD. et TH. DUR.) J. LÉON.	Shoot, petiole	Woody vine	PF, ASF
Hypselodelphys poggeana (K. SCHUM.) MILNE-REDH.	Shoot, petiole	Woody vine	PF, ASF
Hypselodelphys scandens LOUIS et MULLEND.	Shoot, petiole	Woody vine	ASF
Hypselodelphys violacea (RIDL.) MILNE-REDH.	Shoot, petiole	Woody vine	PF, ASF
Sarcophrynium macrostachyum (BENTH.) MILNE-REDH.	Shoot, pulp	Herb	ASF
Trachyphrynium braunianum (K. SCHUM.) BAKER	Pulp	Herb	SB
Melastomataceae			
Memecylon jasminoides GILG.	Pulp	Tall tree	PF
Menispermaceae			
Triclisia dictyophylla DIELS	Pulp	Woody vine	PF
Mimosoideae			
Albizzia gummifera C. A. SM. var. ealaënsis (DE WILD.) BRENAN	Leaf	Tall tree	YSF
Pentaclethra macrophylla BENTH.	Seed, bark	Tall tree	PF
Moraceae			
Antiaris welwitschii ENGL.	Pulp	Tall tree	PF
Ficus ottoniifolia (MIQ.) MIQ.	Pulp	Tall tree (woody vine)	PF
Musanga smithii B. BR.	Pulp	Medium tree	YSF
Myrianthus arboreus P. BEAUV.	Pulp	Medium tree	ASF, YSF
Treculia africana DECNE	Pulp	Tall tree	PF

(continued)

Table I. (*Continued*)

Plant	Part eaten	Life form	Main habitat[a]
Olacaceae			
Ongokea gore (HUA) PIERRE	Pulp	Tall tree	PF
Palmae			
Ancistrophyllum secundiflorum (P. BEAUV.) WENDL.	Pith	Woody vine	ASF
Eremospatha hookeri (MANN et WENDL.) WENDL.	Shoot	Woody vine	PF
Sclerosperma mannii WENDL.	Shoot	Low tree	SWF
Papilionoideae			
Baphia laurifolia BAILL.	Leaf	Tall tree	PF
Dalbergia lactea VATKE	Leaf	Tall tree	YSF
Millettia duchesnei DE WILD.	Seed	Tall tree	PF
Pterocarpus casteelsii DE WILD.	Young leaf	Tall tree	PF
Piperaceae			
Piper guineense SCHUM. et THONN.	Pulp	Woody vine	PF
Polygalaceae			
Carpolobia alba G. DON.	Pulp	Low tree	PF, ASF
Rosaceae			
Parinari excelsa SABINE	Pulp	Tall tree	PF
Sapindaceae			
Allophylus africanus P. BEAUV.	Pulp	Low tree	YSF
Chytranthus carneus RADLK. ex MILDBR.	Young leaf	Low tree	PF
Pancovia laurentii (DE WILD.) GILG ex DE WILD.	Pulp	Medium tree	PF, ASF

	Part eaten	Form	Forest type[a]
Sapotaceae			
Manilkara casteelsii (DE WILD.) EVRARD	Pulp	Tall tree	PF
Pachystela excelsa	Pulp	Tall tree	PF
Synsepalum subcordatum DE WILD.	Pulp	Tall tree	PF, ASF
Sterculiaceae			
Cola bruneelii DE WILD.	Leaf	Shrub	PF, ASF
Cola griseiflora DE WILD.	Pulp	Medium tree	PF
Cola marsupium K. SCHUM.	Pulp, leaf	Low tree	ASF
Leptonychia batangensis (C. H. WRIGHT) BURRET	Pulp	Low tree	PF
Leptonychia tokana R. GERMAIN	Pulp	Low tree	PF
Sterculia bequaertii DE WILD.	Leaf	Tall tree	PF
Tiliaceae			
Grewia malacocarpa MAST.	Pulp	Tall tree	PF
Grewia pinnatifida MAST.	Pulp, young leaf	Medium tree	PF, ASF
Vitaceae			
Cissus aralioides (WELW. ex BAK.) PLANCH.	Pulp	Woody vine	ASF, PF
Cissus dasyantha GILG et BRANDT	Pulp	Woody vine	PF, ASF
Cissus dinklagei GILG et BRANDT	Pulp	Woody vine	PF, ASF
Zingiberaceae			
Aframomum laurentii DE WILD.	Pith	Herb	SWF
Aframomum sp.	Pulp, pith	Herb	SB
Renealmia africana BENTH. ex HOOK. F.	Pith	Herb	SWF
Fungus			
?*Langermania fenzlii*	Whole	Mushroom	PF

Unidentified food types: leaf (five species), fruit (four species), mushrooms (two species), dead branch and petiole (two species)

[a] PF, Primary forest; ASF, aged secondary forest; YSF, young secondary forest; SWF, swamp forest; SB, secondary shrub.

these products the pygmy chimpanzees regularly consume sugar cane sap and pineapple pulp. Banana pith was observed ingested at the artifical feed place. Pulps of coffee (S. Kuroda, personal communication) and papaya were tasted once.

The animal foods recorded were one species of flying squirrel (*Anomalurus erythronotus*), at least two species of earthworms, and one type of Lepidoptera larva, as well as the honey and nests (and probably the larvae and adults) of two species of stingless bee.

As inorganic food, the earth used for nest construction by termites (*Cubitermes* spp.) was recorded. Drinking from standing water was observed once by S. Kuroda (personal communication). It appears that the bonobos obtain practically all the water they require from their foods.

2.2. Dietary Proportions

The relative importance of each food item in the pygmy chimpanzee's diet was assessed by means of direct observation and fecal analysis. Due to the large size of the temporary associations of pygmy chimpanzees at Wamba, it was difficult to record the exact feeding times of each animal by means of the random sampling method employed. Instead, the following method was used. When it was discovered that a certain pygmy chimpanzee was eating a particular food or starting to eat it, that food item was given a feeding score of one, and another point was added after each 30 min that the eating continued from the time the observation was begun. No extra points were added when the continuous or intermittent eating of the same food by the same individual terminated in less than 30 min. When an individual feeding on one food began feeding on a different food within the same 30-min period, these were considered different feeding bouts, and each food was given one point. For example, if an individual that had been eating food item A began to eat food item B and then returned to item A again within the same 30-min period, item A would receive two points and item B would receive one point. We may consider that the proportion of the total number of points represented by the score of a particular item indicates that item's relative importance in the pygmy chimpanzee's diet. This percentage will hereafter be referred to as the feeding proportion. All recordings made in the nonnatural habitat (artificial feeding place) have been omitted. In addition, the data obtained in January 1977, February 11–March 5, 1978, and January–February 1979, when the greater part of the observation was carried out at the feeding place, have not been included in the analysis of feeding proportion. It was judged that the feeding ecology of the pygmy chimpanzees during these periods deviated considerably from the natural state due to the influence of provisioning.

Fecal analysis was carried out on a total of 1001 feces collected between November 1, 1981 and February 28, 1982. Each fecal sample was weighed ($\overline{X}$ = 45 g, range 10–155 g) and its contents identified. (Identification was relatively simple in the case of seeds, but difficult for items such as leaf and shoot fragments). The quantitative ratios of the identified items in each fecal sample were measured from outer appearance. These ratios may be considered indications of feeding proportions based on fecal analysis.

Food types were divided into the following categories.

1. Arboreal fruits. These are tree or vine fruits (pulps) growing in the trees, including those eaten after falling to the ground. They include a very large number of species.
2. Ground fruits. These are mainly herb fruits growing on or near the ground surface. Representative of these are *Sarcophrynium macrostachyum* and *Aframomum* spp.
3. Arboreal leaves. These include plant parts, such as leaves, shoots, piths, stems, petiole, and bark, which are eaten in the trees. Some representative items are the leaves of *Scorodophloeus zenkeri*, *Leonardoxa romii*, and *Manniophyton fulvum*, the petioles and leaf shoots of *Haumania liebrechtsiana*, the pith of *Ancistrophyllum secundiflorum*, and the shoots of *Sclerosperma mannii*.
4. Ground leaves. These are the same food types as mentioned in the previous category, but which grow on the ground. Some of these are leaf shoots of *Sarcophrynium macrostachyum*, the herbaceous stem of *Palisota ambigua*, and the pith of *Aframomum* spp. The shoots and petiole of *Haumania liebrechtsiana* grow both on the ground and in the trees.
5. Seeds. These are various types of beans and seeds.
6. Animal foods. Insect larva, earthworms, etc.
7. Inorganic foods. Earth.

The results of the direct observation and fecal analysis are indicated in Table II. The feeding proportion for arboreal fruits showed the highest values (83.4% and 86.8%, respectively). The consumption of ground fruits was not detected by direct observation, but showed a feeding proportion of 6.3% in fecal analysis. Arboreal and ground leaves accounted for 12.6% and 2.6%, respectively, according to direct observation, but these two foods together accounted for only 5.8% in the fecal analysis. Animal foods showed a feeding proportion of 1.5% in direct observation, but in fecal

Table II. Feeding Proportions (%)

	Direct observation					Fecal analysis
	1975 Nov. 1–1976 Feb. 10	1976 Oct. 9–1976 Dec. 31	1977 Oct. 20–1978 Feb. 10	1981 Oct. 21–1982 Feb. 28	Mean	1981 Nov. 1–1982 Feb. 28
Fruits						
(Arboreal)	84.3	90.7	79.6	79.1	83.4	86.8
(Ground)	0	0	0	0	0	6.3
Total	84.3	90.7	79.6	79.1	83.4	93.1
Leaves						
(Arboreal)	13.4	9.1	15.9	11.8	12.6	3.7
(Ground)	2.3	0.2	1.7	6.0	2.6	2.0
Total	15.7	9.3	17.6	17.8	15.2	5.7
Animal foods	—	—	2.8	3.1	1.5	1.1

analysis this food type appeared only as flying squirrel fur and bone fragments in five fecal samples on three separate occasions, and earthworm remains in three samples on two occasions, accounting for a negligible 1.1% of the total. Only two examples of consumption of inorganic foods were recorded, both by direct observation.

Differences between the results of direct observation and fecal analysis were found in the feeding proportions of items other than tree foods as well. Many ground foods are scattered uniformly in the thick undergrowth. As a result, foraging animals scatter in a corresponding manner, and there are much fewer chances for simultaneous observation of several individuals. In contrast, feeding in trees provides a better view, and it is not uncommon for several individuals to gather in a single fruit tree. Accordingly, the estimated values for consumption of arboreal food components tend to be inflated in direct observation. On the other hand, in fecal analysis, where nothing remains of leaves but a few minute fragments, the data are unfavorably biased with regard to these food types. Thus the disadvantage of ground fruits in direct observation can be balanced by fecal analysis, but it is difficult to determine a correct feeding proportion for terrestrial leaves and shoots through either method. Representative of this type of food are the leaves and shoots of *S. macrostachyum*. Since pygmy chimpanzees make frequent foraging trips in most seasons to the aged secondary forest rich in *S. macrostachyum*, it may be assumed that the actual feeding proportion of this food is quite high, even though direct observation of consumption of *S. macrostachyum* was rare.

In any case, we estimate that the pygmy chimpanzee's diet at Wamba consists of about 80% pulp, 15% fibrous foods such as leaves and shoots, and 5% seeds. Animal foods constitute only a small fraction of the diet.

2.3. Annual and Seasonal Variations in Diet

We shall arbitrarily categorize food items with a feeding proportion of over 30% as staples, those with a value of greater than 10% but less than 30% as substaples, and those under 10% as supplementary foods. In Fig. 3, each month is divided into three periods (1st–10th, 11th–20th, 21st–end), and the feeding proportions of the main food items that appeared at least once as a substaple food are indicated for the respective periods. As a result, we can say that fruits are often overwhelmingly major foods (staples or substaples) with regard to both number of species and frequency. Pulps of *Dialium* spp. were the most stable food item eaten as a staple during several periods every year. No other fruit appeared as a stable annual staple. This indicates that many of the tree species in the forest do not have a fixed annual cycle in fruiting or quantity of production. A typical example of this is seen in the *Landolphia owariensis*. It appeared as an overwhelmingly major food in 1977 and 1981, but did not appear in other years (Fig. 3). The fruits of this woody vine, at least where a large quantity of fruit production is concerned, seem to occur only every few years.

There is little data on differences in diet among the unit groups. Concerning major foods, at least, there was no intergroup difference. With regard to the periods during which these foods were eaten, however, it appeared that there were some time lags among the groups. According to the observations of 1975 and 1976, the *Dialium* season of the B group ended more than 1 month ahead of that of the K group. Another example of this type was seen in the case of *L. owariensis*. According to the fecal analysis carried out in 1981, the end of the *L. owariensis* season for the B group was approaching by the end of November, and the proportion of *L. owariensis* seeds in their diet fell to 3% in the period December 1–10, after which it became 0%. In contrast, for the E group the proportion of this food remained at over 85% until December 10, and after that maintained a high percentage of about 30% until December 31. The fecal analysis carried out by T. Nishida (personal communication) in 1977 at Wamba produced similar results with regard to time lags in feeding season among the groups.

Lags in main food season were seen not only among the different groups but also within the home range of a single unit group. The diets of the E group from November 1981 to February 1982 consisted of an alternation of four major foods: *L. owariensis*, remaining *L. owariensis*

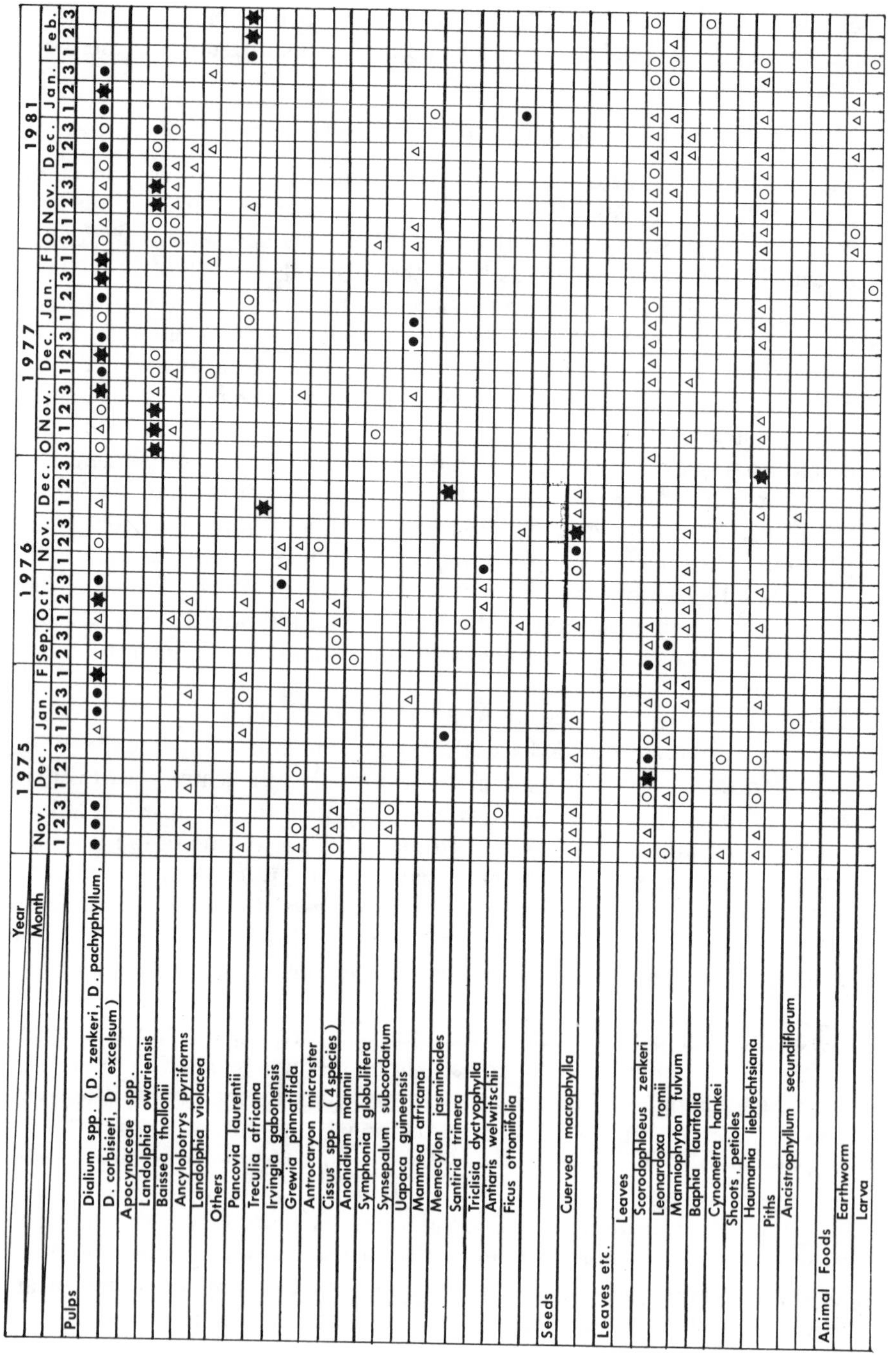

	1975	1976	1977	1981
Year / Month	Nov. Dec. Jan.	F Sep. Oct. Nov. Dec.	O Nov. Dec. Jan. F	O Nov. Dec. Jan. Feb.

Pulps

- Dialium spp. (D. zenkeri, D. pachyphyllum,
- D. corbisieri, D. excelsum)

Apocynaceae spp.
- Landolphia owariensis
- Baissea thollonii
- Ancylobotrys pyriforms
- Landolphia violacea

Others
- Pancovia laurentii
- Treculia africana
- Irvingia gabonensis
- Grewia pinnatifida
- Antrocaryon micraster
- Cissus spp. (4 species)
- Anonidium mannii
- Symphonia globulifera
- Synsepalum subcordatum
- Uapaca guineensis
- Mammea africana
- Memecylon jasminoides
- Santiria trimera
- Tridisia dyctyophylla
- Anttaris welwitschii
- Ficus ottoniifolia

Seeds
- Cuervea macrophylla

Leaves etc.

Leaves
- Scorodophloeus zenkeri
- Leonardoxa romii
- Manniophyton fulvum
- Baphia laurifolia
- Cynometra hankei

Shoots, petioles
- Haumania liebrechtsiana

Piths
- Ancistrophyllum secundiflorum

Animal Foods
- Earthworm
- Larva

Figure 3. Annual changes of major foods. (★) More than 70%, (●) 30–70%, (○) 10–30%, (◁) ...

and other Apocynaceae spp., *Dialium* spp., and *Pancovia laurentii* (Fig. 4). Particularly striking local differences were seen at the beginning, middle, and end of the *L. owariensis* seasons (Fig. 5). In the northern region, the *L. owariensis* season ended in the first part of November, while in the southern region it ended in the middle of November, and in the central region it ended in the middle of December. With the exception of the central region, the season for Apocynaceae, which replaces *L. owariensis*, started around mid-December, when the *L. owariensis* fruit was almost gone; but in the eastern region, the use of *L. owariensis* plus other Apocynaceae fruits began simultaneously in late December and ended by the end of that month. Slight time lags were observed among the regions in the use of other major foods as well. It is difficult to imagine that there would be striking differences in the fruiting season of a particular plant among regions that are continuations of the same forest, having the same principal ecological conditions. It seems, rather, that the direct cause of the regional differences in the period of consumption of major foods is the pygmy chimpanzee's habit of eating until they exhaust the supply of the particular food. In other words, while the principal cause influencing the beginning of the fruiting seasons of the pygmy chimpanzee's major food plants is the phenological factor, the fruiting periods of these plants appear to be brought to an end by their consumption by the animals.

2.4. Food Preferences

The necessary condition for staple and substaple foods is wide availability. It appears, however, that in addition to availability, preference also plays an important role in the pygmy chimpanzees' diet. For example, the pith of *Aframomum* spp. and the herbaceous stem of *Palisota* spp. grow in very large quantities in the herb layers of the secondary forest and primary forest, respectively, but in both direct observation and analysis of feeding remains these foods were found only very rarely. Similarly, the fruits of *Caloncoba welwitschii, S. macrostachyum, Ataenidia conferta, Pachystela excelsa*, etcetera, and the seeds of Caesalpiniaceae spp. were consumed infrequently in relation to their availability. The fruit of *Irvingia gabonensis* was a staple during some periods (Fig. 3), but at other times when more preferred foods were available the pygmy chimpanzees quickly switched from it. Even the *Dialium* spp., which constitute annually the most important food for the pygmy chimpanzees at Wamba, had periods when they were underutilized. This tree species begins its ripening period annually from September to October, at which time the pygmy chimpanzees begin to consume it, but in years when there was a large production of Apocynaceae fruits (represented by *Landolphia owa-*

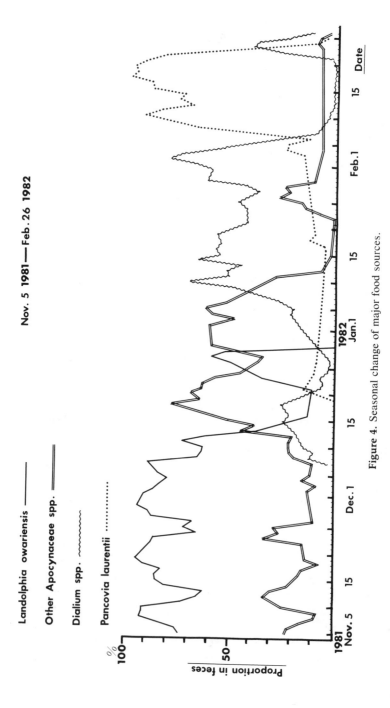

Figure 4. Seasonal change of major food sources.

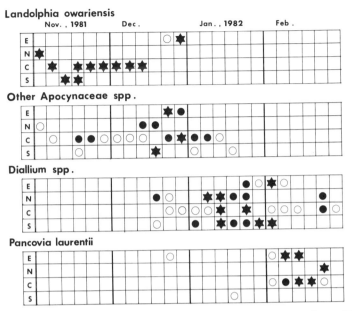

Figure 5. Local difference in major food seasons. (★) More than 70%, (●) 30–70%, (○) less than 30%. E, Easten region; N, Northern region; C, central region; S, southern region.

riensis), the *Dialium* spp. season was always delayed until after the other was exhausted. In addition, according to the data collected in 1982, when the fruit of *Pancovia laurentii* began to ripen at the beginning of February, the pygmy chimpanzees switched from *Dialium* to *P. laurentii*, and returned to *Dialium* in the latter part of February after the *P. laurentii* fruit was depleted (Figs. 4 and 5). A clear difference was seen also between the pygmy chimpanzees' preference for *L. owariensis* and for other Apocynaceae spp. While these were fruiting simultaneously, only the *L. owariensis* was eaten in high percentages. It is probable that this preference is based not only on taste, but on such factors as ease of collecting and processing and feeding efficiency as well. For example, not only is the fruit of *Dialium* much smaller than that of Apocynaceae and *P. laurentii*, but the former requires the operation of peeling the hard, brittle rind off each fruit with the teeth. Accordingly, it may be assumed that the amount of edible parts that can be ingested within the same period of time is much less in the case of *Dialium*.

Upon arriving at a major fruit food source, the pygmy chimpanzees usually fall into a state of excitement. During the earlier part of the feeding period in particular, they eat voraciously, and activities such as chasing, begging, greeting, appeasing, or copulation occur very frequently. The animals incessantly emit feeding grunts, and at times two or more individuals take to whooping loudly at the same time. If we postulate that such behavior reflects the degree of fondness for foods, it would appear that the fruits of *Anonidium mannii, Canarium schweinfurthii, Dacryodes edulis, Cissus* spp., and *Treculia africana*, which have a comparatively scarce availability, are liked by the pygmy chimpanzee, and would become their most important foods if their availability were sufficiently great. It is likely, however, that this kind of judgement on the basis of pygmy chimpanzee behavior is not suitable in the case of fibrous foods. When consuming leaves and shoots (which include a number of important foods), the pygmy chimpanzees generally maintain a calm, lazy manner. Food choice varies even during the course of a single day. Feeding on fruits reaches two peaks: in the early morning and in the afternoon. For feeding on leaves, a prominent peak is observed only once in late afternoon.

2.5. Food Diversity

A determination of daily and monthly dietetic diversity was made based on direct observation and fecal analysis. The average number of food types recorded per day and per party from direct observation was 2.0 (range 0–9, $N = 302$) and from fecal analysis was 6.3 (range 1–12, $N = 140$). A large difference was thus seen between the two methods. In the former case, the number showed a tendency to increase ($r = 0.57$) with the number of hours of observation under natural (unprovisioned) conditions ($\overline{X} = 3.1$ hr, range 0.0–10.3, $N = 303$). This indicates that a large amount of observation time is needed in order to achieve a correct picture of the pygmy chimpanzees' daily diet by means of direct observation. In contrast, there was no high positive correlation ($r = 0.17$) between the number of fecal samples ($\overline{X} = 7.2$, range 1–34, $N = 140$) collected at the same time (in most cases early morning) and the same site (in most cases the sleeping place of the previous night) on the one hand, and the number of food remain types ($\overline{X} = 6.3$, range 1–12, $N = 140$) on the other hand. This indicates that fecal analysis is a suitable method by which to look for a close approximation of the number of food types eaten by a party in one day using a comparatively small amount of data, since feces represent a condensation of feeding and foraging activity carried out over a long period. The above results suggest as well that pygmy chimpanzees from the same party foraging at the same time make common use of the same major food source, and there is little separate exploitation of minor food sources by each individual. It is estimated that

considered it highly likely that these subterranean mushrooms
en (as are earthworms) in the course of digging activity.

imal Items

Mammals. A blue duiker, a chicken, and a squirrel were displayed
gmy chimpanzees together with sugar cane. The number of pygmy
zee individuals that reacted to these items was six of 20 for the
ker, 14 of 25 for the chicken, and two of 17 for the squirrel.
s consisted in behavior such as the following: surprise, detour
dance, charging and displaying while dragging a branch, moving
ation near the animal, arm raising, threatening with wrist shake,
touching the animal, staring at it, and smelling it. Adults became
sinterested after the first one to three reactions, but the reactions
ures tended to be repeated.

Carcasses and Parts of a Carcass. A fresh bat carcass and the
ternal organs of giant rats were offered once each to the pygmy
zees. These were completely ignored by all individuals.

Hen's Eggs. A total of 30 eggs were offered over three occasions.
ere all either ingested or carried away. Individual differences
dent in the consumption of eggs. Of the 44 animals present in
ng place at the time, 12 showed an interest in the eggs. Of these,
were observed to actually eat the eggs, while the others either
them after carrying them away, or took them into the bush
servation was impossible. A certain adolescent female showed
attachment to the eggs; after eating her own eggs, she approached
viduals with begging behavior. Furthermore, in the next several
owing the distribution of the eggs, she would, upon arrival at the
lace, immediately head for the stump where they had first been
see if there were additional eggs. One adult female, upon leaving
g place, discovered an egg discarded by another individual and
up. She then made a 5-m detour to a *Uapaca guineensis* tree to
s buttresses as if to look for some more eggs. From this test, it
educed that in the wild state at least some individuals collect
the nests of birds.

ing Techniques

nt Foods

the animals were searching for food in the trees, quadrupedal
was most usual. On thick horizontal or slightly inclined boughs,

the number of food types consumed by a pygmy chimpanzee party of
Wamba in one day is at most about ten, a figure in accord with results
of fecal analysis.

The number of food types recorded for one month was not correlated
with the number of fecal samples ($\overline{X} = 249$, range 192–328, $N = 4$), falling
between 22 and 27; in direct observation, however, the number bore a
positive relationship to observation time ($\overline{X} = 56.8$ hr, range 4.1–108.7
hr, $N = 18$), varying widely from 9 to 29 (Fig. 6). In each month where
fecal analysis and direct observations were carried out together, a larger
number of food types was detected by the former method than by the
latter, but in months where the number of observation hours exceeded
5000 min, the results of direct observation were close to those of fecal
analysis. From the results of fecal analysis and direct observation, as well
as from a comparison of these with the asymptotic curve obtained through
observation in Gombe National Park, it can be assumed that the number
of food types consumed monthly by the pygmy chimpanzees in probably
around 40 (Fig. 6).

Of the total number of feeding points in a month, 50%, 70%, and 90%
were accounted for by the first 1–3 ($\overline{X} = 1$), 1–5 ($\overline{X} = 3$), and 2–10 ($\overline{X}
= 6$) food types, respectively. Fifty percent of the food types recorded
accounted for 76.0–96.7% ($\overline{X} = 93.1\%$) of the total number of feeding
points.

It would seem that food diversity has a certain degree of relationship
to the types of major foods. The median for food types discovered in one
fecal specimen was less in the *L. owariensis* and *P. laurentii* seasons than
in the Apocynaceae and *Dialium* seasons (two versus 4; Table III, $p
< 0.001$, Mann–Whitney U test). In the Apocynaceae and *Dialium* sea-
sons, four and three species of food plant, respectively, constituted one
major food group. In this test, these were lumped together and counted
as one food type. In transition periods between major food seasons, the
number of food types found in a single fecal specimen was the same as
or greater than the number found during the preceding or following major
food seasons. These results suggest that dietetic diversity is reduced during seasons when preferred foods are abundantly available.

2.6. Food Provisioning

As an aid to understanding the nature of the pygmy chimpanzees'
food habits, several potential food items were distributed experimentally
in the artificial feeding place and the natural habitat.

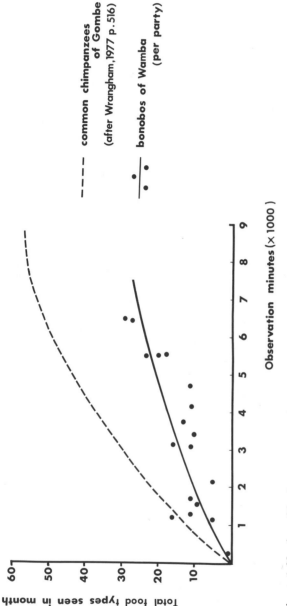

Figure 6. Number of food types per month. Dashed line: common chimpanzees at Gombe (after Wrangham, 1977, p. 516). Dots and solid line: bonobos at Wamba (per party).

Table III. Number of Food Types per Fecal
1982 February

Food season	Number of food types per fecal sample	
	Median	Range
Landolphia owariensis	2	1–5
Transitional	3.5	1–9
Other Apocynaceae spp.	4	1–8
Transitional	4	2–7
Dialium spp.	4	3–8
Transitional	5	3–8
Pancovia laurentii	2	1–6
Transitional	5	3–7

2.6.1. Plant Items

a. Domesticated Plants. Several domestic
to the pygmy chimpanzees. Among them were s
which are principal items of provisioning. Suga
of the E group and some of the members of t
evidence, however, has yet been obtained to i
the B and S groups have eaten sugar cane. Grou
regarding feeding on pineapple as well. The E gr
more than sugar cane when the two were distrib
pineapple was ignored by most members of the P g
of young females who had once belonged to the

Pygmy chimpanzees were indifferent to bana
of four adults smelled or inspected the bananas a
banana from the bunch, only to discard it after sr
broken in half by an adolescent female and throv
In addition to these items, there are a variety
Wamba, but there is no evidence of these havin
bonobos.

b. Subterranean Mushrooms. It was pointed
that the bonobos may dig up and feed on a subt
germania fenzlii) (Kano, 1979, 1983). Four of th
rooms, which occur mainly underground, were o
on February 23, 1981 (the largest of these was al
These were displayed to animals of the E group tog
Direct observation was rendered impossible becau
the sugar cane. When the bonobos left, however,

---

gone. '
were e

2.6.2.

a
to the
chimp
blue
Reac
and a
the v
furtiv
totall
of im

fur a
chimp

The:
were
the
only
disc
whe
the
oth
day
fee
pla
the
pic
ins
ma
eg

2.'

2.

w

knuckle walking was sometimes used. In most arboreal locomotion, however, the palmigrade grasping hand postures were employed. When moving from branch to branch or through a tree crown, scrambling and suspended behavior, including various forms of armswinging, were often seen. Bipedal walking was seen less frequently. In almost all cases it involved movement over short distances, often with the body supported by overhead handholds. During the feeding time, running and leaping were infrequent except in agonistic encounters and when leaving the feeding tree to begin travel.

There were many methods of food gathering, which differed according to the type of food. Comparatively large fruits (with diameters greater than 7–8 cm, e.g., fruits of Apocynaceae) were gathered with the hands and were often carried to a safe place to be eaten. In the case of smaller fruits (e.g., *Dialium* spp.), the fruiting twig was broken off and carried to a safe place. For carrying such foods the hand (58%), mouth (25%), foot (13%), and groin (4%) were used ($N = 24$). When feeding on fruit in the trees, pygmy chimpanzees assumed a sitting posture 90% of the time, while positions such as hanging (5%), quadrupedal standing (2%), and lying (3%) were infrequent ($N = 132$). Leaves, shoots, pith, etc., were not transported in the trees but rather eaten on the spot. The hands were used for harvesting in most instances, but when the mature leaves of *S. zenkeri* were being eaten, the pinnate leaves on the bent twig were pulled off selectively one by one directly with the mouth.

Particular techniques for processing fruit were seen only in the case of *L. owariensis*. The pygmy chimpanzees would hold this spherical fruit in one or both hands, and rotating it with the palms, would peel off the tough rind with their teeth, beginning from the stem. During this procedure, care was taken not to rip the soft, thin membrane that separates the rind from the inner edible parts. When the rind had been peeled off as far as the equator, the membrane, now exposed as a half-sphere, was bitten into and the inside part taken out and eaten. This appeared to be an excellent technique to avoid losing any of the juice. The villagers, too, sometimes eat the *L. owariensis* fruit using this technique. By this technique, the food remains produced from one *L. owariensis* fruit consist of one bowl-shaped rind and several rind fragments. Since other nonhuman primates use a completely different processing technique—making a hole in the rind with the teeth and poking fingers in to remove the insides—it is simple to judge from *L. owariensis* food remains whether or not the fruit was eaten by a pygmy chimpanzee.

The slender shoots of Marantaceae spp., represented by *S. macrostachyum*, were pulled out of the ground with the hands. This operation requires considerable grasping power and muscular strength. Among all the mammals at Wamba, only humans and pygmy chimpanzees have the ability to exploit this plant as food.

2.7.2. Animal Foods

a. Insect Larva. Caterpillars, locally known as *tofilifili* or *tau* (*Andronymus neander* Plötz, Hesteriidae larva) swarm on the giant *Cynometra hankei* trees of the primary forest every year at some time in December or January after dry days have continued for 2 or 3 weeks. When the pygmy chimpanzees encountered these, they consumed them in great number collecting them with their hands or mouths. The feeding rate of one male was four larvae in 16 sec (15 per min), while one female ate 60 in 4 min 43 sec (13 per min). The average length of one feeding bout was 18 min (range 7–34, $N = 6$). It may thus be deduced that the average number of larvae consumed by one pygmy chimpanzee on one occasion was about 200. Judging from the results of nutritional analysis of the larvae, this amount would correspond to 7.5 g of crude protein (R. Asato, personal communication). If there were three feeding sessions per day, then, the average amount of animal protein obtained by one pygmy chimpanzee per day from *tofilifili* would be about 22.5 g. For pygmy chimpanzee nutrition, this is probable quite a meaningful value. The *tofilifili* food season, however, lasts for only about 1 week.

b. Earthworms. The pygmy chimpanzees looked for earthworms almost exclusively on the forest floor of the *Gilbertiodendron* primary forest and in the beds of streams in the swamp forest. The types of earthworm produced in these two different habitats differed from one another, though neither of them is identified as yet. No tools were used for digging. Clearly recognizable holes had been dug in the ground. The pygmy chimpanzees would dig with one or both hands in the earth from a quadrupedal standing position, or, less frequently, in a sitting posture.*

In the swamp forest, the pygmy chimpanzees would forage for earthworms more persistently. When searching for earthworms in the swamp forest, the pygmy chimpanzees did not dig clearly defined holes. Rather, they would move slowly in the shallow streams or adjoining mud beds, pushing the mud aside with their fingers. One party from the E group spent the greater part of three hours in a small stream foraging for earthworms.

The consumption rate for earthworms was very low. During a total

---

*The median size of freshly dug holes was major axis 33 cm (range 12–173 cm), minor axis 23 cm (range 10–70 cm), and depth 16 cm (range 2–55 cm) ($N = 36$). The median number of fresh holes per site was five (range, 1–65, $N = 24$). The median duration of a digging bout performed by one animal was 14 min.

of 41 min of digging activity recorded in the *Gilbertiodendron* forest, food was brought to the mouth only 15 times (0.4 times per min per animal), while in the swamps the rate was four times in 180 min (0.04 times per min per animal).

c. Hunting Behavior. Meat eating by pygmy chimpanzees at Wamba has not been observed directly, but two instances of unsuccessful attempts to catch small mammals were observed. In both instances males from the E group failed in attempts to stalk and catch flying squirrels.

d. Nests of Stingless Bees. Different methods were used for feeding on the nests of the two species of stingless bee, *Apidae* sp. locally known as *belo*, and *Dactyfurina staudingeri* Gribodo, 1893.

On January 26, 1983 between 6:30 and 7:45, a belo nest constructed in and around a hole in a large tree were raided successively by four bonobos (one adult and three immatures). Almost all the parts sticking out around the circumference of the hole, as well as those on the inside near the opening, were demolished and the contents licked away, but no use of tools for poking into the interior of the hole was seen.

On February 19, 1978, an adult male in an E party visiting the feeding place appeared holding in one arm a *D. staudingeri* nest about the size of a papaya fruit. By the time the nest was about one-third consumed, it had passed through the hands of six individuals, but no begging or sharing behavior was observed. Rather, all food transfer took the form of each new owner picking up what the previous owner had discarded.

2.8. Habitat Utilization

Taking into consideration time lags in major food seasons among the four regions of the E group range, and based on direct observation, information derived from observation of vocalizations, study of fresh food remains and tracks, and local information, we calculated the utilization frequencies of vegetation types in each season from October 24, 1981 to February 28, 1982 (Table IV). In all seasons, utilization frequency of the primary forest was the highest (number of days utilized/total number of days was equal to 93.5%; range 77.8–100%, $N = 199$). This situation reflects the fact that the primary forest is the principal habitat of many food plants. In addition, the fact that the primary forest is the location most frequently used as a night-nesting place is probably a contributing factor as well. The next most frequently employed vegetation type was the aged secondary forest (47.2%, range 20.0–63.6%). This result is likely connected with the fact that among the major food plants for this period, *L. owariensis,* other Apocynaceae spp., and *P. laurentii* are found abundantly in both the primary forest and the aged secondary forest. Fur-

Table IV. Utilization of Vegetation Types

Food season	Proportion of utilization,[a] %				
	Dry primary forest	Swamp forest	Aged secondary forest	Young secondary forest and bush	Number of days
Landolphia owariensis	93.5	37.1	54.8	8.1	62
Transitional	100.0	50.0	50.0	0	2
Other Apocynaceae spp.	100.0	41.9	48.4	19.4	31
Transitional	100.0	12.5	43.8	31.2	16
Dialium spp.	97.1	42.9	20.0	31.4	35
Transitional	81.8	18.2	63.6	45.5	11
Pancovia laurentii	84.0	12.0	52.0	20.0	25
Transitional	77.8	11.1	55.6	11.1	9
Undetermined	100.0	25.0	62.5	37.5	8
Total	93.5	31.2	47.2	20.6	199

[a]Proportion = (number of days used by pygmy chimpanzees)/(number of days observed).

thermore, *S. macrostachyum* and *H. liebrechtsiana,* which provided the main ground foods, have the aged secondary forest as their principal habitat. *Dialium,* another staple, is characteristic of the primary forest, and during its season the utilization rate of the aged secondary forest was at its lowest (20.0%). The utilization rate of the swamp forest was 31.2%. It was highest from the *L. owariensis* to the *Dialium* season, and became lower toward the *P. laurentii* season. During this period, foods recorded for the swamp area included earthworms, *Uapaca guineensis* fruit and petioles, *Aframomum* spp. pith and fruit, *Sclerosperma mannii* pith, and *Symphonia globulifera* fruit. In the southern region, the pygmy chimpanzees often ranged into the extensive swamp of the Luo from late morning until late afternoon. It was difficult to follow them there, but from fecal analysis it can be assumed that they were feeding mainly on the fruit of *U. guineensis.*

The young secondary forest and secondary shrub had the lowest utilization rate (20.6%). During the *Dialium* season and the transitional periods before and after it, the utilization rate was relatively high (31.2–45.5%).

While there are differences in utilization rate, we may say that the pygmy chimpanzees make extensive use of their habitat. As the young secondary forest and shrub, with their low utilization rate, produces non-seasonal foods available all the year, such as *Aframomum* spp. fruit and pith and *Musanga smithii* fruit, it seems to be utilized for supplementing shortages in the other vegetation types.

Few data were obtained concerning vertical utilization of the habitat. Judging, however, from the life forms of the recorded food plants as well as the direct observations, the feeding stratum most highly utilized is probably the crowns of emergent trees. Following this in importance is the shrub layer. Not only do many leaf shoots and some fruits occur on the ground, but the fruits of a considerable number of tall and medium tree species are both eaten in the trees and carried down to the ground (e.g., *T. africana, I. gabonensis, A. mannii, Dialium* spp., *Mammea africana, P. laurentii, L. owariensis*). The low and medium tree layers are less important as food sources.

2.9. Routine Activities

The diurnal activities of parties at Wamba from the time of awakening and leaving the sleeping nest (5:00–6:00) to the time of making and settling into a new sleeping nest (17:00–19:00), was classified into five categories:

1. Arboreal feeding. This category includes activity wherein party members are feeding in trees. After reaching the feeding tree, the first several minutes are characterized by excited feeding. After 10–20 min, feeding behavior becomes quieter and eventually activity decreases and animals become restive. The transition from feeding to resting is continuous and proceeds at different rates with different individuals.
2. Arboreal resting. This term refers to the condition where most party members are inactive in the trees, or when they are absorbed in the various activities characteristic of resting, such as allo- or self-grooming and social or individual play. The making of day nests is frequent. All the resting places were at or near the food sources. Even during the resting period, two or three individuals may continue or resume relaxed feeding.
3. Traveling. Traveling usually occurs during the resting time. In general, the ground is used for traveling after the first several tens of meters of arboreal travel. In the case of the unhabituated S group (and in the B and K groups before habituation), there were times when arboreal travel was carried out for 200–500 m.
4. Terrestrial activities. This category includes all activities on the ground other than traveling. As party activities, these are not clearcut. Feeding on the ground is generally neither excited nor simultaneous. While some animals would search for food, others might be grooming, playing, or resting. Group members were separated by relatively long distances when on the ground.
5. Night nesting. The amount of time used by each individual for

night nesting is several minutes, but as this activity is done more or less simultaneously, only about 15 or 20 min elapses from the time the first individual begins to make its nest to the time the last one has finished, party size notwithstanding.

Among the major activities, the usual cycles were feeding–resting–traveling and terrestrial activities followed by more feeding. Omitting the time spent at the artifical feeding place, the diurnal rhythm of routine activities takes the form shown in Fig. 7. Hourly arboreal feeding budgets reach two peaks, one in early morning and the other in late afternoon. Terrestrial activities peak after feeding. Resting budgets have a large peak at midday. The resting peaks in early morning and evening indicate resting in sleeping nests.

At Wamba pygmy chimpanzees have a basic rhythm in their daily activities, which can be summarized as follows: immediately after leaving the nest in early morning, they feed in the trees; they then rest, and later move in a leisurely manner along the ground to the next food tree(s). While moving between fruiting trees they feed on terrestrial plants. They become gradually less active toward midday. In the afternoon the same

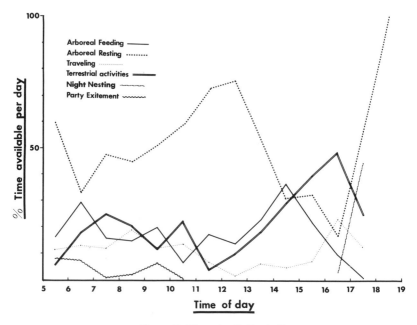

Figure 7. Diurnal activity rhythms.

activities are resumed and the animals eventually nest in trees where they have fed late in the day.

Hourly resting budgets were 43%, accounting for the largest time percentage among the activity patterns. This was followed by terrestrial activities (20%), arboreal feeding (18%), traveling (13%), and others (13%). Because resting was an important factor in terrestrial activities, it may be inferred that the total feeding time (arboreal and terrestrial) does not exceed 30%.

2.10. Daily Range and Party Size

The median length of the day ranges of the E group party during the period from October 26, 1981 to February 28, 1982 was 2.4 km (range 0.4–6.0 km, $N = 91$). Differences between seasons were statistically insignificant (Kruskal—Wallis analysis of variance of ranks, $H = 5.253$, $v = 7, p > 0.05$).

The median party size (counting only independent individuals) for the E group during this period was 15 members (range 1–36, $N = 147$). The median party size was large during the seasons of *P. laurentii* (median $= 18.5$), *L. owariensis* (median $= 17$), Apocynaceae (median $= 15$) and *Dialium* (median $= 13$) in that order; the difference among these was statistically significant ($H = 22.4, v = 7, p < 0.05$). The materials used in this test, however, were biased in a manner that tended to lower the median value of the *L. owariensis* seasons. The following explanation will clarify the reason for this bias. It was difficult to ascertain the size, composition, and membership of large pygmy chimpanzee parties in the forest, even when they were habituated. Mobile provisioning methods were effective for this purpose, but during the *L. owariensis* season the pygmy chimpanzees often formed extremely large parties, and were attracted to the sugar cane we used as a provisioning item (*L. owariensis*, the natural food, was preferred). If the sizes of all parties had been verified, the median value for the *L. owariensis* season would certainly have been the largest. Indirect evidence for this lies in the fact that the E group is composed of two major subgroups (one in the north and one in the south), both of which normally divide up for foraging. Merged subgroups in the E group were encountered 21 times out of the total 49 occasions during the *L. owariensis* season.

The above data suggest that the more abundant a preferred food is in a particular season, the more the pygmy chimpanzees tend to form large parties during that time. This observation coincides with Kuroda's discoveries chiefly concerning the B group (1979).

2.11. Competition

2.11.1. Intragroup Competition

Agonistic interaction occurs with a greater frequency during artificial feeding than in the natural state (3.26 versus 0.70 times per party per hour, Table V). Under natural conditions agonistic behavior is most frequent during periods of group excitement at food sources (3.01 times per hour) and during episodes of arboreal feeding. At times other than these the frequency of occurrence was only 0–0.54 times per hour. Agonistic interactions thus tend to occur in feeding contexts. It appears that in artificial and arboreal feeding, the fact that the food sources are spatially concentrated leads to a shortening of distances between individuals, resulting in aggressive encounters. The form of aggresive behavior in trees was one of an individual displacing another in the food tree. Under artificial feeding conditions, behavior such as threats, attacks, and chasing was so frequent that some party members were excluded altogether from feeding.

Aggression was most frequent between adult males (Kano, 1980). In almost all cases, such behavior was directed by a higher ranking animal toward a lower ranking one. Aggressive encounters between adult females and adult males were infrequent. Females, especially multiparous females, advanced toward food unafraid of males. Neither did males monopolize the food. At the approach of these females, the males (even high-ranking males) would become disconcerted and avoid the females, carrying food with them as they left. As a result, males were thus always supplanted

Table V. Agonistic Incidents

	Number of agonistic interactions recorded	Observation hours	Number per hour
Group activity in natural habitat			
Arboreal feeding	68	54.4	1.25
Arboreal resting	53	124.8	0.42
Traveling	17	32.1	0.53
Terrestrial activities	25	46.0	0.54
Night nesting	0	6.1	0
Group excitements	28	9.3	3.01
Total	191	272.7	0.70
Under artificial feeding			
Mobile provisioning	236	118.4	1.99
Fixed provisioning	307	48.0	6.40
Total	543	166.4	3.26

by females, and the latter had free access to food. Young nulliparous females, while rarely threatened by males, tended to be hesitant about approaching them. They were always sexually receptive, however, and were able to obtain food through the agency of copulatory interaction with males (Kano, 1984). Interactions between males and females during feeding suggest endogenous inhibition of feeding-related aggression by males toward females. Agonistic interactions between females were also rare. Interactions were peaceful between females during feeding, and their interindividual distances were much smaller than those between males. Females engaged in genital rubbing, which can be thought of as mutual appeasement behavior (Kano, 1980; Handler *et al.,* this volume), and it is conceivable that this practice plays an important role in preventing disputes among females regarding feeding.

It is possible that intragroup competition for food exists among the pygmy chimpanzees with regard to arboreal foods (fruits). If this is the case, such competition probably reaches a significant degree only among adult males. The patterns of inter- and intraclass interactions during feeding indicate that for the pygmy chimpanzees the social factor is more important for food getting than is locomotor or food-searching and processing ability.

2.11.2. Intergroup Competition

Many main food sources are included in the overlapping parts of the home ranges of neighboring unit groups. Two instances were recorded in which two unit groups (K and E) sharing common territory at their peripheries used the same food source quite peacefully except for a few minor skirmishes. On the other hand, a case of intergroup encounter leading to a violent intergroup fight and resulting in serious injuries to several individuals, has been recorded between the E and P groups (K. Kitamura, personal communication). In general, however, group avoidance prevents direct intergroup encounters (Kano, 1982*a*). Observations seem to indicate that party size is the most important factor for displacement-avoidance relationships between unit groups. Accordingly, while the formation of large parties increases intragroup competition, thereby lowering individual feeding efficiency, large parties are advantageous for intergroup competition.

Upon arriving at a preferred food source, large parties emit frequent simultaneous food calls, but food calls are rare in the case of small parties. In addition to informing the members of the party of the location of the food source, food calls may serve to inform other groups of the size of one's own party.

2.11.3. Interspecies Competition

Many of the fruits eaten by pygmy chimpanzees at Wamba are also consumed by red-tail monkeys (*Cercopithecus ascanius*) and mona monkeys (*Ceropithecus mona*). The forest in the Wamba region is inhabited also by *Cercopithecus salongo, Cercopithecus neglectus, Cercocebus atterrimus, Colobus angolensis, Colobus badius,* and *Allenopithecus nigroviridis.* Little is known of the feeding behavior of these primates. Due to the fact that villagers kill these animals with guns and traditional weapons, the population density of the monkeys of Wamba has been drastically reduced, and encounters with them were rare. Even when such encounters did take place, there was practically no opportunity to make useful observations, because of their great fear of humans. Among all the monkeys of Wamba, *Cercopithecus ascanius* and *C. mona* were the most common, but in comparison with places such as Yalosidi, where there are fewer hunters of monkeys, the population densities of these two species were also much lower.

Certain of the pygmy chimpanzees' foods are eaten by nonprimate mammals and birds. The impact of these animals on the major pygmy chimpanzee foods is slight due to the low numbers of these animals also as a result of hunting.

Among the pygmy chimpanzee foods, a considerable number are eaten regularly by humans. Most of these are gathered opportunistically, but some are systematically exploited. These include *S. macrostachyum* shoots and fruits of *L. owariensis, C. schweinfurthii,* and *T. africana.* Some species are cut down in order to obtain their fruits. These have become very sparse in the Wamba area. The *S. macrostachyum* shoots have commercial value, and each day women gather this plant in aged secondary forest in the western part of the E group range. Since each woman can gather several kilograms of edible shoots in a single day, their gathering activity has considerable implications for pygmy chimpanzee feeding activity.

3. Discussion

In all localities where intensive study has been carried out the diet of pygmy chimpanzees consists of both plant and animal foods [in addition to this study, see Horn (1980); Badrian *et al.* (1981), Badrian and Malenky (this volume), and Kano (1983)]. Thus, like the common chimpanzee (*Pan troglodytes*), *Pan paniscus* is omnivorous. At Gombe, 10 g of meat is consumed per animal per day, and insects are said to be an important part of the chimpanzee diet (Hladik, 1977). Judging from the data presented by Badrian *et al.* (1981) and Badrian and Malenky (this volume), it is clear that the proportion of animal foods eaten by the pygmy chimpanzees in the Lomako is higher than at Wamba and Yalosidi. But even in the Lomako, it is doubtful whether the animal foods in the diet reach the level observed in common chimpanzees. For pygmy chimpanzees, plant foods appear to be more important than for common chimpanzees. Among the edible plants recorded in Yalosidi (Kano, 1983), Lomako (Badrian *et al.*, 1981; Badrian and Malenky, this volume), and Lake Tumba (Horn, 1980), species (genera) common to Wamba constitute 50% (71%), 47%* (75%) and 50% (58%), respectively, of the total number recorded there. Considering differences in food species resulting from vegetation differences, variances in period and conditions of study, etcetera, we may say that the above rate of correspondence is high.

Fruit (pulp and seed) is the most important food category in the pygmy chimpanzee's diet. It constitutes 59% of the total number of wild plant food types recorded at Wamba, 63% at Yalosidi (Kano, 1983), and 58% (Badrian *et al.*, 1981) and 42% at Tumba (Horn, 1980). The feeding proportion of fruit in the diet has not yet been calculated for the study sites other than Wamba, where it is estimated at about 80%. When the study sites are compared on the basis of indications such as recorded observation time in the early stage of study, it seems that their population density is greatest at Wamba, followed in order by Yalosidi, Lomako, and Lake Tumba. The fact that the percentage of fruits in their diet is high at Wamba, where they live in the most favorable environment, indicates that, like the common chimpanzees, pygmy chimpanzees are basically frugivorous.

Among the various ecological factors that influence the density of pygmy chimpanzees, the most important factors are probably vegetation and humans. The pygmy chimpanzees have the ability to use their habitat comprehensively [this study and Kano (1983], but they are best adapted to the dry primary forest. Horn (1975) also pointed out that the pygmy chimpanzees of Lake Tumba depend on areas of dry forest for a majority of their diet. Wamba and Yalosidi most likely have drier vegetation than either the Lomako or Lake Tumba. At Lake Tumba, where the habitat is largely swampy, the density of pygmy chimpanzees is lowest. A much more important factor, however, in determining the distribution of pygmy chimpanzees is human interference. According to an extensive survey (T.

---

*Anthonotha fragrans and *Megaphrynium macrostachum* in the Lomako list (Badrian *et al.*, 1981) are synonymous with *Macrolobium fragrans* and *Sacrophrynium macrostachyum* in the Wamba list (Table I), respectively.

Kano, unpublished data), the distribution of pygmy chimpanzees is extremely irregular and discontinuous, with large differences in density as well. Areas in which pygmy chimpanzees are in low density within this region are those in which there is heavy predation by humans.

Among the four pygmy chimpanzee study sites, Lake Tumba is the area where hunting of pygmy chimpanzees is most prevalent (Nishida, 1972; Horn, 1980). This must certainly exert a strong influence on the density and ecology of the pygmy chimpanzees in this region. At Yalosidi, also, some people still eat pygmy chimpanzees (S. Uehara, personal communication). In the area from Boende to Befale, pygmy chimpanzees have been widely preyed upon. In the territory of the Ngandu people, which includes Wamba, however, there is a long-standing proscription against hunting and eating bonobos. It is conceivable that this situation has resulted in the high density of Wamba pygmy chimpanzees, and that due to the absence of human persecution, the original lifeways of the pygmy chimpanzees have been preserved.

Pygmy chimpanzees share many ecological traits with common chimpanzees. Both have wide niche breadth, using different strata of a variety of forest types. Both travel on the ground rather than in trees from one food source to another, which enables them to have wide group (or individual) ranges, and to exploit many plant resources. In both species, a few food types comprise the bulk of the daily diet (see Section 2.5), and in common chimpanzees of Gombe, 50% of the feeding time is spent on the first two to five food types according to Wrangham (1977); nevertheless, they have broad diet breadth. The unit groups of both species are flexible, dividing into a number of temporary parties, by which feeding efficiency is maximized (Wrangham, 1979). All of these traits common to the two species of *Pan* appear to be related to their frugivorous habits. Fruit is often an unreliable and irregular food source.

Comparisons of more detailed aspects of ecology suggest that there are subtle differences between the two species. Pygmy chimpanzees inhabit lowland forests exclusively, but common chimpanzees range over a much wider spectrum of habitats, from lowland to montane area, and from moist forest to arid open savanna. In common chimpanzees, population density appears to be largely affected by the vegetation. In some densely populated areas, their density (2.6–6 per km^2) is higher than that at Wamba (1.7 per km^2), which is one of the highest density areas for pygmy chimpanzees (Table VI). In those areas, common chimpanzees live in smaller group ranges (5–26 km^2) than the pygmy chimpanzees (58 km^2; Table VI). In an opposite extreme case, in some arid areas, the former's range is estimated to be as large as 450 km^2 (Kano, 1971), or even to reach 700–750 km^2, with very low density (Itani, 1979). Such a

Table VI. Group Ranges and Densities at the Study Sites of *Pan paniscus* and *Pan troglodytes*

	Group range, km²	Group size	Size/ range	Population density per km²	Reference
Pan paniscus					
Wamba	58	65	1.1	1.7[a]	—
Pan troglodytes					
Gombe	26	—	—	2.6–3.9	Teleki (1973)
Mahale (K)	10.4	27	2.6⎱	5.7	Nishida (1979)
Mahale (M)	13.4	80	6.0⎰		
Budongo	20	85	4.3	3.9,[b] 6[c]	Suzuki (1977)
Bossou	5–6	21	3.5–4.2	3.5–4.2	Sugiyama and Koman (1979a)

[a]Kano (1982a) formerly estimated the pygmy chimpanzee density in the Wamba study area at about three animals per km². Since it has been ascertained, however, that the K, E, and P groups travel beyond the former boundary of the study area, that figure was an overestimate. At present, reliable measurements of group size and range are available for the E group only. The density for the E group is 1.7 animals per km² (see Section 1.1).
[b]Reynolds and Reynolds (1965).
[c]Sugiyama (1977).

broad home range allows them to survive in the extremely severe habitat (Itani, 1979). No information has been obtained regarding pygmy chimpanzee group range in a low-density area. However, as the pygmy chimpanzees' low density is mainly the result of thinning out by humans, it may be assumed that local variations in group ranges are much smaller in pygmy chimpanzees, and it is unlikely that their largest group range attains that of the common chimpanzees.

At Gombe, the feeding budget accounts for 42.8% of the total available time on a daily basis (Teleki, 1981). This value is larger than that of Wamba pygmy chimpanzees (30% at most), which indicates that common chimpanzees spend more time feeding than do the pygmy chimpanzees. The average number of food types eaten by Gombe chimpanzees per day and per month is 14.6 and 60, respectively (Wrangham, 1977), The values for daily and monthly food types eaten are lower at Wamba, 10 and 40, respectively. This may indicate that the supply of major foods is more stable at Wamba than at Gombe. Wrangham (1977) pointed out that diet breadth increases as overall food availability decreases. The number of food types recorded for common chimpanzees is 140 at Gombe (Wrangham, 1977), 205 at Mahale (Nishida, 1974), and 285 at Ipassa [141 identified and 144 unidentified types (Hladik, 1977)], each of which is greater than that at Wamba. However, due to the fact that the observation time at Wamba was less, and that no observations were made during half of the year, such a simple comparison does not give a proper indication of differences in diet breadth between the two species (but see Badrian and Malenky, this volume). As the Wamba pygmy chimpanzees live near the village, many potential foods that are cultivated by humans are available to them, but among these, only pineapples and sugar cane are regularly eaten by pygmy chimpanzees. Moreover, the food lists from the other three study sites do not contain domesticated plants, and in spite of extensive inquiry in the pygmy chimpanzees range (T. Kano, unpublished data), only few pieces of information were positive concerning the eating by bonobos of any of the cultivated items. In West Africa, many domesticated plants (e.g., maize, sugar cane, pineapples, mangos, oil palms, avocados, pears, grapefruits) are regularly eaten by wild common chimpanzees everywhere (Kortlandt, 1967). Bananas, which are consistently ignored by pygmy chimpanzees at Wamba, are accepted by common chimpanzees of Gombe (Wrangham, 1974), Mahale (Nishida, 1974), and Beni (Kortlandt, 1967).

The use of tools for obtaining or processing food is widely seen among common chimpanzees (e.g., Beatty, 1951; Goodall, 1965; Teleki, 1974; McGrew, 1974, 1979; Sugiyama and Koman, 1979b; Nishida, 1973; Nishida and Hiraiwa, 1982; Wrangham, 1977). Among the pygmy chimpanzees, while the use of tools for improvement of the environment has been seen (Kano, 1982b), it has not been observed at Wamba for food acquisition. There is, however, indirect evidence of termiting from the Lomako (Badrian et al., 1981).

Copulatory behavior and other behavior deriving from it are much richer in variety and higher in frequency in the pygmy chimpanzees (Kano, 1980, 1984; Kitamura, 1984; Handler et al., this volume). It is likely that the development of such behavior is made possible by the freedom resulting from the low amount of energy put into subsistence activities. From the point of view of behavior categorization, these behaviors could be classified as sexual behavior, but functionally they constitute social behavior, and serve to heighten the affiliation between individuals (Kano, 1980, 1982a). Mounting and rump contact between pygmy chimpanzee males seem to be reassurance behavior (Kano, 1980), aiding in the establishment of male bonds. Females are sexually receptive almost continually except for a short anestrous period (within a year) after parturition, and maintain close associations with males (Kano, 1984; Handler et al., this volume). In addition, females exhibit genital rubbing, which strengthens affiliation with each other (Kano, 1980; Handler et al., this volume). As a result, group integrity is strengthened and it is infrequent for the group to split into two solitary or small unisexual parties (Kano, 1982*a*). This social pattern is advantageous in competition with other groups, and may be an adaptation to abundant local food sources. On the other hand, among the common chimpanzees, the male–male bond is extremely strong, and serves as the framework of the group (Nishida, 1979). Postpartum anestrus is much longer [3–4 years (Nishida, 1977)], and during this period parous females lead a rather solitary life with their dependent offspring (Wrangham, 1979). This social relation constitutes a better feeding strategy in the sense that the animals are able to exploit scattered minor food sources as well (Wrangham, 1979).

Based on the above comparisons, we may say that the pygmy chimpanzees historically have existed in a stable environment rich in sources of food. Pygmy chimpanzees appear conservative in their food habits and unlike common chimpanzees have developed a more cohesive social structure and elaborate inventory of sociosexual behavior. In contrast, common chimpanzees have gone further in developing their resource-exploiting techniques and strategy, and have the ability to survive in more varied environments. These differences suggest that the environments occupied by the two species since their separation by the Zaire River has differed for some time. The vegetation to the south of the Zaire River, where *Pan paniscus* is found, has been less influenced by changes in climate and geography than the range of the common chimpanzee to the north. Prior to the Bantu (Mongo) agriculturists' invasion into the central Zaire basin, the pygmy chimpanzees may have led a carefree life in a comparatively stable environment.

4. Summary

1. One hundred and thirty-three types (113 species) of wild plants, five types of domesticated plants, six animal species (five wild and one domestic), and one inorganic source were recorded as pygmy chimpanzee foods at Wamba.
2. Fruit (pulp and seed) constituted about 80% of the total feeding score. Animal foods formed only a small part of the total intake.
3. In spite of the broad diet breadth, the major share of the diet consisted of a small number of foods.
4. There are large seasonal and annual variations in the availability of the major foods, which occur in asynchronous fruiting cycles.
5. Changes in intake of the foods appeared to be related to both availability and relative preference.

6. Foraging party size varied with food type. Large parties, which appear to be advantageous in competition with other unit groups, were formed when preferred foods were abundant.
7. Pygmy chimpanzees utilized all of the vegetation types and forest strata, but the most frequently exploited strata were the crowns and shrub strata of the dry primary forest.
8. The people of Wamba do not hunt pygmy chimpanzees, but they have decimated the monkey fauna and other potential pygmy chimpanzee competitors, to the advantage of the pygmy chimpanzees.
9. The average number of food types eaten per day and per month were estimated to be 10 and 40, respectively. These figures are lower than those calculated by Wrangham (1977) for common chimpanzees at Gombe. The feeding budget of pygmy chimpanzees at Wamba was estimated to be at most 30% of the total diurnal time, which is also lower than that recorded at Gombe (Teleki, 1981). These comparisons suggest that pygmy chimpanzees at Wamba are living in an environment that is richer in food species than at Gombe and other common chimpanzee study sites.
10. From the conservative diet and highly sophisticated sociosexual behavior of pygmy chimpanzees and the common chimpanzee's more elaborate feeding techniques and strategies, it is inferred that the pygmy chimpanzee may have existed in a more stable environment than that of the common chimpanzee since their divergence from a common ancestor.

ACKNOWLEDGMENTS. This study was supported financially by 1973–1978 and 1981 Grants-in-Aid for Scientific Research (Grant-in-Aid for Overseas Field Research) from Monbusho, the Ministry of Education, Science and Culture, Japan. Acknowledgment is due to Dr. Iteke Bochoa, then Délégue Général à la Recherche Scientifique, l'Institut de Recherche Scientifique du Zaire, for research permission. We also press our gratitude to our field colleagues Dr. Suehisa Kuroda and Dr. Koji Kitamura for useful discussions, and to Noel Badrian and Nancy Thompson-Handler, who offered important information from Lomako. Botanical specimens were kindly identified by Dr. H. Breyne, Prof. C. Everard of INERA, and members of the Botanical Garden of Mbandaka. We are also grateful to Miriam Eguchi for help in translating the original draft of this manuscript into English. We especially thank Dr. Randall L. Susman for the opportunity to publish this work.

References

Altmann, J., 1974, Observational study of behaviour: Sampling methods, *Behaviour* **49**:227–267.

Badrian, A., and Badrian, N., 1977, Pygmy chimpanzees, *Oryx* **12**:463–468.

Badrian, N., Badrian, A., and Susman, R. L., 1981, Preliminary observations on the feeding behavior of *Pan paniscus* in the Lomako forest of central Zaire, *Primates* **22**(2):173–181.

Beatty, E. H., 1951, A note on the behaviour of the chimpanzee, *J. Mammal.* **32**:118.

Coolidge, H. J., 1933, *Pan paniscus:* Pygmy chimpanzee from south of the Congo River, *Am. J. Phys. Anthropol.* **18**(1):1–57.

Goodall, J., 1965, Chimpanzees of the Gombe Stream Reserve, in: *Primate Behavior: Field Studies of Monkeys and Apes* (I. Devore, ed.), Holt, Rinehart and Winston, New York, pp. 425–473.

Hladik, C. M., 1977, Chimpanzees of Gabon and chimpanzees of Gombe: Some comparative data on the diet, in: *Primate Ecology* (T. H. Clutton-Brock, ed.), Academic Press, London, pp. 481–501.

Horn, A., 1975, Adaptations of pygmy chimpanzee (*Pan paniscus*) to forests of Zaire basin, *Am. J. Phys. Anthropol.* **42**:307.

Horn, A., 1980, Some observations on the ecology of the bonobo chimpanzee (*Pan paniscus* Schwarz, 1929) near Lake Tumba, Zaire, *Folia Primatol.* **34**:145–169.

Itani, J., 1979, Distribution and adaptation of chimpanzees in an arid area, in: *The Great Apes* (D. A. Hamburg and E. R. McCown, eds.), Benjain/Cummings, Menlo Park, California, pp. 55–71.

Kano, T., 1971, The chimpanzee of Filabanga, western Tanzania, *Primates* **12**:229–246.

Kano, T., 1979, A pilot study on the ecology of pygmy chimpanzees, *Pan paniscus:* in *The Great Apes* (D. A. Hamburg and E. R. McCown, eds.), Benjamin/Cummings, Menlo Park, California, pp. 123–135.

Kano, T., 1980, Social behavior of wild pygmy chimpanzees (*Pan paniscus*) of Wamba: A preliminary report, *J. Hum. Evol.* **9**:243–260.

Kano, T., 1982a, The social group of pygmy chimpanzees (*Pan paniscus*) of Wamba, *Primates* **23**(2):171–188.

Kano, T., 1982b, The use of the leafy twigs for rain cover by the pygmy chimpanzees of Wamba, *Primates* **23**(3):453–457.

Kano, T., 1983, An ecological study of the pygmy chimpanzees (*Pan paniscus*) of Yalosidi, Republic of Zaire, *Int. J. Primatol.* **4**(1):1–31.

Kano, T., 1984, Reproductive behavior of the pygmy chimpanzees (*Pan paniscus*) of Wamba, Republic du Zaire, in: *Sexuality of the Primates* (T. Maple and R. D. Nadler, eds.), Van Nostrand Reinhold Co., New York, (in press).

Kitamura, K., 1983, Pygmy chimpanzee association patterns in ranging, *Primates* **24**(1):1–12.

Kitamura, K., 1984, Genito-genital contacts in the pygmy chimpanzee (*Pan paniscus*), in: *Primate Sexuality* (T. Maple and R. D. Nadler, eds.), Van Nostrand Reinhold Co., New York (in press).

Kortlandt, A., 1967, Experimentation with chimpanzees in the wild, in: *Neue Ergebnisse der Primatologie* (D. Starck, R. Schneider, and H.-J. Kuhn, eds.), Gustav Fischer, Stuttgart, pp. 208–224.

Kuroda, S., 1979, Grouping of the pygmy chimpanzees, *Primates* **20**:161–183.

Kuroda, S., 1980, Social behavior of the pygmy chimpanzees, *Primates* **21**(2):181–197.

McGrew, W. C., 1974, Tool-use by wild chimpanzees in feeding upon driver ants, *J. Hum. Evol.* **3**:501–508.

McGrew, W. C., 1979, Evolutionary implication of sex differences in chimpanzee predation and tool use, in: *The Great Apes* (D. A. Hamburg and E. R. McCown, eds.), Benjamin/Cummings, Menlo Park, California, pp. 440–463.

Nishida, T., 1972, Preliminary information of the pygmy chimpanzees (*Pan paniscus*) of the Congo Basin, *Primates* 13:415–425.

Nishida, T., 1973, The ant-gathering behavior by the use of tools among wild chimpanzees of the Mahali Mountains, *J. Hum. Evol.* 2:357–370.

Nishida, T., 1974, Ecology of wild chimpanzees, in: *Human Ecology* (R. Otsuka, J. Tanaka, and T. Nishida, eds.), Kyoritsu Shuppan, Tokyo, pp. 15–60.

Nishida, T., 1977, Chimpanzees of the Mahale Mountains (I) Ecology and social structure of a unit-group, in: *The Chimpanzees* (J. Itani, ed.), Kodansha, Tokyo, pp. 543–638.

Nishida, T., 1979, Social structure among wild chimpanzees of the Mahale Mountains, in: *The Great Apes* (D. A. Hamburg and E. R. McCown, eds.), Benjamin/Cummings, Menlo Park, California, pp. 72–121.

Nishida, T., and Hiraiwa, M., 1982, Natural history of a tool-using behavior by wild chimpanzees in feeding upon wood-boring ants, *J. Hum. Evol.* 11:73–99.

Reynolds, V., and Reynolds, F., 1965, Chimpanzees of the Budongo forest, in: *Primate Behavior: Field studies of Monkeys and Apes* (I. Devore, ed.), Holt, Rinehart and Winston, New York, pp. 366–424.

Sugiyama, Y., 1977, The chimpanzees of Budongo forest, in: *The Chimpanzees* (J. Itani, ed.), Kodansha, Tokyo, pp. 473–542.

Sugiyama, T., and Koman, J., 1979a, Social structure and dynamics of wild chimpanzees at Bossou, Guinea, *Primates* 20:323–339.

Sugiyama, Y., and Koman, J., 1979b, Tool-using and -making behavior in wild chimpanzees at Bossou, Guinea, *Primates* 20(4):513–524.

Suzuki, A., 1977, Society and adaptation of chimpanzees, in: *The Chimpanzees* (J. Itani, ed.), Kodansha, Tokyo, pp. 251–336.

Teleki, G., 1973, *The Predatory Behavior of Wild Chimpanzees*, Bucknell University Press, Lewisburg.

Teleki, G., 1974, Chimpanzee subsistence technology: materials and skills, *J. Hum. Evol.* 3:575–594.

Teleki, G., 1981, The omnivorous diet and ecletic feeding habit of chimpanzees in Gombe National Park, Tanzania, in: *Omnivorous Primates: Gathering and Hunting in Human Evolution* (R. S. O. Harding and G. Teleki, eds.), Columbia University Press, New York, pp. 303–343.

Vuanza, P. N., and Crabbe, M., 1975, *Les Regimes Moyens et Extremes des Climats Principaux du Zaire*, Centre Meteorologique de Kinshasa.

Wrangham, R. W., 1974, Artificial feeding of chimpanzees and baboons in their natural habitat, *Anim. Behav.* 22:83–93.

Wrangham, R. W., 1977, Feeding behaviour of chimpanzees in Gombe National Park, Tanzania, in: *Primate Ecology* (T. H. Clutton–Brock, ed.), Academic Press, London, pp. 503–538.

Wrangham, R. W., 1979, Sex differences in chimpanzee dispersion, in: *The Great Apes* (D. A. Hamburg and E. R. McCown, eds.), Benjamin/Cummings, Menlo Park, California, pp. 481–489.

CHAPTER 11

Feeding Ecology of Pan paniscus in the Lomako Forest, Zaire

NOEL BADRIAN AND RICHARD K. MALENKY

1. Introduction

Studies pertaining to the feeding ecology of free-ranging common chimpanzees (*Pan troglodytes*) are numerous, and have been carried out in a wide range of habitats (Reynolds and Reynolds, 1965; Goodall, 1968; Nishida, 1968; Suzuki, 1969; Jones and Sabater Pi, 1971; Hladik, 1977; Wrangham, 1977; McGrew *et al.*, 1981). But field studies on the pygmy chimpanzee (*Pan paniscus*) began only in the early 1970s, and are still far less numerous than those on common chimpanzees (Nishida, 1972; Badrian and Badrian, 1977; Horn, 1977; Kano, 1979; Kuroda, 1979; Susman *et al.*, 1980).

This study explores the habitat and diet of unhabituated, unprovisioned pygmy chimpanzees in the primary rain forest of the Lomako River in central Zaire. Data on the composition and seasonal changes of the forest are presented, and an attempt is made to relate the behavior of the pygmy chimpanzees to selected environmental factors.

2. Study Area

The study area is comprised of approximately 35 km² of undisturbed forest north of the Lomako River in Befale Zone, Province of Equateur,

NOEL BADRIAN • Department of Anthropology, State University of New York at Stony Brook, Stony Brook, New York 11794. *RICHARD MALENKY* • Department of Ecology and Evolution, State University of New York at Stony Brook, Stony Brook, New York 11794.

Zaire (0° 51' north, 21° 5' east; Fig. 1). The study site is isolated, with the nearest permanent village located at Bokoli, 35 km to the south. The forest block between the Lomako and Yekokora Rivers has been uninhabited since the 1920s, when villages were relocated along roads for administrative purposes.

The Lomako is predominantly *terra firma* forest dissected by four or five small streams and their aquifers. The area surrounding these streams is characterized by small valleys with occasionally innundated terrain, supporting swamplike vegetation. Secondary forest (on the sites of abandoned villages) accounts for less than 5 km² in approximately three separate blocks within the study area. Places adjacent to the disturbed areas are characterized by the virtual absence of trees in the 12- to 20-cm-diameter range. These disturbed areas are generally adjacent to secondary forest and probably the result of exploitation for building materials with specific size requirements. Generally, this type of forest is indistinguishable, on the basis of factors other than trunk size, from primary forest. The remainder is primary and disturbed primary forest.

The fauna of the study area is varied, and typical of tropical rain forest (Moreau, 1966). Over 100 bird species were provisionally identified

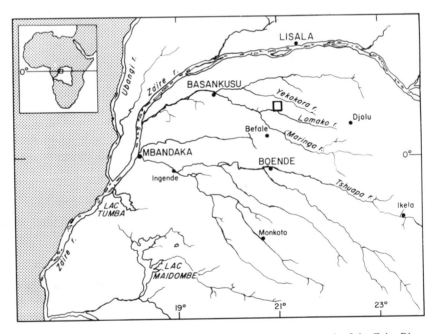

Figure 1. Map of Equateur, Zaire. The range of *Pan paniscus* is south of the Zaire River.

during a 6-day census in February 1981; the total species list is likely to be much longer. Of the birds, three are thought to be important with respect to the Lomako primates. The crowned hawk eagle (*Stephanoaetus coronatus*) was observed to prey on monkeys (including confirmed predations of adult *Cercopithecus mona* and *Colobus angolensis*) and may be a predator on infant pygmy chimpanzees as well. Two other important birds are the hornbills *Ceratogymna atrata* and *Bycanistes albotibialis*, which are potential food competitors with the pygmy chimpanzees and other primates. Numerous primate species also inhabit the area. The most common are *Galago demidovii*, *Colobus angolensis*, *Cercopithecus mona*, and *Cercocebus atterimus*. *Allenopithecus nigroviridus* and *Cercopithecus neglectus* are also present along the Lomako but are rarely observed in the study area.

Animals of importance as potential predators on pygmy chimpanzees are leopards (*Panthera pardus*) and pythons (*Python sebae*). Elephants (*Loxodonta africana*), which had been common during an earlier study (Badrian and Badrian, 1977), are no longer present in significant numbers. Bush pigs (*Potamochoerus porcus*) are common, and four species of duiker (genus *Cephalophus*) are the most frequently encountered ungulates.

3. Study Population

The study population of bonobos includes members of two different communities [cf. Nishida's (1968) "unit groups"]. These are referred to as the Bakumba and Eyengo communities and are named after the streams that drain their respective ranges. As familiarity with the two communities has increased (see Susman, this volume), it has become apparent that the study area includes most of the range utilized by the Bakumba community and a smaller proportion of the Eyengo community's range (Badrian and Badrian, this volume). The Bakumba community occupies roughly 22 km² of the study area and is composed of approximately 50 individuals, classified as infants, juveniles, adolescents, or adults (Badrian and Badrian, this volume). At the outset of the study it was difficult to identify individuals, but we have since been able to identify 17 animals. Individuals are recognized by distinguishing characteristics such as deformed hands, malformed lips, facial scarring, and other features.

4. Methods

Pygmy chimpanzees were often difficult to locate. A series of numbered trails were laid out utilizing existing animal paths, and main trails were joined when necessary for comprehensive coverage of the area. Each day observers with guides systematically walked sections of the study area. Trees that had recently been visited by pygmy chimpanzees and trees containing ripening foods were regularly checked. Other clues to the whereabouts of pygmy chimpanzees were the activities of hornbills, and often black mangabeys.

Once the pygmy chimpanzees were located, behavioral data were recorded using a combination of scan sampling and *ad libitum* sampling methods (Altmann, 1974). Tree heights, dense foliage, and the difficulty of identifying individuals make a focal animal technique impractical. Locomotor, feeding, and interactive behaviors were recorded for those animals in sight. We stayed with the chimpanzees for as long as they remained in the area, being careful not to behave in a way threatening to them. When they descended from the trees to travel along the ground, or left the area through the tree canopy, we generally did not follow, so as to minimize the threatening effect of our presence. On some occasions in which we tried to follow animals on the ground and in trees, they moved too quickly for us to maintain contact. At other times, when we *were* able to follow, they kept too great a distance to allow detailed observations to be made.

All indirect data pertaining to pygmy chimpanzees, such as vocalizations, feeding remains, fecal samples, nests, and disturbed vegetation, were recorded. A phenological study was carried out on ten tree species determined to be important food sources for *Pan paniscus* during earlier feeding studies at this site (Badrian *et al.*, 1981). Meteorological data, including daily rainfall, temperature, and humidity, were collected (Fig. 2). Plant identifications were made by Roger DeChamps of the Tervuren Museum in Belgium from wood samples and leaves.

Forest surveys were conducted in nine randomly chosen 20 × 100 m plots. All trees with a diameter at breast height (DBH) greater than 8 cm were identified and measured (height and DBH). Computations were made of (1) relative frequencies, (2) relative dominance, and (3) relative density. An importance value (the sum of 1, 2, and 3) was then calculated for each species. Table I shows the results for the ten species with the highest importance values (out of a total of 115 recorded). Seven other species known to be important foods are included for comparison.

5. Plant Foods

One hundred and thirteen plant foods have been recorded for the Lomako, representing at least 81 species from 26 families (Table II). Eight

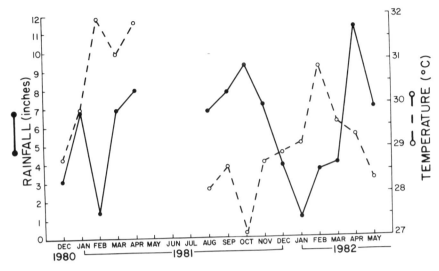

Figure 2. Graph of rainfall and temperature.

seeds and one flower collected from fecal samples but not identified have been included in this table. In addition, we suspect that the three *Dialium* species shown in Table II actually represent five species. Of the 113 foods, 54% (N = 62) were fruit, 21.2% (N = 24) were leaves, 7.1% (N = 8) were seeds, 6.2% (N = 7) were leaf petioles, 6.2% were pith (N = 7), and 4.4% (N = 5) were flowers. A preliminary analysis suggests that fruit is the most important element in the pygmy chimpanzee diet (importance being determined by the frequency with which the items were consumed). The order of importance of plant foods other than fruits is: pith, leaves, leaf petioles, seeds, and flowers. The first four categories accounted for foods eaten during more than 90% of feeding episodes.

Within the fruit category, the ten most important were those of *Dialium* (three species), *Uapaca guineensis*, *Ficus* (two species), *Antiaris toxicaria*, *Pancovia laurentii*, *Polyalthia suaveolens*, and *Anonidium mannii*. These ten sources accounted for 70% of all fruit feeding episodes. The *A. toxicaria*, *Ficus* spp., *Dialium corbisieri*, and *Dialium* spp. are all very large trees (up to 60 m in height); *Dialium pachyphyllum* and *U. guineensis*, although somewhat smaller, have large crowns. Thus all of these species can provide feeding sites for relatively large groups of pygmy chimpanzees (as many as 17 individuals) in one tree. Usually, however, they were found in *U. guineensis* in smaller groups than might be expected from the concentration of food and the availability of feeding sites in this tree species. *Dialium* spp. and *U. guineensis* are often found in "groves"

Table I.   Tree Survey[a] Data Showing Ten Species with the Highest
Importance Values, Together with Other Important Food Trees

Family Genus and species	F	RF, %	D	RD, %	d	rd, %	IV
Cesalpiniaceae							
Crudia laurentii	8/9	3.10	61,980.34	11.4	46	5.8	20.30
Annonaceae							
Polyalthia suaveolens	9/9	3.48	24,358.91	4.5	88	11.2	19.18
Olacaceae							
Dioga zenkeri	9/9	3.48	28,335.23	5.2	65	8.2	16.88
Cesalpiniaceae							
Scorodophloeus zenkeri	9/9	3.48	33,788.50	6.2	56	6.8	16.48
Olacaceae							
Strombosiopsis tetrandra	7/9	2.71	31,662.27	5.7	23	2.8	11.12
Sterculiaceae							
Cola griseiflora	8/9	3.10	10,607.22	2.0	39	4.9	10.00
Guttiferaceae							
Garcinia punctata	7/9	2.71	6,336.71	1.2	31	3.9	7.81
Cesalpiniaceae							
Dialium zenkeri	8/9	3.10	7,939.16	1.5	25	3.2	7.80
Ebenaceae							
Diospyros alboflavescens	6/9	2.33	4,963.35	0.9	28	3.6	6.83
Moraceae							
Chlorophora excelsa	6/9	2.33	6,581.25	1.2	18	2.3	5.83
Euphorbiaceae							
Uapaca guineensis	2/9	0.76	21,775.17	4.0	2	0.2	5.06
Cesalpiniaceae							
Dialium corbisieri	3/9	1.15	4,050.30	0.7	3	0.4	2.25
Dialium pachyphyllum	1/9	0.38	2,827.30	0.5	1	0.1	0.98
Moraceae							
Antiaris toxcaria	1/9	0.38	1,963.50	0.4	1	0.1	0.88
Annonaceae							
Anonidium mannii	2/9	0.76	5,150.84	0.9	2	0.3	2.72
Sapindaceae							
Pancovia laurentii	1/9	0.38	1,075.20	0.2	1	0.1	0.68
Moraceae							
Ficus spp.	—	—	—	—	—	—	—

[a]Based on nine 0.2-ha plots. Total area surveyed was 1.8 ha. *F*, Frequency (number of plots in which the species appears). RF, Relative frequency (*F*/total frequency for all species × 100). *D*, Dominance (total basal area). RD, Relative Dominance (*D*/total dominance for all species × 100). *d*, Density (total number of trees sampled). rd, Relative density (*d*/total densities for all species × 100). IV, Importance value (sum of RF, RD, and rd)

Table II. Plant Foods in the Diet of *Pan paniscus* in the Lomako Forest

Scientific name	Item[a]	Dec 1980	Jan 1981	Feb	Mar	Apr	May	June	July	Aug	Sep	Oct	Nov	Dec	Jan 1982	Feb	Mar	Apr	May	June
Annonaceae																				
Anonidium mannii	Fr							—	1	7	4									
	Fl							—										1		
Monodora angolense	Fr							—				1								
Polyalthia suaveolens	Fr		6			1		—		2						2			3	
Apocynaceae																				
Anthoclitandra robustior	Fr							—		2	3		1							
Carpodinus gentilii	Fr							—	1	1										
Landolphia congolensis	Fr							—					4	1						
Landolphia jumellei	Fr							—			1			1						
Landolphia sp. (Boseja)	Fr	1974–1975[b]						—					1							
?Landolphia sp. (Bonsele)	Fr		1			1		—			1			1						
Cesalpiniaceae																				
Anthonota fragrans	Sd	1974–1975[b]	1					—												
Anthonota macrophylla	Sd	1974–1975[b]	5	1	2			—												
Crudia laurentii	Lv		1		2			—		1	1		1	1	2		2			1
Cynometra sessiliflora	Fr		1					—												
Dialium corbisieri	Fr		4		1	3		—			1				1	4	2	2		2
	Lv					2		—			1									
	Fl	1979[c]						—												
Dialium pachyphyllum	Fr							—		2	2		7	11	8	2	5	5		
Dialium sp. (Loleka)	Fr							—						7	4	2				
Gilbertiodendron dewevrei	Lv			1				—									2	5	1	1
Monopetalanthus microphyllus	Fr							—				1								
Oxystigma oxyphyllum	Fr							—				2								
	Lv							—				1								
Pachyelasma tessmannii	Fr							—												

(continued)

Table II. (*Continued*)

Scientific name	Item[a]	Dec 1980	Jan 1981	Feb	Mar	Apr	May	June	July	Aug	Sep	Oct	Nov	Dec	Jan 1982	Feb	Mar	Apr	May	June
Paramacrolobium coeruleum	Fr	1						—												
	Lv							—												
Scorodophloeus zenkeri	Lv	6	4	2	1	2		—	1					6		1	1	2	1	
Ebenaceae																				
Diospyros alboflavescens	Fr							—				2								
Diospyros hoyleana	Lv							—				1								
Euphorbiaceae																				
Hymenocardia ulmoides	Lv	1																		
Uapaca guineensis	Fr	2	18	2				—									1	2		5
	Lp							—		1										
Flacourtiaceae																				
Caloncoba welwitschii	Fr	1974–1975[b]																		
Paropsia sp. (Bontongentonge)	Fr				1			—												
Guttiferaceae																				
Garcinia ovalifolia	Fr	1						—												
Garcinia punctata	Fr							—		1	2									
Garcinia sp. (Tatolongo moke)	Fr							—				1								3
Mammea africana	Fr							—										1		
Irvingiaceae																				
Irvingia gabonensis	Sd[d]																			
Irvingia wombolu	Fr							—	2									1		
Klainedoxa gabonensis	Fr					1		—										2		
	Lv					1		—												
Lauraceae																				
Beilschmiedia corbisieri	Fr	1974–1975[b]																		
Meliaceae																				
Guareae cedrata	Fr							—						1						
Lovoa trichilioides	Lv		1					—												

Taxon														
Moraceae														
Antiaris toxicaria	Fr	6	—	1			1				1	5		1	2	8
	Lv	2	—												2
	Fl	1974–1975[b]													
Ficus exasperata	Lv	1974–1975[b]													
Ficus sp. (Lokumo)	Fr	2	3	—						2			2	2	
Ficus sp. (Lokumo moke)	Fr	1	3	—								1		6	
Treculia africana	Fr[a]	1	1	—	1	2	6			1	1		2	6	3
	Sd		1												
Myristicaceae														
Pycnanthus angolensis	Fr	1974–1975[b]		—			1								
Staudtia stipitata	Fr	1		—											
Olacaceae														
Onkokea gore	Fr			—									1		
Strombosia glaucescens	Sd	1974–1975[b]		—							1				
Strombosia grandifolia	Lv			—											
Strombosiopsis tetrandra	Fr			—			1								
Strombosiopsis zenkeri	Sd	1974–1975[b]		—											
Rosaceae														
Parinari excelsa	Fr	1	1	—				3							
Rubiaceae														
Canthium sp. (Botitsuampono)	Fl	1		—											1
Leptactinia seretii	Fr[c]			—										1	
Nauclea diderichii	Fr			—											
Nauclea sp. (Likangu)	Lv	1													
Sapindaceae														
Chytranthus sp. (Kulutende)	Fr			—			1	1					3		
Pancovia laurentii	Fr	2	1	2	6		1	2					3	4	
	Sd	2	1	2	6		1	2					3	4	
Sapotaceae														
Chrysophyllum lacourtianum	Fr			—				1					2	4	1
Synsepalum subcordatum	Fr		1	—	1	2								1	

(continued)

Table II. (*Continued*)

Scientific name	Item[a]	Dec 1980	Jan 1981	Feb	Mar	Apr	May	June	July	Aug	Sep	Oct	Nov	Dec	Jan 1982	Feb	Mar	Apr	May	June
Simarubiaceae																				
Quassia africana	Fr							—					1							
Sterculiaceae																				
Cola griseiflora	Fr			1				—			2									
	Lv			1				—												
Pterygota beguaertii	Sd	1			1			—						1					1	
Tiliaceae																				
Grewia louisii	Fr							—												
Ulmaceae																				
Celtis mildbraedii	Fr	2		2				—												
	Lv	2		1				—												
	Fr	1		1				—												
Celtis tessmannii	Fr							—												
Verbenaceae																				
Vitex sp.? (Bonsanga)	Fr				1			—												
Unidentified																				
Sp. indet. (Bontio)	Fr							—				1								
Sp. indet. (Boseta)	Lv							—					1							
Sp. indet. (Lonkunga)	Fr							—												
Sp. indet. (?)	Fr							—						3	1					
Herbaceous and other plants																				
Commelinaceae																				
Palisota sp. (Lintentele)	Pth	4	2	3		1		—					2	2	2		1			1
	Lv		2	2				—					4							

		C1	C2	C3	C4	C5	C6	C7	C8	C9	C10	C11	C12	C13	C14	C15	C16	C17
Marantaceae																		
Haumania liebrechtsiana	Pth	8	17	7	4	8	—	4	1	2	8	9	13	1	3	9	7	2
	Lp	3	13	6	3	7	—	4	1	2	6	8	12	3	4	8	6	1
	Lv	4	8	4	3	5	—	1	1	2	3	3	11	2		6	3	
Megaphrynium macrostachum	Lp		1	1		1	—											
	Lv			1		1	—											
Sarcophrynium schweinfurthii	Lp	2	5	3		2	—	1			2	3		4	3	2		
	Lv	2	5	3		2	—	1				2	3		4	3		
Trachyphrynium braunianum?	Pth			1			—	1				1						
	Lp			1			—	1				1						
	Lv			1			—	1				1						
Palmaceae																		
Sclerospermum mannii	Pth							1										
	Lp							1										
Zingiberaceae																		
Afromomum sp. (Mbole)	Fr	2				2												
	Pth											1						
Costus afer	Pth		1															
Costus sp. (Besomboko)	Pth			3														
	Fr					1												
Renealmia africana?	Fr																	
	Pth																	
	Lp		1															
Unidentified																		
Semiaquatic herb (Tokondoko)	All											1						
Unidentified from fecal samples: eight fruits (seeds of), one flower																		

Fr, fruit; Fl, flowers; Sd, seeds; Lv, leaves; Lp, leaf petioles; Pth, pith.

[a] Not recorded in this study (1980–1982).

[b] Not recorded in 1974–1975 or 1980–1982.

[c] Jelly in center of seed consumed.

[d] Immature fruit eaten.

(where a number of trees of the same species are found together in a relatively restricted area), whereas *A. toxicaria* and the *Ficus* spp. are more widely dispersed. The fruits of all these species, with the exception of one *Ficus* (Lokumo, Table II) are small (maximum 2 cm in diameter), and grow in clusters on terminal branches. Pygmy chimpanzees spend considerable periods of time feeding and resting in all of these trees. The most prolonged periods are spent in *Dialium* spp. (see Fig. 3) and the small-fruited *Ficus*—5–7 hr/day.

Pancovia laurentii is a tree that produces medium-sized, succulent fruits (up to 4 cm in diameter), which usually contain three edible seeds (1–1.5 cm in diameter). The fruits of *P. laurentii* grow in clusters on stalks found in groups on terminal branches. Individual trees are widely separated, fairly small, and seldom provide feeding sites for more than six adult pygmy chimpanzees. The animals did not spend extended periods of time feeding and resting in these trees, as they did, for example, in *Dialium* spp. and *Ficus* sp. (Lokumo moke, Table II).

Polyalthia suaveolens is a very common, emergent tree with a small crown, and consequently provides a restricted number of possible feeding

Figure 3. Adult female feeding on *Dialium* sp. fruit.

sites at any one time. It produces abundant small, round fruits, but pygmy chimpanzees do not spend long periods feeding on them. In addition to being a major food source, *P. suaveolens* is also a favorite nesting tree.

Anonidium mannii, another common tree, produces huge ovoid, compound fruits up to 50 cm long. Upon ripening, the fruits often fall to the ground, where the majority of pygmy chimpanzee feeding bouts for this species took place. Some of these fruits, however, were also consumed in the trees. Individual *A. mannii* trees produce fruit asynchronously between the end of May and the end of August. Individual fruits of a single tree may also ripen asynchronously. Although the trees are relatively large, the fact that few fruits are ripe at any one time means that there are few feeding sites associated with individual trees, and this, in turn, does not permit large groups of animals to aggregate.

Pith from seven herbaceous species comprise the second most frequently consumed food type. For most of these species, immature leaves and/or the pithy ends of leaf petioles were also consumed, often during the same feeding bout (Table II). These ground-dwelling plants were sometimes consumed during travel, as pygmy chimpanzees moved from place to place. In these instances, animals would pause momentarily to snack on leaf petioles or shoots and then move on. Of the herbaceous group, the most important species by far is *H. liebrechtsiana*. It is, in fact, the food most frequently utilized by pygmy chimpanzees in the Lomako forest (as determined from observation, fecal analysis, and the frequency of feeding remains recorded). This species is prevalent throughout the study area in all forest zones. It usually occurs at low density, but it forms dense thickets where tree falls have allowed sunlight to penetrate to the forest floor. Although this species is heavily utilized, it replaces itself rapidly, shoots growing up to 8 cm/day.

Another herbaceous plant prominent in the diet of pygmy chimpanzees is *Sarcophrynium schweinfurthii*. These plants produce very large leaves at the terminus of tall (1 m) petioles that grow up from the ground. They are found in scattered patches throughout the study area, and in some places form almost pure stands 20 m or more in diameter. Pygmy chimpanzees consume the immature unopened leaves of *S. schweinfurthii* and the pithy ends of their leaf petioles. The distribution of these and other herbaceous species, together with the fact that each plant provides only a small amount of food, precludes the formation of large feeding groups of pygmy chimpanzees at any one site; groups of up to ten animals were occasionally observed to constitute scattered foraging groups. These groups often moved slowly along while feeding on *Haumania* and other ground-level foods.

In addition to the immature leaves of herbaceous species mentioned above, pygmy chimpanzees eat the leaves of 18 tree species. *Crudia laurentii* and *Scorodophleous zenkeri* are the two most important leaf sources. The trees of both species are large and provide feeding sites for large numbers of pygmy chimpanzees. These two species are among the most common and evenly distributed trees in the forest (Table II). The leaves of these (and other) trees were often consumed by some members of a group who were feeding in an adjacent fruit tree. In general, however, larger groups were not associated with leaf-eating even in these large trees, and the time spent in them was never as long as in fruiting trees such as *Dialium* and *Ficus*.

The seeds of *Pancovia laurentii* (considered here as a separate food since both fruit and seeds were eaten and digested) and *Treculia africana* were the most important foods in this category. The spherical fruits of the latter tree are very large (up to 50 cm in diameter) and contain thousands of small seeds about the size and shape of sunflower seeds, embedded in a fibrous matrix. As with *A. mannii*, these large fruits often fall when ripe and are eaten on the ground. The *A. mannii* trees fruit asynchronously and only a few of the fruits are ripe at any one time, thus precluding the formation of very large foraging groups.

6. Seasonal Patterns

No clearcut pattern of seasonal variation in the production of pygmy chimpanzee plant foods is evident in the Lomako area. Fruit, leaves, and pith are all available throughout the year. Of the fruits listed in Table II as pygmy chimpanzee foods, no less than seven were available in any one month, as determined by a random survey of available fruits, which was conducted bimonthly (Fig. 4).

While some fruit was always available, the pattern of fruit production varied greatly from species to species, and even within species. For example, *Dialium*, one of the most important pygmy chimpanzee foods, was in almost continuous production between December 1980 and the following April, and from July 1981 to May 1982. *Ficus* spp. fruited asynchronously throughout the study period. *Pancovia laurentii* also lacked a fixed fruiting pattern, some individual trees producing fruit up to four times in a year and others not fruiting at all during the study period. *Polyalthia suaveolens* fruit is produced sporadically between January and October, with great individual variation from tree to tree. The fruits of *Antiaris toxicaria* are abundant but ripen asynchronously between November and April. *Uapaca guineensis* produces abundant fruit, but its fruiting pattern is more synchronous. In 1980–1981, the fruit of this species lasted for 3

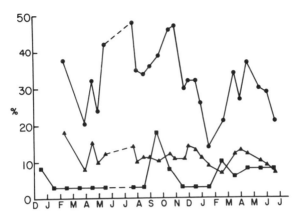

Figure 4. Seasonal variation in group size, food availability, and total fruits available. (●) Total fruits, (▲) known foods, (■) modal group size.

months (December through February), but fruit was not produced again until the latter part of May 1982. *Scorodophloeus zenkeri* and *Crudia laurentil* produce new leaves throughout the year. *Haumania liebrechtsiana*, the major source of pith in the pygmy chimpanzee diet, was also available throughout the year. Thus, no strictly seasonal pattern of plant food availability can yet be discerned among the widely varying production cycles of the main food plants in the Lomako area.

7. Feeding Patterns and Group Size

Between March and October of 1981, pygmy chimpanzees were more difficult to locate, and for significant periods of time they were neither seen nor heard. Their presence at the study site, however, was evidenced by frequently encountered fresh feeding remains. We believe this is due to the existence of smaller, widely dispersed foraging groups during this period. In fact, when groups were contacted they were generally small (2–5 animals), and, as noted by Badrian and Badrian (this volume) and Kuroda (1979), small groups are generally less vocal than larger ones.

The major foods being exploited during these months were from *Dialium corbisieri, Polyalthia suaveolens, Garcinia punctata, Anonidium mannii,* and *Haumania liebrechtsiana.* These are all common, evenly dispersed species. In addition, the less common, more scattered *Treculia africana* and *Anthoclitandra robustior* (a liana) were being utilized (see Table II). None of these species, with the exception of *D. corbisieri,* provide the opportunity for large groups to gather and feed in one place, due to the smaller number of feeding sites available in any one tree. The larger groups encountered during this time were either associated with *D. corbisieri* (which possesses a large crown with evenly distributed fruits) or were feeding in several different food trees in close proximity to one another.

Between March and July 1982, larger groups were encountered with greater frequency. The major foods being eaten at this time were from *P. suaveolens, H. liebrechtsiana, U. guineensis, Dialium* spp., *Ficus* sp., (a species with small, abundant, evenly dispersed fruits in a widespread canopy), *A. toxicaria,* and *Chrysophyllum lacourtianum* (a low-density, large-crowned tree, with abundant fruit), as is shown in Table II. With the exception of *P. suaveolens, U. guineensis,* and *H. liebrechtsiana,* these species can all provide sufficient feeding sites for larger groups. During the same months of 1981, as noted above, the group sizes recorded were smaller, and the foods being eaten were different.

During January 1981, *U. guineensis* was being heavily utilized, accounting for 95% of all fruit-eating episodes. These trees provide large numbers of feeding sites. In addition, they often exhibit a clumped distribution in the Lomako, and fruit synchronously. The most efficient method of exploiting this resource would be to utilize the trees or groves more intensively but in smaller groups. Due to the distribution of these trees a smaller group can find sufficient food over successive days without having to move very far. At the same time, a greater proportion of the available fruit would be utilized during the relatively short fruiting period. Pygmy chimpanzees of the Lomako were observed in smaller groups at the time during which *U. guineensis* was a particularly important food source (Badrian and Badrian, this volume).

8. Animal Foods

Table III lists all the animal foods recorded for *Pan paniscus* in the Lomako forest, including those reported by Badrian et al. (1981). Some of these foods were recovered from fecal analysis and were not fully identifiable. During the 1980–1982 study period, the columnar nests of two termite species (possibly *Microtermes* sp. and *Trinervitermes* sp.) were occasionally found to have been disturbed and broken open [for description, see Horn (1980) and Badrian et al. (1981)]. These disturbed mounds presented a different appearance from those that had been utilized by pangolins (*Manis* spp.), and were probably opened by pygmy chimpanzees. Remains from the large grubs of a wood-boring beetle were recovered from three fecal samples in January and February 1981. On

Table III.   Animal Foods in the Diet of the Lomako Forest
Pygmy Chimpanzees

Identification	Description
Invertebrates	
Coleoptera	Adults of two species of bettles; larvae of one species of wood-boring beetle
Hymenoptera	Adults of two species; larvae of *Pachysima aethiops*
Apoidea	Adult melliponid bees
Isoptera	Adult ?*Microtermes* sp., adult ?*Trinervitermes* sp.
Lepidoptera	Larvae of three species of Hesperiid butterflies
Orthoptera	Adults of two species
Diplopoda	Adult giant millipede
Gastropoda	Adults of one species of land snail and two species of aquatic snails
Vertebrates	
Osteichthyes?	(See text)
Reptilia	Snake
Insectivora	Adult shrew
Rodentia?	—
Artiodactyla	Infant *Cephalophus dorsalis*, infant *Cephalophus nigrifrons*

one occasion in April 1981, an old, rotting tree stump, which had been broken open by what could only have been pygmy chimpanzees, was found. The 2-m-tall stump (which had been intact the previous day) was also surrounded by fresh *H. liebrechtsiana* feeding remains.

In January and February 1981, many fecal samples studied contained the remains of caterpillars. During January 1982, *C. laurentii* trees were observed to be infested with caterpillars of at least three species of hesperid butterflies. Pygmy chimpanzees were seen to feed on both the leaves and the caterpillars during this period. On other occasions, small groups were observed on the ground collecting and eating caterpillars that had fallen out of the trees to pupate in the vegetation of the shrub layer. This caterpillar "harvest" lasted for only a few weeks and was repeated again, although on a smaller scale, in April 1982.

Between the end of December and the beginning of April of both years (1981–1982), the water level in the small streams and the swamplike areas surrounding them was low. At these times we often found feeding remains in and around the streams. In addition, the vegetation and mud at the edge of the streams were often disturbed. From footprints and plant feeding remains, this appeared to be the result of pygmy chimpanzee activity.

On two occasions small groups were observed actually foraging in the shallow water of small streams. On one of these occasions only plant foods were observed being eaten, including an aquatic or semiaquatic plant as yet unidentified (tokondoko, Table II). The second instance involved 6 min of direct observation. The animals were feeding in the leaf litter and mud at the edge of the stream. Rapid movements by the pygmy chimpanzees suggested that they were feeding on small animals (fish or shrimp) that are abundant in the shallows.

Three instances of predation were recorded during the study period. All involved infant forest duikers (*Cephalophus* spp.) weighing no more than 5 kg. Pursuit and capture of the prey was not seen, but the killing of one animal and the "sharing" and eating of meat was observed in two cases.

The first episode occurred in January 1981 at 11:33. Alerted by the distress calls of an infant duiker (*Cephalophus dorsalis*), we encountered a large male running tripedally along the ground dragging the infant duiker by one forelimb (Fig. 5). We followed him quietly for the next 21 min, during which we made brief contact three times. He fed from both the body and severed leg in the company of 3–6 other animals. At one point two of the other adults pulled legs off the kill and ran off with them.

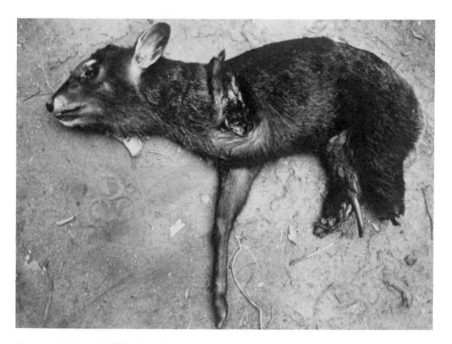

Figure 5. Infant *Cephalophus dorsalis* partially consumed by pygmy chimpanzees.

The second predatory episode took place in the same general area of the study site at 08:36 in August 1981. The events surrounding this incident were very similar to the previous one, but the prey this time was an infant black-fronted duiker (*Cephalophus nigrifrons*). This second group of pygmy chimpanzees was also disturbed by the presence of observers, and abandoned the duiker while it was alive. On both occasions the pygmy chimpanzees were totally silent throughout the episodes. We would not have discovered them but for the distress calls of the infant duikers.

The third case of predation occurred in December 1981 at 14:45. It involved a larger, very vocal group. On following the vocalizations of both the pygmy chimpanzees and the infant duiker, a small adult male was located in a tree holding a struggling infant *Cephalophus nigrifrons*. The male was surrounded by a group of excited, vocalizing group members, who were watching him intently from the nearby trees. Shortly after the observation began the duiker was killed (possibly strangled). Throughout the sighting (which lasted for nearly 80 min), the carcass was in the possession of the same individual, although he was chased by another male on one occasion.

The male in possession of the duiker was constantly approached and surrounded by up to four adult females plus infants and juveniles, whom he constantly tried to avoid. They were observed to make begging gestures and vocalizations similar to those described by Goodall (1968) and Kano (1980) for common chimpanzees.

9. Discussion

Seasonal variations in the Lomako forest are very complex, and hence difficult to define. Local patterns of rainfall, and precipitation in distant areas of the catchment basin (which affects local levels of ground water by draining into the Lomako River), plus other factors, such as temperature, humidity, and plant competition, acts simultaneously to create the unpredictable and complex phenological patterns observed (see also Richards, 1966; Fleming, 1979).

Fruit and most other food types are available throughout the year. Some species fruited at widely different times from one year to the next. Some species were synchronous in their production of fruit, and others were highly asynchronous. Thus, one could not be certain that a given month in one year would yield the same array of foods as the same month of the previous year (see also Strusaker, 1975; Hladik, 1977). We are therefore forced to examine factors other than strictly seasonal or more regular climatological ones in an attempt to explain variations in feeding strategies.

Wrangham (1977) found that chimpanzee party size increased in relation to the availability of more abundant food at a given source. The analysis of our data suggests that group size in the Lomako tended to vary in relation to the particular structural and spatial characteristics of food sources. Small groups were associated with foods that offered fewer feeding sites, even if the food available was sufficient to support larger sized groups (i.e., *Polyalthia suaveolens, Pancovia laurentii, Treculia africana, Anonidium mannii*), or (as in the case of *Uapaca guineensis*) when abundant but patchily distributed food was available within a relatively short period of time. Large foraging groups were associated with foods that were distributed so as to offer a greater number of feeding sites and were widely distributed. These food items were also often produced asynchronously. We also observed that pygmy chimpanzees spent longer periods of time feeding in trees of these species (i.e., *Dialium* and *Ficus* spp.) than they did in the other trees mentioned above. Social interactions were more frequent under the latter conditions, i.e., in the larger groups, and may have encouraged pygmy chimpanzees to remain for longer periods of time (Badrian and Badrian, this volume).

In this regard Nishida (1968) also reports large groups (of common chimpanzees) associated with concentrated fruit sources (e.g., *Saba florida* and *Garcinia huillensis*). Suzuki (1969) reports observing a group of more than 17 common chimpanzees in a large trees (*Saba* spp.), feeding on ripe fruit. Finally, Sugiyama (1968) reports large groups of *P. troglodytes* frequently moving from tree to tree when the trees are tall, have large canopies, and produce abundant fruits.

We had predicted a reciprocal relationship between the abundance of ground feeding remains, particularly of *H. liebrechtsiana*, and the amount of time pygmy chimpanzees spent feeding in major fruit trees. Our data show, however, that ground feeding remains were not less abundant at times when arboreal feeding was most frequent. As Wrangham (1977) has suggested for common chimpanzees, pygmy chimpanzees may be seeking a diversified diet, and, while they are feeding on a single type of fruit for extended periods, they may be supplementing this high-carbohydrate intake with easily available, ubiquitous, herbaceous foods, which are likely to contain significant amounts of water, cellulose, protein in immature leaves, and possibly minerals in pith and petioles (Hladik, 1977).

Some ecological differences are apparent when the two major *Pan paniscus* study sites in the Lomako and at Wamba are compared. This is particularly evident in the major food items recorded for the two populations. For example, Kuroda (1979) reports that *H. liebrechtsiana* is a "complementary" food at Wamba; it is not eaten throughout the year, as it is in the Lomako. Two of the foods shown as staples for a major portion of the year at Wamba, *Megaphrynium macrostachum* and *Afromomum* sp., are found only in very low densities at our site and are eaten infrequently. Both *Megaphyrnium* and *Afromomum* occur in secondary growth. *Sarcophrynium schweinfurthii*, which is an important food in the Lomako, but not at Wamba (Kuroda, 1979), is a closely related but smaller forest equivalent of the species *M. macrostachum*.

Another staple food at Wamba was *Musanga cercropioides* fruit (Kuroda, 1979). This species is rare in the Lomako and has not been recorded as a pygmy chimpanzee food species to date. *Musanga* is another secondary forest species, and strengthens our impression that there are considerable habitat differences between the two sites. Both Kuroda (1979) and Kano (1980) report large areas of secondary forest at Wamba, and Kano states that the primary forest of their study site is "more or less depleted" due to human activity. Indeed many of the major pygmy chimpanzee foods are species that are heavily utilized by the indigenous human population.

One of the more clearly defined differences between the *Pan paniscus* populations of the Lomako and Wamba is the larger group sizes at the latter site. This can, perhaps, be partially explained by the greater utilization of abandoned fields and secondary forest that results from shifting root-crop agriculture. It is in these areas that some of the important staple foods of the Wamba pygmy chimpanzees are to be found. *Afromomum* sp. is abundant in these clearings (Kano, 1980) and presents the pygmy chimpanzees with the same situation as that described above for very large, widely dispersed fruit trees—food is abundant and evenly dispersed within a prescribed area, allowing the formation of larger feeding groups. A similar situation is found for *Megaphyrnium macrostachum*. Kuroda (1979) reports small group sizes in early March 1975, when *Afromomum* fruit was a major source of food, but he states that the fruit was already scarce by that time (other major foods being consumed at this time were fruit from *P. suaveolens, P. laurentii,* and *M. cercropioides,* a small secondary forest tree). Kuroda goes on to state that in November 1975 *Afromomum* and *M. macrostachum* were abundant, and notes that, in fact, group sizes were larger at this time. He also notes that smaller groups were associated with *U. guineensis* and larger ones with *Dialium* and other sparsely distributed, abundantly fruiting species (e.g., *Saba florida* and *Landolphia owariensis*).

On the whole, the composition of the pygmy chimpanzee diet and that of the common chimpanzee are quite similar (Goodall, 1965; Reynolds and Reynolds, 1965; Suzuki, 1969; Hladik, 1977; Wrangham, 1977). Both species consume fruits with greater frequency, and in greater quantities, than any other food. Both exhibit a flexibility in the diversity of their diet that is more extensive than any of the other apes. The common chimpanzee, in its exploitation of a greater variety of habitats (Reynolds and Reynolds, 1965; Goodall, 1968; Hladik, 1977; McGrew et al., 1981), could be said to be more flexible in its feeding and ranging habits than are pygmy chimpanzees, since the latter are confined to lowland tropical forest. Pygmy chimpanzees were, however, found to utilize almost all food categories recorded for common chimpanzees, i.e., fruit, leaves, flowers, pith, insects, and small mammals. Insects, particularly ants and termites, are consumed in the Lomako, but not with the same regularity as that reported for common chimpanzees (Hladik, 1977). Some items (bark, resin, other primates, and soil) have not been reported as pygmy chimpanzee foods so far in the Lomako. The consumption of bark by P. troglodytes is reported as intense by Nishida (1968). This behavior is directly linked to the seasonal fluctuations recorded at Mahali, and occurs when fruit is scarcest—toward the end of the rainy season. Monkeys are not hunted by pygmy chimpanzees in the Lomako as they are at Gombe (Goodall, 1968; Teleki, 1973; Wrangham, 1975). This may be related to the virtually continuous forest canopy in the Lomako, which provides ready escape routs for potential arboreal prey and could confound even cooperative hunting by pygmy chimpanzees. We have not observed soil eating (geophagy) in pygmy chimpanzees, but this may be due to the fact that this population is still difficult to observe on the ground. Similarly, the frequency with which ants and termites are consumed may be found to be considerably higher once all-day "follows" are possible.

Some herbaceous foods are very important in the pygmy chimpanzee diet, and may be correlated with their molar structure and dental wear patterns (Kinzey, this volume). The proportion of pith from herbaceous species in the pygmy chimpanzee diet is much higher than that recorded for common chimpanzees studied by Reynolds and Reynolds (1965), Sugiyama (1968), Kano (1971), and Wrangham (1977). Jones and Sabater Pi (1971), however, report that pith is a food eaten commonly by the chimpanzees of Rio Muni. Hladik (1977) reports that the pith of a marantaceous herb is eaten frequently and throughout the year by chimpanzees in Gabon. These differences could be related to forest habitats, but the situation is confusing and will not be resolved until comparative data are available from all sites.

The predatory episodes described above are very similar to those described for Pan troglodytes (Goodall, 1963; Teleki, 1973; Wrangham, 1975; Nishida et al., 1979). Voluntary sharing of meat, however, was not actually observed in the Lomako. The low frequency with which pygmy chimpanzees have so far been recorded to hunt, together with the intense social activity surrounding the events, supports Teleki's suggestion that social factors may be as important as nutritional ones in governing this behavior (Teleki, 1973). Some researchers have viewed predation by common chimpanzees as atypical or an artifact of habituation and provisioning (Gaulin and Kurland, 1976). The present data will, perhaps, help to clarify this issue, since among pygmy chimpanzees, the species most closely related to the common chimpanzee, predation is now confirmed in an unhabituated and unprovisioned population. It is interesting to note that at Wamba, where some groups are habituated and provisioned, no predation has been reported to date (Kuroda, 1980; Kano, 1982). This, however, could simply be due to the proximity of the latter site to a permanent village (Kano, 1980) and the resulting depletion of the local fauna.

The population density of pygmy chimpanzees is higher at Wamba (3/km²) than in the Lomako (2/km²), and group sizes are significantly larger at the former site (Kuroda, 1979; Kano, 1982; Badrian and Badrian, this volume). Local agricultural practices do not seem to be detrimental to the existence of *Pan paniscus,* since these animals are able to exploit secondary forest and abandoned fields successfully, as evidenced at Wamba. West African gorillas exploit and even prefer similar areas (Groves, 1971; Jones and Sabater Pi, 1971). Among pygmy chimpanzees the association of foraging groups with secondary forest areas may not be as obligatory, but could still be significant in relation to distributional patterns. The patchy distribution of *Pan paniscus* within the Congo Basin (Badrian and Badrian, 1977), therefore, may not be due primarily to habitat modification by the local populace, and is perhaps more directly related to human predation. In the two areas in which substantial pygmy chimpanzee populations are known to exist (Lomako and Wamba), the local villagers do not hunt or eat these animals (Kuroda, 1980; Badrian and Badrian, 1977), as is the more general custom throughout the area.

ACKNOWLEDGMENTS. We wish to thank the Zairean Government for permission to conduct this research in the Lomako forest. Dr. Kankwenda M'Baya and other members of IRS-Kinshasa were especially helpful and supportive of our work.

Our research would not have been possible without the friendship, hospitality, and help that we received from numerous people, particularly members of Cultures Zairoises; the British and American Embassies in Kinshasa; the Peace Corps; and the missionaries of Boende, Befale, and Mbandaka. We will always be grateful for the help and cooperation of the people of the Lomako area, and especially our guides and helpers from Bokoli.

We would like to thank Dr. John Fleagle and Alison Badrian for commenting on earlier drafts of this manuscript, and Dr. Gretchen Gwynne for editorial assistance. The research was supported by grants from NSF and the National Geographic Society.

References

Altman, J., 1974, Observational study of behavior: Sampling methods, *Behaviour* **49**:227–267.

Badrian, A., and Badrian, N., 1977, Pygmy chimpanzees, *Oryx* **13**:463–472.

Badrian, A., and Badrian, N., 1978, Wild bonobos of Zaire, *Wildl. News* **13**(2):12–16.

Badrian, N., Badrian, A., and Susman, R. L. S., 1981, Preliminary observations on the feeding behavior of *Pan paniscus* in the Lomako forest of central Zaire, *Primates* **22**(2):173–181.

Fleming, T. H., 1979, Do tropical frugivores compete for food?, *Am. Zool.* **19**:1157–1172.

Gaulin, S. J. C., and Kurland, J. A., 1976, Primate predation and bioenergetics, *Science* **191**:314–315.

Goodall, J., 1963, Feeding behaviour of wild chimpanzees, *Symp. Zool. Soc. Lond.* **10**:39–48.

Goodall, J., 1965, Chimpanzees of the Gombe Stream Reserve, in: *Primate Behavior* (I. DeVore, ed.), Holt, Rinehart and Winston, New York, pp. 425–473.

Goodall, J., 1968, The behaviour of free-living chimpanzees in the Gombe Stream Reserve, *Anim. Behav. Monogr.* **1**:161–311.

Groves, C. P., 1971, Distribution and place of origin of the gorilla, *Man* **6**(1):44–51.

Hladik, C. M., 1977, Chimpanzees of Gabon and chimpanzees of Gombe: Some comparative data on the diet, in: *Primate Ecology* (T. H. Clutton-Brock, ed.), Academic Press, London, pp. 481–501.

Horn, A. D., 1977, A Preliminary Report on the Ecology and Behavior of the Bonobo Chimpanzee (*Pan paniscus*, Schwarz 1929) and a Reconsideration of the Evolution of the Chimpanzees, Ph.D. Dissertation, Yale University, New Haven, Connecticut.

Horn, A. D., 1980, Some observations on the ecology of the bonobo chimpanzee (*Pan paniscus*, Schwarz 1929) near Lake Tumba, Zaire, *Folia Primatol.* **34**:145–169.

Jones, C., and Sabater Pi, J., 1971, Comparative ecology of *Gorilla gorilla* (Savage and Wyman) and *Pan troglodytes* (Blumenbach) in Rio Muni, West Africa, *Bibl. Primatol.* **13**:1–96.

Kano, T., 1971, The chimpanzees of Filabanga, western Tanzania, *Primates* **12**:229–246.

Kano, T., 1979, A pilot study on the ecology of pygmy chimpanzees, *Pan paniscus*, in: *The Great Apes* (D. A. Hamburg and E. R. McCown, eds.), Benjamin/Cummings, Menlo Park, California, pp. 128–135.

Kano, T., 1980, Social behavior of wild pygmy chimpanzees (*Pan paniscus*) of Wamba: A preliminary report, *J. Hum. Evol.* **9**:243–260.

Kano, T., 1982, The social group of pygmy chimpanzees *Pan paniscus* of Wamba, *Primates* **23**(2):171–188.

Kuroda, S., 1979, Grouping of the pygmy chimpanzee, *Primates* **20**(2):161–183.

Kuroda, S., 1980, Social behavior of the pygmy chimpanzees, *Primates* **21**(2):181–197.

McGrew, W., Baldwin, P. J., and Tutin, C. E. J., 1981, Chimpanzees in a hot, dry habitat Mount Asserik, Senegal, West Africa, *J. Hum. Evol.* **10**:227–244.

Moreau, P., 1966, *The Bird Fauna of Africa and Its Islands*, Academic Press, New York.

Nishida, T., 1968, The social group of wild chimpanzees in the Mahali Mountains, *Primates* **9**:1167–1224.

Nishida, T., 1972, Preliminary information on the pygmy chimpanzee (*Pan paniscus*) of the Congo Basin, *Primates* **13**:415–425.

Nishida, T., Uehara, S., and Nyundo, R., 1979, Predatory behavior of wild chimpanzees of the Mahale Mountains, *Primates* **20**(1):1–20.

Reynolds, V., and Reynolds, F., 1965, Chimpanzees in the Budongo Forest, in: *Primate Behavior* (I. DeVore, ed.), Holt, Rinehart and Winston, New York, pp. 368–424.

Richards, P. W., 1966, *The Tropical Rain Forest*, Cambridge, England.

Strusaker, T. T., 1975, *The Red Colobus Monkey*, University of Chicago Press, Chicago.

Sugiyama, Y., 1968, Social organization of chimpanzees in Budongo Forest, Uganda, *Primates* **9**:225–258.

Susman, R. L. S., Badrian, N., and Badrian, A., 1980, Locomotor behavior of *Pan paniscus* in Zaire, *Am. J. Phys. Anthropol.* **53**:69–80.

Suzuki, A., 1969, An ecological study of chimpanzees in a savanna woodland, *Primates* **10**:103–148.

Teleki, G., 1973, *The Predatory Behavior of Wild Chimpanzees*, Bucknell University Press, Lewisburg.

Wrangham, R. W., 1975, The Behavioural Ecology of Chimpanzees in Gombe National Park, Tanzania, Ph.D. Dissertation, Cambridge University.

Wrangham, R. W., 1977, Feeding behavior of chimpanzees in Gombe National Park, Tanzania, in: *Primate Ecology* (T. H. Clutton-Brock, ed.), Academic Press, London, pp. 503–538.

CHAPTER 12

Interaction over Food among Pygmy Chimpanzees

SUEHISA KURODA

1. Introduction

Many authors have suggested that food sharing behavior arose in connection with the emergence of hunting in the hominization process (e.g., Pilbeam, 1972; Washburn and Lancaster, 1968). The presence of predatory and meat sharing behavior in the common chimpanzee (*Pan troglodytes*) has been thought to support this theory. In recent studies of the pygmy chimpanzee (*Pan paniscus*), however, it has become clear that these animals frequently share plant food, and the process of sharing large fruits resembles that of meat sharing in the common chimpanzee (Kano, 1980, Kuroda, 1980). Therefore, a more precise comparison of this behavior between species is needed to understand not only their phylogenetic relationship, but also the evolutionary development of food sharing, which was a necessary condition for the development of the sexual division of labor in human society.

Pygmy chimpanzees' food sharing often occurs in a complex context in connection with sexual or other behavior. Animals sometimes share abundantly available food. This chapter describes interaction over food among pygmy chimpanzees, with particular attention to the context surrounding food sharing.

SUEHISA KURODA • Laboratory of Physical Anthropology, Faculty of Science, Kyoto University, Sakyo-ku, Kyoto, 606 Japan.

2. Subjects and Methods

2.1. Subjects and Data Collection

Most of the data were collected from E group [previously called L group by Kuroda (1979)] at Wamba, in central Zaire [see Kuroda (1979) and Kano (1980, and this volume) for details of the study area]. Observations were made from October 1978 to February 1979 and from October 1979 to December 1979. Total direct observation time was 427 hr in the forest and 165 hr at the artificial feeding place.

During the study E group was provisioned with sugar cane [as it had been since 1976 (Kano, 1980, 1982)] and it consisted of 59 individuals (Fig. 1). The mothers of young adult males and adolescents were all known. Interactions and behavior were recorded by the *ad libitum* sampling method (Altmann, 1974) both at the artificial feeding place (AFP) and in the forest. Usually, a chimpanzee party (temporary association) visited the AFP in the morning and left before noon. The size of a party varied from two to nearly 59, but in most cases parties contained 10–35 animals. The adult sex ratio (male to female) was almost equal in most parties. When chimpanzees visited the AFP, we provided them with sugar cane intermittently 4–5 times a day. Sugar cane was cut into 30- to 40-cm pieces and placed against fallen trees at regular intervals so that all chimpanzees might obtain at least a few pieces (Fig. 2). We also occasionally provided pineapples.

2.2. Terminology

The following terms have been employed to describe various food-related behaviors:

1. Possession of food: physical contact with food or keeping food within hand's reach.
2. Food sharing: food transfer between chimpanzees without antagonistic interaction.
3. Food-taking bout (FTB): a bout or effort to take another's food or begging. An FTB is scored whether or not the taker or beggar is successful in the attempt to take food.
4. Food interaction unit (FIU): a series of any interactions including approach between a possessor and a nonpossessor where more than one FTB is involved. When a nonpossessor is eating shared food or has less than a 3-min break of interaction during which it is not eating, the FIU is considered as continuous. An FIU is

counted for each possessor–nonpossessor dyad when more than two chimpanzees interact over food at one time.

3. Results

3.1. Interaction over Naturally Occurring Food

3.1.1. Food Items

Pygmy chimpanzees at Wamba occasionally eat small mammals [flying squirrels (T. Kano, personal communication)], but interaction over them has not been observed. Fourteen items of naturally occurring food were observed to be begged for or transferred among pygmy chimpanzees (Table I). Forty FIUs and 60 FTBs were recorded for these foods (Table II). Two scrambles over honey were seen, but they were not recorded as FIUs because both cases were of short duration in the trees.

3.1.2. Interaction over Preferred Foods

Kano (1980) classified foods shared by pygmy chimpanzees into three types according to age–sex classes of the donors and recipients, abundance of food, and ease of preparation. Type food I (Table I) can be said to be highly valued (preferred) and is shared mostly among adults. The interaction pattern over type I food resembles that over meat among common chimpanzees.

Type I food attracted more chimpanzees for a longer time than either type II or type III foods. It was consumed by two or more adults in most cases [78.6%, $N = 14$ (Kano, 1980; Kuroda, 1980)]. The largest feeding–begging cluster over a *Treculia africana* fruit consisted of nine chimpanzees (Kuroda, 1980), and that over an *Anonidium mannii* fruit consisted of five individuals. Foods of types II and III did not attract more than two adults at a time, and even this was very rare. Sharing between adult males and a scramble by more than two chimps were observed only over type I food. These characteristics were also observed over pineapples, which were artificially provided (and were type I food) (Tables III and IV).

The differences in interactive behavior in pygmy chimpanzees over type I foods on the one hand and type II and type III foods on the other were similar to those that distinguish interaction over meat and over plant food among common chimpanzees (Goodall, 1968; Teleki, 1973). There

Adult males (16 chimps)
KUMA[1]', YASU[2], UDE[3], HACHI[4], YUBI[5], IBO[6], KURO[7], KAKE[8], HATA, IKA, JESS, HANAJIRO, MOPAYA, KURODASHI, KOSHI, MASU.

Adult females (14 chimps) and their offspring

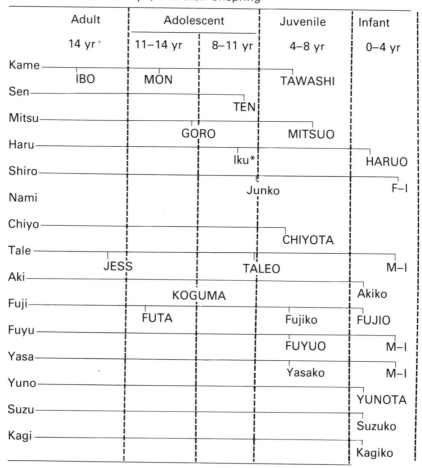

Young females (5 chimps)
Mugi, Mayu, Bihi, Lan, Ayei.**

Figure 1. Members of E group (December 1979). Numerals following names of males indicate dominance ranking order. The female adolescent Iku emigrated out in October 1979. The young female Ayei immigrated into the group in November 1979. FI, Female infant; MI, male infant. The total number of members is 59. (*) Emigrated out in Oct. 1979; (**) immigrated into E-group in Nov. 1979. (!) dominance ranking order among males.

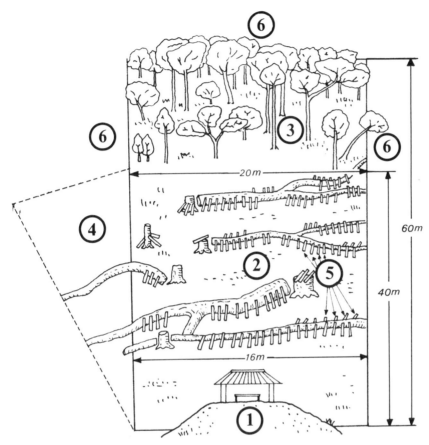

Figure 2. The artificial feeding place (AFP). (1) Observation hut; (2) feeding place; (3) area shrub-cleaned; (4) area expanded in November 1979; (5) sugar cane; (6) forest.

are also similarities in the distribution pattern: type I food (in *Pan paniscus*) and meat (in *Pan troglodytes*) are often divided into large portions forcibly at first; only later is it shared in small pieces with begging animals (Nishida *et al.*, 1979; Teleki, 1973; Kuroda, 1980). Thus type I food for pygmy chimpanzees corresponds to meat for common chimpanzees. Such characteristics of interaction seem to come from the high value placed on specific foods by individuals of each species.

3.1.3. Sharing of Abundant Food

Pygmy chimpanzees sometimes share abundant food among adults. This has not been observed in common chimpanzees.

Table I. Food Items Shared by Pygmy Chimpanzees

Item	Part eaten	Size	Availability
Type I			
Anonidium mannii	Fruit	Large	Rare
Pineapple[a]	Fruit	Large	Rare
Honey	—	—	Rare
Type II			
Sugar cane[a]	Pith	Medium	Rare
Type III			
Dialium corbisieri	Fruit	Small	Abundant
Dialium excelsum	Fruit	Small	Abundant
Dialium zenkeri	Fruit	Small	Abundant
	Leaf	Small	Abundant
Ficus sp.	Fruit	Small	Abundant
Gilbertiodendron dewevrei	Bean	Small	Abundant
Landorphia owariensis	Fruit	Medium	Abundant
Manniophyton fulvum	Leaf	Medium	Abundant
Musanga cercropioides	Fruit	Medium	Abundant
Synsepalum subcordatum	Fruit	Medium	Abundant
Tabernaemontana crassa	Fruit	Medium	Abundant
? (boaton)	Fruit	Large	Rare[b]

Fifteen species, 16 items

[a]Artificially provided.
[b]Chimpanzee's preference low.

In two cases foods from *Dialium* were shared. In the first, a young female with a *Dialium* twig was approached by a male who also had a *Dialium* twig. The male shared *Dialium* fruits from the female's twig; later 5 dyads of adults shared *Dialium* leaves. In both instances food was abundant and concentrated. It may be better to call these behaviors co-feeding rather than food sharing. Similar cases were observed where a male and female or females cofed on leaves of trees or vines.

Since low-value (over-ripe or unripe) sugar cane is readily shared among pygmy chimpanzees [as it is among common chimpanzees (Nishida, 1970; McGrew, 1975)], it might be expected that abundant food is also readily shared if a chimpanzee can get access to a possessor. The high interindividual tolerance between both females and between male and female pygmy chimpanzees (Kuroda, 1979, 1980) facilitates food sharing. It is unlikely that all FIU dyads are kin related, due to the fact that

Table II. Food Interactions over Naturally Occurring Food[a]

P	N	FIU	FIU[+]	FTB	FTB[+]	Agonistic FIU
M	M	1 (1)	1 (1)	4 (4)	4 (4)	0
M	F	1 (1)	1 (1)	1 (1)	1 (1)	0
F	M	4	2	4	2	0
F	F	11 (3)	6 (2)	13 (3)	7 (2)	0
M	I,J	1	0	1	0	0
F	I,J	12 (6)	1 (1)	18 (9)	1 (1)	2 (2)
I,J	I,J	3 (1)	2	7	3	0
Mo	Off	7	6	12	9	0
Total		40 (12)	19 (5)	60 (17)	27 (8)	2 (2)

[a]P, Possessor; N, nonpossessor; FIU[+], FIU (food interaction units) containing FTB[+] (successful food-taking bout); M, male; F, female; I,J, infant and juvenile; Mo, mother; Off, offspring. Number in parentheses is the value for type I food.

females emigrate from their natal unit groups when they are young (Kano, 1982; Kuroda, 1982).

It was often observed that chimpanzees begged or took other's food even when food was plentiful and also in the AFP.

3.2. Interaction over Artificially Provided Food

To analyze patterns of interaction over food, I rely mainly on data from provisioned animals. I observed 608 FIUs in which sugar cane and pineapple were offered to the subject (Table V).

Table III. Food Interaction over Pineapple[a]

P	N	FIU	FIU[+]	Percent	FTB	FTB[+]	Percent	Agonistic FIU	Food sharing FTB	Food sharing FTB[+]	Food sharing Percent
M	M	8	4	50.0	8	4	50.0	6	2	0	0
M	F	12	5	41.7	27	18	66.7	0	27	18	66.7
F	M	3	3	100.0	10	7	70.0	0	10	7	70.0
F	F	4	3	75.0	10	8	80.0	0	10	8	80.0
M	I,J	11	4	36.4	19	12	63.2	0	19	12	63.2
F	I,J	7	3	42.9	12	4	33.3	0	12	4	33.3
AdM	I,J	3	0	0	4	0	0	1	0	0	—
Total		48	22	45.8	90	53	58.9	7	79	49	62.0

[a]Key as in Table II. Ad, Adolescent.

Table IV. Comparison between Food Interactions over Pineapples and
Sugar Cane

Food	Agonistic FIU	Scramble by more than two	Food distribution	Begging–feeding cluster	Preference
Pineapple (type I)	+ + Among males only	+	+ + Among adults + mother–infant	1–6[a]	+ + Attracts chimps any time
Sugar cane (type II)	+ Rare	−	+ + Among adults + + mother–infant	1–3[b]	+ Ignored in fruitful season

[a]Mean 3.1; N = 20.
[b]Mostly one.

Interactions were classified according to their complexity and whether
they were antagonistic or peaceful. The simplest interaction was that in
which no attempt was made to either take food or beg on the part of the
animal. Sixty-seven of the FIUs involved this category. Interaction was
ended by the possessor's negative response to the approacher, in most
cases with the possessor moving away. In four cases the possessor left
one sugar cane for the approaching animal. The ordinary interaction consisted of an approach of one animal and an attempt to take food (469
FIU). More complex interactions involved sexual behavior, grooming, or
"baby sitting" by the approaching animal of the possessor's infant (72
FIUs).

3.2.1. Agonistic Interaction

In 66 FIUs a threat, physical attack, forcible taking of food, or other
aggressive behavior was observed. These FIUs were classified as agonistic
encounters. Slight grimaces or avoidance itself is not considered to be a
sign of agonistic interaction, because even a dominant possessor sometimes displayed grimaces and often avoided a subordinate's approach.

Agonistic FIUs were infrequent (10.9%, Table V). This frequency is
similar to that reported for common chimpanzees (Nishida, 1970; Teleki,
1973). Such FIUs consisted of almost one FTB alone. Agonistic interaction seldom occurred when the nonpossessor did the begging or was

Table V. Food Interactions over Artificially Provided Food[a]

P	N	Agonistic								Food sharing		
		FIU	FIU+	Percent	FIU	Percent[b]	FTB	FTB+	Percent	FTB	FTB+	Percent
M	M	39	23	59.0	12	30.0	41	23	56.3	33	17	51.5
M	F	134	77	57.5	6	4.5	211	128	60.7	210	125	59.5
F	M	21	13	61.9	1	4.8	73	62	84.9	72	62	86.1
F	F	101	52	51.5	8	7.9	158	83	52.5	152	81	53.3
M	AdM	26	11	42.3	7	26.9	28	11	39.3	25	8	32.0
AdM	M	2	1	50.0	1	50.0	2	1	50.0	1	0	0.0
F	AdM	11	2	18.2	0	0.0	19	3	15.8	19	3	15.8
AdM	F	2	1	50.0	0	0.0	2	1	50.0	2	1	50.0
M	IJ	77	33	42.9	5	6.5	126	50	39.7	123	48	39.0
IJ	M	5	3	60.0	2	40.0	6	3	50.0	4	2	50.0
F	IJ	46	13	26.0	2	4.3	73	22	28.6	73	22	30.1
IJ	F	12	5	41.7	3	25.0	13	5	38.5	12	3	25.0
AdM	IJ	8	0	0.0	2	25.0	10	0	0.0	10	0	0.0
IJ	IJ	16	2	12.5	7	43.7	19	2	10.5	15	2	13.3
Mo	Off	87	64	73.6	0	0.0	180	127	70.6	180	127	70.6
Off	Mo	12	10	83.3	4	33.3	12	10	83.3	8	6	75.0
Sib	Sib	10	6	60.0	6	60.0	10	6	60.0	8	2	25.0
Total		608	315	51.8	66	10.9	983	537	54.6	947	509	53.7

[a]Key as in Table II. Ad, Adolescent. Sib, Sibling.

[b]Agonistic FIUs as percent of total FIUs.

cautious and submissive. Food transfer in agonistic FIUs was observed 28 times, which accounts for only 5.2% (N = 537) of all food transfer. The most common forms of agonistic interactions were threatening and tug of war over food (28 and 22 FIUs, respectively). Though two exceptional FIUs included biting between females, physical attack was mild and rare (15 FIUs).

In some instances a subordinate animal took food from a dominant animal even while the latter was threatening it (11 FIUs). On the other hand, attempts by dominant individuals to snatch a subordinate's food often failed (14 FIUs). These instances also indicate a level of tolerance by dominant individuals and reduced intragroup aggression.

The presence of pineapples incited agonistic interactions among males more frequently than did sugar cane (Fisher's exact test, p = 0.006). It appeared that highly prized foods attracted many chimpanzees and significantly increased levels of tension. Participants, including dominant males, often engaged in a tug of war over pineapples wherein there was grimacing and screaming.

3.2.2. Food Taking and Begging

Attempts to get another's food vary from grabbing and snatching to begging with grimaces or whimpers. These attempts were classified into four types: (1) snatching or forcible taking, (2) cautious taking, (3) pilfering, (4) begging (Table VI).

a. Forcible Taking. This was infrequent (25 FTBs) and was performed by a dominant approacher except in 2 FTBs. The attempt to take food forcibly was often foiled by the possessor's resistance (7 FTBs).

b. Cautious Taking. This occurs slowly in front of the possessor. This behavior was frequently seen when a female took sugar cane from a male (50 FTBs). Cautious taking was extremely successful (88.2%, N = 127) if the desired food was not actually held by the possessor. Though an animal with more sugar canes than it could hold would consume those around it if another animal did not take them, surplus seems the main reason for high frequency of food sharing in these cases.

Animals who engaged in cautious taking (and begging) were subordinate in most cases (91.3%, N = 127). This behavior is considered near to begging because of the relative passivity of the subordinate individual when approaching its more dominant counterpart.

c. Pilfering. This was performed by young adults or adolescent males (7 FTBs). In this case a youngster approached a dominant male possessor from behind to take sugar cane, resulting in an attack by the possessor. However, usually the attack was brief.

d. Begging. This refers to a group of gestures that has been reported previously (Kano, 1980; Kuroda, 1980). It is similar to the begging reported in common chimpanzees (Goodall, 1968; Nishida, 1970; Teleki, 1973), except for the rare gesture of mimicking the eating of food (2 FIUs, 2 FTBs). Only rarely (2 FIUs) did staring result in food sharing. Scavenging scraps never elicited sharing of food.

The most frequent begging gesture is extending the hand in front of the possessor toward his food or mouth, and/or touching the possessor (70.2%, $N = 692$). Touching the possessor's mouth is the most frequent of these (75.4%, $N = 487$), but it was not seen when a male begged from another male. The typical "give me some" gesture of the common chimpanzee (Goodall, 1968; Nishida, 1970; Teleki, 1973), i.e., extending the hand with palm upward, was rare. Fingers are usually bent slightly and the palm was not necessarily upward. "Kissing" the mouth or food is rare in pygmy chimpanzees even in infants (17 FTBs).

Begging was most frequently seen in an offspring begging from its mother (Table VI). The age of begging offspring varies from 1 to 16 years of age (Table VII). Youngsters are more successful than older animals in begging. Infants younger than 2 years rarely beg, but 3-year-olds actively beg. Begging is most frequent in the 4- to 5-year-old group. Begging occurs at a much earlier age in common chimpanzees. Begging has been observed in 3-month-old infant common chimpanzees and most begging for natural food is seen in 18- to 24-month-old animals. Modal age for begging of provisioned food is 30–36 months (Silk, 1978). Pygmy chimpanzee infants appear to have a longer period of dependence on their mothers than do common chimpanzees. Long-term dependence is also seen in the fact that young adults and adolescents were almost always near their mothers. When attached, mothers protected their young adult and adolescent offspring (Kano, 1982; Kuroda, 1982) as well as their infants.

Most beggars were subordinate to possessors except for male–female combinations, in which there is no clear dominance relationship. Dominant individuals did not behave aggressively to beggars except in 4 FIUs. Even in those cases where possessors reacted aggressively they eventually shared food. In only one instance did a dominant individual act aggressively after a begging episode.

e. Taking and Begging of Abundant Food. Pygmy chimpanzees sometimes begged or took another's food despite the presence of abundant food. On several occasions an adolescent male stared at an adult male who was eating even when the adolescent had his own sugar cane (5 FIUs). Young females tended to beg from dominant males even when sugar cane was available to them. We interpret a number of such events as behaviors that reinforce social bonds between individuals.

Table VI. Nonpossessor's Behavior and Possessor's Response[a]

P	N	A	Ag-1	Ag-2	Ag-3	P-1	P-2	P-3	P-4	P-5	Total	Ag-1
					Nonpossessor's behavior[b]							
M	M	5	2	5	1	17	6	2	2	1	41	5
M	F	21			1	50	10	24	10	95	211	6
F	M			1		1	1	7		63	73	1
F	F	10		6		11	17	33	7	74	158	2
M	AdM	5			3	11		9			28	7
AdM	M			1						1	2	
F	AdM	1				1		13	1	3	19	
AdM	F	1				1					2	
M	I,J	4		1	2	16	6	20	17	60	126	3
I,J	M			2			2		2		6	
F	I,J	1				1	3	26	5	37	73	2
I,J	F	6	1				3	1	1	1	13	2
AdM	I,J	7						2	1		10	2
I,J	I,J	4	1	3		3	2	3	2	1	19	6
Mo	Off	1				6	7	10	5	151	180	
Off	Mo			4		4	4				12	
Sib	Sib	1		2		5			1	1	10	4
Total		67	4	25	7	127	61	150	55	487	983	40

[a] Key as in Table II. Sib, Sibling. Ad, Adolescent.
[b] A, Approach only; Ag-1, attack or threat; Ag-2, snatch or tug; Ag-3, pilfer; P-1, cautious taking of food at possessor's front; P-2, cautious taking of food held by the possessor; P-3, staring; P-4, scavenging of scrap; P-5, active begging.
[c] T, Tug food; R-1, pull food; R-2, touch, grasp, or draw food; R-3, touch beggar; R-4, give scrap, R-5, turn to back or side; R-6, move away; R-7, ignore begging.

3.2.3. The Possessor's Response

a. Rejection Response. The possessor never approached a nonpossessor except in the case of mother and infants. The possessor's responses are listed in Tabe VI. The term "leaving food" was applied to the cases where a possessor left one cane for the beggar despite the former's capacity to remove all the cane. Females frequently gave chewed, sugar-depleted, scrap to beggars, which usually discouraged begging.

Dominant possessors occasionally rejected beggars with mildly aggressive behavior (37 FTBs), but the most frequent method of rejection was simply ignoring the approach. Begging by immatures often tended to be ignored by adults. Female begging was often simply ignored by other

Table VI. *(Continued)*

T	Possessor's rejecting response[c]								Possessor's sharing response[d]					
	R-1	R-2	R-3	R-4	R-5	R-6	R-7	Total	S-1	S-2	S-3	S-4	S-5	Total
5	1	4			2	3	4	24	15	1			1	17
	3	6		4	3	34	35	91	46	7	69	1	2	125
					2		8	11	1	1	60			62
6	3			16		4	46	77	11	13	47	4	6	81
		3			3	1	6	20	8					8
1							1	2						0
					2	4	10	16	1		2			3
						1		1	1					1
1	4	7	5	6	3	4	49	82	14	5	29			48
2						1	1	4		2				2
		1		6	3	3	36	51	1	2	19			22
					6	2		10		3				3
					6	2		10						0
1					1	6	3	17	2					2
	1	4		4	3	1	40	53	6	6	115			127
4	2							6	4	2				6
2					1		1	8	2					2
22	15	24	5	36	23	74	244	483	112	42	341	5	9	509

[d]S-1, Allow cautious taking of food not physically contacted; S-2, allow cautious taking of food in physical contact; S-3, allow beggar to take food or share food; S-4, break food into two; S-5, leave food.

females rather than eliciting aggressive responses. Males, however, showed rejection as frequently by moving away as by simply ignoring beggars.

b. *Food Sharing.* Food sharing occurred in 509 (53.7%) of 947 FTBs. Sharing of food that corresponds to the "food transfer" of McGrew (1975) occurred in 397 (52.%) of 753 FTBs. These are classified according to the age–sex class of the participants in Table V.

There is a clear contrast between the food sharing patterns of the two species of *Pan.* Whereas female common chimpanzees seldom share either plant food or meat with others except their infants (McGrew, 1975; Teleki, 1973), pygmy chimpanzee females frequently share food with unrelated animals as well as with their infants. This indicates their high degree of sociality, which also appears in other social interactions, such as grooming, coattack by females, and aggregation patterns (Kano, 1980; Kuroda, 1980; Handler et al., this volume).

Table VII. Begging/Taking and Sharing between Mother–Offspring Pairs

Offspring's age, years	Number of beggars	FTB	FTB+	FTB+/FTB
0–1	0	0	0	—
1–2	3	6	5	0.83
2–3	6	41	36	0.88
3–4	3	21	19	0.90
4–5	2	42	30	0.71
5–6	1	19	9	0.47
6–7	1	7	3	0.43
7–8	3	15	7	0.47
8–11	2	3	2	0.66
11–14	1	1	0	0.00
14–17	1	24	16	0.67
Total	23	179	127	0.71

Most plant food sharing in common chimpanzees occurs in the direction of mothers to infants. At Gombe Stream, 79% of banana sharing took this direction (McGrew, 1975). However, food sharing in this direction amounts to only 25.0% of all sharing among pygmy chimpanzees. If we consider only the sharing corresponding to McGrew's "food transfer," the index rises to 30.5%, which is still well below that of common chimpanzees.

The frequency of each individual's food sharing varied greatly. There was, however, a clear tendency for dominant males to share food more frequently. Most often females approached alpha and beta males to take or beg food (Table VIII). This share accounted for 52.0% of all sharing by males with females.* Immatures frequently took or begged food from dominant males (Table IX). This accounted for 66.7% of all sharing by males with immatures.

Frequent food sharing by dominant males is due not only to the fact that they obtain more food, but also because they are the center of attention. When low-ranking males obtained sugar cane, they scattered and sat on the periphery of the AFP or in nearby trees, whereas high-ranking males tended to remain with females at the center of the party, in the center of the AFP. If a beggar approached a low-ranking male who was eating food, the latter moved away. A typical example could be seen in subadult males who do not have a position in the dominance hierarchy.

*With the value for sharing by one individual IBO, this accounted for 83.2% of all sharing by males with females. IBO was the sixth-ranking male, but was the most dominant male in the subgroup that formed the nucleus of the parties that visited the AFP frequently.

Table VIII. Food Sharing by Possessor Males with Nonpossessor Females[a]

	KUMA	YASU	HACHI	YUBI	IBO	KURO	KAKE	JESS	IKA	Total
Kame	2/3	15/19	2/4	0/1	—	0/1	—	—	0/1	19/29
Sen	—	0/1	—	1/2	—	1/1	—	—	—	2/4
Haru	2/2	—	—	—	1/3	—	1/2	—	—	4/7
Mitsu	0/4	0/8	—	—	1/1	—	0/4	—	—	1/17
Shiro	2/2	2/8	1/3	—	1/1	—	—	—	—	6/14
Nami	2/2	8/9	—	—	0/1	—	—	—	—	10/12
Chiyo	2/3	2/3	—	3/3	—	—	—	—	—	7/9
Tale	1/1	—	—	0/1	—	—	—	—	—	1/2
Yasa	0/4	—	—	—	—	—	—	—	—	0/4
Suzu	10/11	1/2	—	1/1	5/8	—	0/3	—	—	17/25
Kagi	—	—	—	0/1	—	—	—	2/2	—	2/3
Mayu	1/2	4/8	1/5	—	11/12	—	—	1/1	—	18/27
Bihi	3/4	6/8	1/1	—	17/19	—	1/1	—	—	28/33
Mugi	0/1	0/1	—	—	—	—	—	1/2	—	1/4
Lan	—	—	—	—	—	—	—	3/10	—	3/10
Ayei	—	2/3	—	—	3/5	1/1	—	—	—	6/9
Total	25/38	40/70	5/13	5/9	39/50	2/3	2/10	7/16	0/1	125/210

[a]Names of males are given in capital letters. Values are ratios of successful FTBs to total peaceful FTBs.

Table IX. Food Sharing by Possessor Males with Nonpossessor Immatures[a]

	KUMA	YASU	HACHI	YUBI	IBO	KURO	KAKE	HATA	JESS	Others	Total
Iku	0/1	1/2	—	0/1	—	—	—	—	—	—	1/4
Junko	0/2	—	—	0/3	1/2	—	—	—	0/1	—	1/8
TEN	1/3	2/5	—	—	1/2	0/1	—	0/3	—	—	4/14
TALEO	0/1	6/8	—	0/2	—	—	—	—	0/1	—	6/12
Yasako	—	—	—	—	—	—	—	—	0/3	—	0/3
CHIYOTA	1/1	2/13	—	—	—	—	—	—	—	0/1	3/15
TAWASHI	8/15	6/8	0/3	—	1/1	1/1	1/1	1/12	—	—	18/41
Fujiko	—	—	—	—	—	—	—	—	0/2	—	0/2
MITSUO	0/1	0/1	—	—	—	—	2/3	—	—	—	2/5
Kagiko	—	—	—	—	—	—	—	—	5/10	1/1	6/11
Suzuko	5/5	—	—	0/1	—	—	—	—	—	—	5/6
HARUO	—	—	—	—	—	—	1/1	—	—	1/1	2/2
Total	15/29	17/37	0/3	0/7	3/5	1/2	4/5	1/15	5/17	2/3	48/123

[a]Key as in Table VIII.

When they obtained a pineapple, they would rapidly run away to a place where others would not bother them. By contrast, high-ranking males did not leave the central area, though they did avoid beggars by moving short distances. If they possessed prized foods, they were often surrounded by beggars and compelled to share their food.

   c. Voluntary Food Sharing. Although most food sharing occurred as a result of allowing the approacher to take food, there were several cases in which the possessor volunteered to share food (7 FIUs).

   In 1 FIU, a female possessor broke her sugar cane into two and shared it with another female who was simply staring at her (but not actively begging). In 1 FIU, a young female put a bit of pineapple on a male's hand after she copulated with him.

   Other cases were observed in mother–infant pairs. In one a mother put a bit of *Musanga cercropioides* fruit into the mouth of an 18-month-old infant in response to its incipient begging. The other 2 FIUs were by the same mother–infant pair. The mother responded to the infant's whimpering in one case and to its approaching in another case after which the mother offered a piece of sugar cane to the infant. In the final case a mother held out an unpeeled cane for her 3-year-old infant, which had burst into a temper tantrum.

   It seems that a mother's active response to her infant's incipient begging prompts infants to beg more actively. This hastens the weaning of infants.

3.2.4. Interactions over Food with Genitogenital Contact

   a. Copulation. A unique aspect of pygmy chimpanzee behavior is genitogenital contact, which appears in various contexts of social interaction and various age–sex class combinations (Savage and Bakeman, 1977; Kano, 1980, 1984; Kuroda, 1982; Kitamura, 1983, and 1984; Handler *et al.*, this volume).

   Genitogenital contact between a male and a female was observed in 26 FIUs (Table X), in which copulation and food sharing occurred successively. One example occurred on January 3, 1979. A young female approached a male, who was eating sugar cane. They copulated in short order, whereupon she took one of the two canes held by him and left. In another case a young female persistently presented to a male possessor, who ignored her at first, but then copulated with her and shared his sugar cane.

   Most of these interactions were observed between a male possessor and approaching estrous females (22 FIUs). In one case the solicitor was an anestrous female. There was only one case in which a female possessor

Table X. Food Interactions with Genitogenital Contact

A. Copulation

P	N	FIU	FIU+	Percent	FTB	FTB+	Percent	During/after copulation		
								FTB	FTB+	Percent
M	F	23	17	73.9	58	43	74.1	46	37	80.4
F	M	1	1	100.0	4	4	100.0	4	4	100.0
AdM	F	2	1	50.0	2	1	50.0	2	1	50.0

B. Genital rubbing

P	N	FIU	FIU+	Percent	FTB	FTB+	Percent	During/after genital rubbing		
								FTB	STB+	Percent
F	F	24	21	87.5	51	33	64.7	46	35	76.1

C. Other

Behavior	P	N^b	FIU	FIU+	Agonistic FIU	FTB	FTB+
Immature copulation	F	I,JM	3	2	0	9	7
Genital rubbing	F	IF	1	0	0	1	0
Mounting with penile erection	M	IF,JF	4	2	0	8	4
Mounting	M	AdM	2	1	1	2	1
	M	JF	2	2	1	2	2
	F	F	1	1	1	1	1
Rump–rump contact[c]	M	M	1	1	1	1	1
	M	AdM	1	1	0	1	1
Total			15	10	4	25	15

[a]Key as in Table II. Ad, Adolescent.
[b]IM, JM, IF, JF: male infant, male juvenile, female infant, and female juvenile.
[c]Kano (1980); Kuroda (1980).

copulated with an approaching male after which she shared her food with him. It appears that females are more successful in getting food from males if they copulate first. The proportion of successful FTBs after copulation (80.4%) is significantly higher than that of others between a male possessor and a female beggar ($\chi^2 = 13.8$, df $= 1$, $p < 0.001$).

In addition to these cases, there were similar ones involving both estrous and anestrous females, but in which copulation did not occur (4 FIUs). Numerous observations suggest that in these instances the female's purpose was to obtain food, not simply to copulate. Presenting and copulation buffer the food interaction and render the males more tolerant. In common chimpanzees, estrous females often gain meat from males more frequently than anestrous females (Teleki, 1973), but instances of actual copulation during food sharing are very few (Goodall, 1972). However, according to data from common chimpanzees in the Mahale Mountains, females that copulated were more successful in obtaining meat because they remained near the male possessors for a longer period (Takahata *et al.*, 1982).

b. Genital Rubbing. Genitogenital rubbing (Kano, 1980; Kuroda, 1980) is commonly seen in interactions over food (24 FIUs, Table X). FTBs after or during this behavior and FTBs during courtship were more successful than others between females (χ^2 = 13.8, df = 1, p < 0.001). The solicitor of genital rubbing was the beggar in 12 of 15 instances. In three instances genitogenital rubbing was ignored by an animal trying to obtain food from another. In two cases females who were soliciting genital rubbing succeeded in acquiring the second female's sugar cane without further interaction.

As genital rubbing serves to ease interindividual tension (Kuroda, 1980), it works also to lessen tension during food transfer between females. This is important in their food distribution, because food sharing in FIUs with genital rubbing accounted for nearly half of all sharing between females (44.4%).

c. Other Genitogenital Contact. Genitogenital contact was observed in FIUs of other age–sex combinations (Table X). Interaction between a male possessor and a female infant or juvenile was observed in 4 FIUs. Conversely, immature copulation by a male infant or juvenile with a female possessor was observed in 3 FIUs. Mounting by a possessor without penile erection was also seen before allowing an approacher to take food, and additionally in one FIU a juvenile took sugar cane from a male who was copulating with his mother.

3.2.5. Other Interactions

a. Grooming. In addition to genitogenital contact, grooming, "baby sitting," and temper tantrum were observed during interactions over food (Table XI). Chimpanzees strategically, though infrequently, groomed food possessors, which gave them access to the counter parts' food.

b. "Baby Sitting." The pygmy chimpanzees sometimes groomed or briefly played with a possessor's infant before or during begging. Three adults seemed successful in getting food from mothers with this behavior, while subadult males' begging was ignored in spite of their baby sitting behavior. In addition to brief grooming behavior as greeting or appease-

Table XI. Food Interactions with Other Behavior[a]

	P	N	FIU	FIU[+]	FTB	FTB[+]
Grooming	M	F	2	2	3	2
	M	JM	1	1	12	1
Baby sitting	F	M	2	2	15	15
	F	AdM,JM	2	0	3	0
	F	F	1	1	5	3
Temper tantrum	Mo	Off	8	5	33	10
	M	JM	1	0	1	0
Total			17	11	72	31

[a]Key as in Table II. Ad, Adolescent.

ment when a subordinate attempted to gain access to a dominant (Kuroda, 1980), baby sitting can be used to appease a mother. These behaviors in FIUs function to increase a possessor's tolerance and to allow the beggar to remain in the vicinity in order to beg for food.

c. *Temper Tantrums.* Pygmy chimpanzee infants or juveniles burst into temper tantrums when they were strongly rejected in their attempts to gain food. This was also observed in common chimpanzees (Goodall, 1968). Temper tantrums were seen in 3- to 5-year-old immatures during 8 FIUs between mother and offspring, and in 1 FIU between a male possessor and a juvenile. Temper tantrums did not seem more effective for food acquisition than ordinary begging (Table XI).

4. Discussion

4.1. Food Sharing Behavior in Pygmy and Common Chimpanzees

The food sharing behavior of pygmy chimpanzees differs from that of common chimpanzees in the following ways: (1) adult pygmy chimpanzees often share plant food; (2) females share food with individuals other than their own infants; (3) males seldom share food among themselves; (4) adults often share food that is in large supply; (5) sexual behavior often occurs in connection with food sharing. However, in both pygmy and common chimpanzees interaction patterns over highly prized foods are similar. Several common tendencies include: (1) increased frequency of food sharing by high-ranking males; (2) uneasiness of the individual in possession of the highly valued food; (3) rare occurrence of physical attack brought on by food interaction; (4) importance of females' sexual behavior in obtaining food from a male.

It is likely that regular hunting brought rapid progress in the food sharing of early hominids. However, it is also likely that behavioral and sociological preadaptations for sharing behavior had been acquired before this, as suggested by the fact that most elements of meat sharing in the common chimpanzee can be found in the plant food sharing of the pygmy chimpanzee. We will discuss such preadaptive features or rudiments for systematic food sharing in *Pan* behavior along three important food distribution patterns in the pygmy chimpanzee, i.e., food division (by tugging), begging-sharing, and cofeeding.

4.2. Food Division

Food division occurs mostly among males for highly valued food. As each male's force and ability are nearly equal, it seems beneficial for the possessor to divide his food rather than to struggle or run away and risk losing all of the food (Nishida, 1981). Dominance ranking order does not function well in the scramble over food. This distribution is the product of balance of power among males.

The fact that older males of the common chimpanzee are more successful in obtaining meat irrespective of their involvement in the kill, can be related to their increased size and strength. We postulate that the possessor and others surrounding him are uneasy about the potential for a struggle over the kill. In such a situation it is possible that the more experienced and confident animals will obtain the food and be the center of the feeding cluster. As McGrew (1979) points out, this fact might constitute a preadaptation for the development of a distributive role.

Among common chimpanzees in the Mahale Mountains, an advanced form of food division was observed. A lone chimpanzee had not been observed to break his sugar cane before eating it. However, when other chimpanzees approached him, he sometimes divided his large piece of sugar cane into two pieces before eating it. Nishida (1981) postulated that this was done in order to avoid conflict with others who approached him. Such conflict risks not only his food but also male bonding which is essential in maintaining his unit-group against the severe inter-unit-group antagonism (Goodall *et al.*, 1979; Nishida, 1981).

Male pygmy chimpanzees have never been observed to divide their food before it is necessary, as required by others. Lack of advanced food division may be due to their short and mildly aggressive interactions over food, in which the possessor is less likely to lose it, and due to relatively weak bonding among males (Kano, 1980; Kuroda, 1979, 1980, 1982). As intergroup relations are not as agonistic as in the case of common chimpanzees and as pygmy chimpanzees are almost free from predators, strong male bonding is not necessary (Kuroda, 1982). It is possible that food distribution of this kind will evolve into even division for participants with an increase in their strength and foresight, and the increasing importance of male bonding.

4.3. Begging and Sharing

Begging seems most likely to be derived from an infant's behavior towards its mother. It indicates a kind of submissive behavior which inhibits the possessor from attacking the beggar. In relation to this, one might hypothesize that the common chimpanzee's appeasement or greeting behavior, seen in extending the hand, is a ritualized form of begging (Nishida, 1981).

Though many questions remain unanswered, some begging and taking of otherwise abundant food by pygmy chimpanzees can be interpreted as attempts to improve social relationships. This behavior may play an important role because there is a possibility that food sharing occurs more frequently between more familiar pairs. Thus, such begging-sharing interaction might contain rudiments of the evolution of hominid habits to exchange food or co-feed which promote social integration.

Food sharing among adult chimpanzees may occur because persistent begging annoys the possessor, leading him to share his food in order to alleviate the annoyance (Nishida, 1981), or because a female's selection of a mating partner has shaped such altruistic behavior in males (Tutin, 1979).

Additional factors should be considered in a discussion of food sharing. One is the inhibition of aggression produced by the action of begging. Another is the sociological one that high-ranking males are not able to escape from the party. They are the focus of social attention, a fact that accrues to and reinforces their high social status. It may be very difficult for them to feed without some approach by other group members. If they cannot escape or drive away others, they must share their food. Thus, food sharing may have developed in situations where sociality supersedes strict feeding considerations. It may be possible that food sharing will acquire the additional meaning of displaying high status and sociality in such situations.

4.4. Cofeeding

Food sharing by male and female pygmy chimpanzees seems to contain rudiments of reciprocity between the two sexes. Highly prized foods tends to be acquired by males and shared with females. Possession of less highly valued food does not seem to cause tension between the possessor and the nonpossessor, therefore it can be shared by females with males or fed upon by both. Higher interindividual tolerance (Kuroda, 1979, 1980, 1982) and females' high sexual activity smooth these distribution patterns.

4.5. Hypothesis of Preadaptation to Systematic Food Sharing

Hypothesizing from the food sharing behavior of pygmy chimpanzees, which are dense forest dwellers, it may be that early forest-dwelling hominids already engaged in a certain degree of food sharing.

If we consider that some surplus is required for food sharing in *Pan*, the generally rich food resources of the forest seem a likely factor to shape food sharing in its elementary form such as that involves the begging–sharing interaction and high interindividual tolerance. Prolongation of estrus in hominid females, as in pygmy chimpanzees (Kano, 1984; Kuroda, 1982), may also have occurred in the rich forest environment. This may have contributed to formation of a cohesive society, which was later advantageous for protection against predators in more open country. If males hunt, the estrous females could attract them back and perhaps share in the meat. Daily food exchange and a high degree of cooperation between males and females should have developed on the basis of such preadaptation in a savanna life.

ACKNOWLEDGMENTS. This study was conducted as part of the Ryukyus University African Primatological Expedition in cooperation with l'Institut de Recherche Scientifique du Zaire. I would like to express my hearty thanks to all those who made this study possible. I thank Prof. T. Kano for his leadership and great help; Prof. J. Itani, J. Ikeda, and T. Nishida for their guidance and support during the course of this study; Drs. A. Mori, K. Kitamura, and E. O. Vinberg for useful discussions and kind help; and Dr. P. J. Asquith for her critical comments on an early version of this chapter. I am greatly indebted to l'Institut de Recherche Scientifique. I am also indebted to my assistants. N. Batolumbo, B. Bokonda, B. Mbale, and I. Lokake, and all the people of Wamba. I thank the Fathers of Yalisele and Boende Missions, R. Smith, members of the American Peace Corps at Mbandaka and Boende, M. Ito, and K. Nakatsuka for their encouragement and kind help, and T. N. Robb for his help in translating this chapter into English.

References

Altmann, J., 1974, Observational study of behavior: Sampling method, *Behavior* **49**:227–267.

Goodall, J., 1968, The behavior of free-living chimpanzees in the Gombe Stream Reserve, *Anim. Behav. Monogr.* **1**:161–311.

Goodall, J., 1972, *In the Shadow of Man*, Houghton-Mifflin, Boston.

Goodall, J., Bandora, A., Bergmann, E., Busse, C., Matama, H., Mpongo, E., Pierce, A., and Riss, D., 1979, Intercommunity interactions in the chimpanzee population of the Gombe National Park, in: *The Great Apes* (D. A. Hamburg and E. R. McCown, eds.), Benjamin/Cummings, Menlo Park, California, pp. 13–53.

Kano, T., 1980, Social behavior of wild pygmy chimpanzees (*Pan paniscus*) of Wamba: A preliminary report, *J. Hum. Evol.* **9**:243–260.

Kano, T., 1982, The social group of pygmy chimpanzees (*Pan paniscus*) of Wamba, *Primates* **23**(2):171–188.

Kano, T., 1984, Reproductive behavior of the pygmy chimpanzees (*Pan paniscus*) of Wamba, Republic du Zaire, in: *Sexuality of the Primates* (T. Maple and R. D. Nalder, eds.), Van Nostrand Reinhold Co., New York (in press).

Kitamura, K., 1983, Pygmy chimpanzee association patterns in ranging, *Primates* **24**(1):1–12.

Kitamura, K., 1984, Genito-genital contacts in the pygmy chimpanzee (*Pan paniscus*), in: *Primate Sexuality* (T. Maple and R. D. Nalder, eds.), Van Nostrand Reinhold Co., New York (in press).

Kuroda, S., 1979, Grouping of the pygmy chimpanzees, *Primates* **20**(2):161–183.

Kuroda, S., 1980, Social behavior of the pygmy chimpanzees, *Primates* **21**(2):181–197.

Kuroda, S., 1982, *The pygmy chimpanzee: Its secret life*, Chikuma-shobo, Tokyo (in Japanese).

McGrew, W. C., 1975, Evolutionary implications of sex differences in chimpanzee predation and tool use, in: *The Great Apes* (D. A. Hamburg and E. R. McCown, eds.), Benjamin/Cummings, Menlo Park, California, pp. 441–463.

McGrew, W. C., 1979, Patterns of plant food sharing by wild chimpanzees, in: *Proceeding of the 5th Congress of the International Primatological Society, Nagoya, Japan*, Karger, Basel, pp. 304–309.

Nishida, T., 1970, Social behavior and relationship among wild chimpanzees of the Mahali Mountains, *Primates* **11**:47–87.

Nishida, T., 1981, *The World of the Wild Chimpanzee*, Chuokoronsha, Tokyo (in Japanese).

Nishida, T., Uehara, S., and Nyundo, R., 1979, Predatory behavior among wild chimpanzees of the Mahali Mountains, *Primates* **20**(1):1–20.

Pilbeam, D., 1972, *The Ascent of Man*, Macmillan, New York.

Savage, E. S., and Bakeman, 1976, Sexual morphology and behavior in *Pan paniscus*, in: *Recent Advances in Primatology* (D. J. Chivers and J. Herbert, eds.), Academic Press, New York, pp. 613–616.

Silk, J. B., 1978, Patterns of food sharing among mother and infant chimpanzees at Gombe National Park, Tanzania, *Folia Primatol.* **29**:129–141.

Takahata, Y., Nishida, T., Hayaki, H., and Takasaki, H., 1982, Mahale Mountains Chimpanzee Project Ecol. Rep. No. 2.

Teleki, G., 1973, *The Predatory Behavior of Wild Chimpanzees*, Bucknell University Press, Lewisburg.

Tutin, C. E. G., 1979, Mating patterns and reproductive strategies in a community of wild chimpanzees (*Pan troglodytes*), *Behav. Ecol. Sociobiol.* **6**:29–38.

Washburn, S. L., and Lancaster, C. S., 1968, The evolution of hunting, in: *Man the Hunter* (R. B. Lee and I. DeVore, eds.), Aldine-Atherton, Chicago, pp. 293–303.

CHAPTER 13

Social Organization of Pan paniscus in the Lomako Forest, Zaire

ALISON BADRIAN AND NOEL BADRIAN

1. Introduction

Of the four great apes—the pygmy chimpanzee (*Pan paniscus*), common chimpanzee (*Pan troglodytes*), gorilla (*Gorilla gorilla*), and orangutan (*Pongo pygmaeus*)—the latter three are known to exhibit marked differences in the way their respective societies are organized and structured. Common chimpanzee society consists of a flexible community of strongly bonded adult males and more solitary females (Wrangham, 1975; Tutin and McGinnis, 1981). Within chimpanzee society, the size and composition of groups ["temporary associations," "bands," "parties" (Goodall, 1968; Reynolds and Reynolds, 1965)] are unstable. The only enduring bonds within a community are those between mothers and their offspring (Goodall, 1968). The social structure of the gorilla consists of cohesive groups comprising a dominant adult male, a variable number of other males, adult females, and young (Schaller, 1963). Solitary males may be found on the fringes of the group and will associate with its members on occasion (Harcourt, 1979). Orangutans are largely solitary except for mothers with dependent offspring and temporary male–female consortships (Mac-Kinnon, 1971, 1974; Rijksen, 1978).

The fourth and least studied of all the great apes, *Pan paniscus,* has a social organization that is only beginning to be understood from data accumulated in the last 10 years of field studies (Horn, 1977, 1980; Kano,

ALISON BADRIAN AND NOEL BADRIAN ● Department of Athropology, State University of New York at Stony Brook, Stony Brook, New York 11794.

1979, 1980, 1982*a,b;* Kuroda, 1979, 1980; Badrian and Badrian, 1977, 1978, 1980; Badrian *et al.,* 1981; Susman *et al.,* 1980).

The data presented in this report were gathered at the Lomako Forest study site over a period of 18 months between November 1980 and June 1982. The results indicate that pygmy chimpanzee society consists of relatively stable geographic communities within which predominantly mixed bisexual foraging groups are found. The groups change in size and composition with time, and the Lomako pygmy chimpanzees manifest several behavioral strategies which can be compared both with other populations of *Pan paniscus* and with common chimpanzees from various locales.

2. Methods

The study site is in the Lomako Forest, Zaire, and has been described by Badrian and Badrian (1977) and Badrian and Malenky (this volume).

Although pygmy chimpanzees were not hunted for meat in the area of the study site (Badrian and Badrian, 1977; Susman *et al.,* 1981), they were extremely timid and difficult to observe, especially in the early part of the study period. No attempt was made to provision the animals, as we were concerned that this might have a deleterious effect on their regular patterns of ranging and their social behavior, as has been discussed by Wrangham (1974) for common chimpanzees.

Old elephant trails traversing the study site, primarily from north to south, were modified and extended. Additional trails were cut to link the existing ones; wherever possible, existing animal paths were used. The resulting network of trails was then marked, numbered, and mapped. These trails were patrolled routinely in order to locate pygmy chimpanzees and to collect data on food remains, nests, and other indirect evidence related to pygmy chimpanzee activity and ranging patterns. Trees bearing ripe fruit from known food sources (Badrian *et al.,* 1981; Badrian and Malenky, this volume) were visited regularly.

This study population included members of at least two distinct communities ["unit groups" (Nishida, 1968)]. The home range of these two communities overlapped to some extent in the middle of the study site (Fig. 1). These ranges were estimated from the distribution of observations, feeding remains, and nest sites within the areas. As it was only toward the end of the study period that positive identication of individuals could be made, data from both communities have been combined unless otherwise specified.

Data were gathered on group size and composition and on all social interactions between individuals of both sexes and all age classes. Age

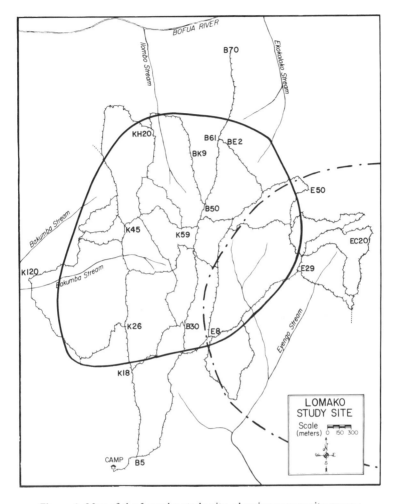

Figure 1. Map of the Lomako study site, showing community ranges.

and sex of individuals were determined on the basis of size and behavioral characteristics. Four age categories were determined:

1. Infants: those young still unweaned and requiring transport from their mothers during travel.
2. Juveniles: those young who were independent of their mothers for transport but always found in proximity to her. Juveniles were occasionally carried by their mothers in threatening situations or during social excitement.

3. Adolescents: those subadults of both sexes that were independent, and might be observed alone or in association with other adults as well as their mothers. Adolescent females exhibited small, erratic estrous swellings. Adolescents of both sexes still showed marked circumanal tail tufts, as did many adult males and some adult females (the presence or absence of tail-tuft was difficult to see in females with large estrous swellings).
4. Adults: full-grown individuals, with the females exhibiting fully developed sexual swellings.

There were certain problems with this broad classification system. As Coolidge (1933) has noted, young pygmy chimpanzees are precocious, and thus the distinction between infant and juvenile was often difficult to make. Very "independent" juveniles had on occasion to be reclassified as infants when they were subsequently seen to suckle and be transported by their mothers.

3. Results

3.1. Population Density and Group Size

The Bakumba community (one of the two communities regularly utilizing the study area) was estimated to contain approximately 50 individuals. This community ranged over an area of 22 km^2 or more, resulting in an estimated population density of 2/km^2. This figure is provisional, and may need to be revised when identified individuals can be followed for extended periods. The size of pygmy chimpanzee groups ["bands" (Reynolds and Reynolds, 1965) and "parties" (Wrangham, 1975)] encountered in our study area between November 1980 and June 1982 varied from one to approximately 50 ($\bar{X} = 7.6$). Figure 2 depicts group size data for our study site (as well as for Wamba; see discussion). The information for the Lomako study area presented in Fig. 2 utilizes only sightings when accurate counts could be made ($N = 268$). The most frequently encountered group in the Lomako forest (the modal group size) contained 2–5 individuals. The second most commonly observed grouping was in the 6–10 category.

3.1.1. Seasonal Variation in Group Size

There was a tendency for group size in the Lomako to fluctuate with time. Table I shows the frequency distribution of group size during each

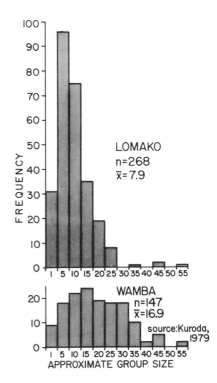

Figure 2. Frequency of *Pan paniscus* group
sizes at Lomako and Wamba.

month of the study period. Variations in group size appeared to be related
to the production cycles of particular food plants rather than to seasonality
(see Badrian and Malenky, this volume).

3.1.2. Group Size and Activity Level

During the initial part of the research period, we observed that the
reaction of pygmy chimpanzees to observers was related to group size.
Small groups (2–5) would usually leave quickly and quietly when encountered. Medium-sized groups vocalized and displayed toward the observers (shaking or dropping branches, slapping or stamping on boughs,
running bipedally or arm swinging directly above our heads, and urinating
or defecating). The group would then resume normal activity or leave
noisily. Large groups (over 20 individuals) were found in only 4% of
encounters. These big groups not only vocalized and displayed in the
trees but sometimes actually advanced toward the observers. Occasionally adult males descended to the ground to display within a few feet of
us. In 30 min or so, however, the members of these groups would settle

Table I. Frequency Distribution of Group Sizes

Group size	Dec 1980	Jan 1981	Feb	Mar	Apr	May	June	July	Aug	Sep	Oct	Nov	Dec	Jan 1982	Feb	Mar	Apr	May	June	Total
1	—	8	2	2	2	—	×	1	1	—	—	3	2	—	1	3	2	3	2	31
2–5[a]	5	18	1	3	7	1	×	3	4	—	—	7	7	16	4	6	5	6	3	96
6–10	6	12	2	2	3	—	×	1	1	—	4	5	5	4	5	6	9	2	8	75
11–15	1	2	1	—	—	—	×	1	—	—	2	3	—	8	5	1	3	6	2	35
16–20	—	1	—	—	—	—	×	—	—	2	—	2	1	3	3	—	3	3	—	19
21–25	—	—	—	—	—	—	×	—	2	—	—	—	1	2	1	—	1	1	—	8
26–30	—	—	—	—	—	—	×	—	—	—	—	—	—	—	—	—	—	—	—	0
31–35	—	—	—	—	—	—	×	—	—	—	—	—	1	—	—	—	—	—	—	1
36–40	—	—	—	—	—	—	×	—	—	—	—	—	—	—	—	—	—	—	—	0
41–45	—	—	—	—	—	—	×	—	—	—	—	—	1	1	—	—	—	—	—	2
46–50	—	—	—	—	—	—	×	—	—	—	—	—	—	—	—	—	—	—	—	0
51–55	—	—	—	—	—	—	×	—	—	—	—	—	—	—	1	—	—	—	—	1
Total	12	41	6	7	12	1	×	6	8	2	6	20	18	34	20	16	23	21	15	268
Mean $\bar{X}$	9.6	4.4	5.2	4.0	3.3	3.5	×	7.2	8.4	16.5	9.5	6.6	9.2	9.8	12.5	5.3	8.9	9.2	6.8	7.9

[a]Female alone, but with dependent offspring included here.

down and resume feeding, etc. Toward the end of the study period the medium and large groups tended to become less agonistic in their reaction to the observers.

3.2. Group Composition

Groups of common or pygmy chimpanzees can be categorized as mixed groups (containing individuals of both sexes and any age class), bisexual groups (adult individuals of both sexes), and unisexual groups (adults of one sex only, with or without offspring). The distinction between bisexual and mixed groups is made to facilitate comparisons with common chimpanzee populations (see discussion). Lone individuals of both sexes are also encountered in both species of *Pan*, although their solitary state is presumed to be temporary.

The exact group composition of pygmy chimpanzees in the Lomako study area was determined on 191 occasions between December 1980 and June 1982. Table II shows the frequency of each category, and compares them with forest-dwelling common chimpanzees from Budongo (Reynolds and Reynolds, 1965) and with *Pan paniscus* at Wamba (Kuroda, 1979) (see discussion).

The majority of groups encountered in the Lomako (68%) were composed of adult and immature individuals of both sexes (mixed groups).

Bisexual groups, containing only adults of both sexes ["adult groups" (Kuroda, 1979; Reynolds and Reynolds, 1965)] were observed on 15 occasions in the Lomako study site. Table III gives the detailed composition

Table II. Group Composition

Group type	*Pan paniscus* This study	*Pan paniscus* Wamba[a]	*Pan troglodytes* Budongo[b]
Mixed	68.4%	74.2%	37.2%
Bisexual (adult)	7.9%	2.5%	30.2%
Females/females with infants	5.2%	4.9%[c]	16.8%[c]
Males	4.7%	2.5%	15.8%
Solitary	18.9%[c]	6.1%	—[d]
Unknown	—	9.1%	—
Groups observed N	191	163	215

[a]From Kuroda (1979).
[b]From Reynolds and Reynolds (1965).
[c]Figure includes solitary female with dependent offspring.
[d]Solitary individuals not analyzed.

Table III.   Composition of Bisexual Groups

Date of observation	Males	Estrous females	Anestrous females	Total group size
December 4, 1980	2	—	1	3
January 7, 1981	1	—	1	2
January 13	1	1	—	2
January 15	1	1	—	2
January 22	1	1	—	2
January 23	1	1	—	2
April 9	2	1	—	3
April 10	2	1	—	3
April 10	2	2	—	4
April 13	1	—	1	2
August 28	2	3	—	5
January 21, 1982	1	1	—	2
January 21	1	1	—	2
February 20	1	—	2	3
February 27	1	1	—	2

of these groups. The majority of the bisexual groups (60%) consisted of one male and one female (seven of whom were in estrus). Although mating between these couples was observed only once, the observation periods were usually brief (as noted above, small groups tended to be more timid); thus, the possibility of their being consortships cannot be excluded. The bisexual groups as a whole, however, are not to be considered "sexual groups" (Kortland, 1962), as the majority of observed matings occurred in mixed groups [see discussion; also Handler *et al.* (this volume)].

The frequency of groups containing individuals of one sex only was low. All-male groups were encountered on nine occasions; seven of these groups contained two males, and two others were made up of three males. All-female groups with or without offspring were also infrequent and accounted for only 5.2% of all groups observed. Lone females with dependent offspring were classified as solitary in Table II, but if these were included in the all-female groups, it would increase the figure to 7.5%, which is still comparatively low. Females were found in unisexual groups containing 2–5 members.

Lone individuals were encountered on 36 occasions; 24 were adult males and 12 were adult females. Of these "lone" females, five were mothers accompanied by dependent offspring. In order to elucidate interindividual affinities both between and within sexes, "alone" is defined here as "with no other adults present" [as in Wrangham and Smuts (1980)]. Thus, in considering group composition (Table II), a mother with an infant was classed as "alone," but with regard to group size (Fig. 2) such a pair was counted as two individuals.

3.2.1. Intragroup Adult Sex Ratio

The predominance of mixed groups (Table II) indicates that interindividual affinity among pygmy chimpanzees is high. Table IV illustrates the various combinations of adult males and adult females for all groups where the sex and number of individuals was known for certain. As can be seen from the table, there is a tendency for groups to be composed of fairly equal numbers of males and females. As group size increases, however, a tendency toward groups containing more females than males becomes evident (socionomic sex ratio 0.69). This may be related to the high level of interfemale affinity found in *Pan paniscus* (see discussion).

The most frequently observed groups, other than solitary individuals, were those containing one male and one female (with or without attendant

Table IV. Combinations of Males and Females in All Groups[a]

Number of females	Number of males												Total
	0	1	2	3	4	5	6	7	8	9	10	11	
0		24	7	2	—	—	—	—	—	—	—	—	33
1	12	21	8	2	1	—	—	—	—	—	—	—	44
2	2	12	4	8	3	—	—	—	—	—	—	—	29
3	6	4	5	6	5	1	—	—	—	—	—	—	27
4	1	2	1	2	1	—	—	—	—	—	—	—	6
5	1	4	4	5	4	1	—	—	—	—	—	—	19
6	—	—	2	3	1	3		—	—	—	—	—	9
7	—	—	1	3	2	1	1		—	—	—	—	8
8	—	1	1	1	—	2	—	—		—	—	—	5
9	—	—	—	—	—	4	1	—	—		—	—	5
10	—	—	—	1	—	—	—	—	—	—		—	1
11	—	—	—	—	1	—	—	—	—	—	—		1
12	—	—	1	—	—	—	—	—	—	—	—	—	1
13	—	—	—	—	—	—	—	—	—	—	—	—	0
14	—	—	—	—	—	—	—	—	—	—	—	—	0
15	—	—	—	—	—	—	—	—	—	—	—	—	0
16	—	—	—	—	—	—	—	—	—	—	—	—	0
17	—	—	—	—	—	—	—	—	—	—	—	—	0
18	—	—	—	—	—	—	—	—	—	—	—	1	1
Total	22	68	34	33	18	12	2	0	0	0	0	1	190

Males 378, females 545, total 923, sex ratio 0.69
[a]Diagonal line indicates socionomic sex ratio of 1.0.

offspring). The discrepancy in the totals between Tables II and IV is occasioned by one solitary adolescent not represented in Table IV. The high incidence of solitary individuals could be, at least in part, an artifact of the tendency for single individuals (particularly males) to remain conspicuous when the rest of the group moved away upon being disturbed by observers.

3.3. Agonistic Behavior

3.3.1. Intracommunity Interactions

When two groups from the same community met at a food source, the level of social excitement and agonistic behavior engendered by the encounter varied with the size of the two groups concerned. When small groups joined up at a food tree there was little or no vocalizing or displaying. Individuals from the two groups were only occasionally seen to display greeting behavior, such as embracing or brief mutual grooming. Encounters between larger groups might involve loud vocalizing, displaying, and general social excitement, including increased sexual behavior. Agonistic interactions were infrequently seen between males or between males and females. When they did occur, they often involved an adult male chasing a subadult animal. The tendency seemed to be for males to avoid confrontations with one another.

3.3.2. Intercommunity Interactions

Intercommunity interactions were infrequently observed, but appeared to be characterized by vocal "contests" and avoidance of confrontations between the groups concerned. Two particularly favored fruit species, *Pancovia laurentii* and *Uapaca guineensis,* occur in the overlapping area of the two communities in the Lomako. In fact, at one particular place four *P. laurentii* trees grow within an area of less than 1 ha (the only place in the study site where this concentration was observed). These trees were used by members of both communities at different times. One instance of intercommunity interaction was witnessed in this overlapping zone.

A large group (>15) from the Eyengo community was in the area of the *P. laurentii* grove, and some members of the group were feeding on fruit in at least one of the trees. A smaller group from the Bakumba community, who were traveling toward the site, began to vocalize some distance from it. The members of the Eyengo community immediately answered and showed evidence of great agitation. Both groups continued vocalizing for about 20 min, but the Bakumba group came no closer, finally became silent, and was presumed to have left the area. The Eyengo group continued vocalizing long after the others became silent and still showed signs of agitation 30 min later. Other very vocal exchanges were recorded between groups in the overlap zone, but it was not usually possible to determine if the groups belonged to different communities.

3.4. Grooming Interactions

Both mutual and self-grooming took place mainly during rest periods. Participants included both sexes and all age classes. In addition, mutual grooming was often associated with sexual behavior and social excitement, and formed an important part of mother–offspring interactions. Grooming sessions did not invariably occur when two groups met, but most of the recorded grooming sessions took place in larger groups containing 11–20 members. Pygmy chimpanzee grooming involves the same techniques as those described for common chimpanzees by Goodall (1968).

During the study period, 72 mutual grooming sessions were recorded in which the sex and age class of the participants was known (Table V). Grooming sessions lasted from 1 to 125 min. The number of bouts per session ranged from one to 18 ($\overline{X}$ = 2.5). The most frequently observed grooming partnership was that of mother and infant (40.2% of grooming pairs). Mothers groomed infants twice as often as they were groomed and for longer periods. A mother often groomed her suckling infant, and grooming was frequently integrated with play sessions between them.

Male–female grooming partnerships accounted for 25% of the grooming sessions. Males groomed females more often than females groomed males, and these sessions were occasionally associated with copulations between the two partners. Interfemale partnerships accounted for 16% of the observed grooming sessions. One such session lasted for 88 min and involved an old female with a small estrous swelling and a younger female with a maximal estrous swelling. The young female was the more active of the two, grooming the older one more frequently and for longer periods.

Table V. Grooming Partnerships as Percent of Total Bouts (N = 72)[a]

M–F	F–F	M–M	M–I	F–I	F–J
25.0%	16.6%	6.9%	1.3%	40.2%	9.7%

[a]M, male; F, female; I, infant; J, juvenile.

Males groomed males in only 6% of the sessions, but the longest grooming session observed in the Lomako (125 min) involved two adult males.

3.5. Food Sharing

The "sharing" of meat was observed on two of the three occasions when predatory behavior was recorded (Badrian and Malenky, this volume). On one occasion, when an infant black-fronted duiker (*Cephalophus nigrifrons*) had been killed by a small adult male a number of estrous females were seen to approach the male and to make repeated and persistent begging gestures and vocalizations. The infants and juveniles in the group also made occasional begging noises and gestures. Some of the individuals who had been persistent beggars were subsequently seen to be in possession of pieces of meat. It is presumed that the male had given it to them or permitted them to take it. The most persistent of the female beggars were seen to copulate with the male, and were also observed to participate in homosexual behavior.

The other instance of "sharing" meat involved no voluntary giving by the male possessor. In this incident, two males snatched at and tore off pieces of meat from the carcass of an infant bay duiker (*Cephalophus dorsalis*) shortly before it was abandoned by the whole group.

3.6. Sociosexual Behavior

Few anestrous females (defined by the absence of a sexual swelling) were encountered during the study period. The main exceptions were females with very small infants who had probably given birth within a short time of the observations. Other mothers with small infants were seen to be in estrus and sexually active. In addition, pygmy chimpanzees were observed copulating throughout the females' estrous cycle. In fact, the majority of copulations that were observed involved females with submaximal swellings (Handler *et al.*, this volume). This is in marked contrast to the behavior of common chimpanzees, where copulation tends to be concentrated in the period when the female is at maximal tumescence (Tutin and McGrew, 1973). It thus appears that female pygmy chimpanzees are sexually receptive over much longer periods than are common chimpanzee females (Handler *et al.*, this volume).

Homosexual behavior between females [hereafter referred to as GG—genitogenital—rubbing (Handler *et al.*, this volume)] was seen in the Lomako population on 25 occasions and involved females of all ages and in all stages of estrus. Most of the GG rubbing incidents took place between mature females, and often occurred before or after heterosexual mating of at least one of the females involved. GG rubbing occurred most often however, during feeding sessions (16 times). Some possible homosexual behavior was also observed between males (Handler *et al.*, this volume).

4. Discussion

Some interesting comparisons can be made between the social organization of the pygmy chimpanzees in the Lomako and those studied by Kano, Kuroda, and others at Wamba. Intraspecific comparisons can also be made on socioecological differences between the two species of *Pan*.

The population density of pygmy chimpanzees is higher at Wamba than it is in the Lomako, i.e., 3/km² (Kano, 1982*a*) versus 2/km². This difference in population density between the two study sites may be related to the plant species composition and other habitat differentials at the two sites (Badrian and Malenky, this volume). Group sizes at both study sites are flexible, but there is a distinct difference in modal group size between the two populations, i.e., 2–5 in the Lomako and 11–15 at Wamba (see Fig. 2).

At Gombe National Park, the modal group size for *Pan troglodytes* was 2–4 (Goodall, 1968) and in the Budongo forest it was 2–6 individuals (Reynolds and Reynolds, 1965). It appears from this that pygmy chimpanzees in the Lomako show greater similarity, as regards modal group size, to common chimpanzees than to the pygmy chimpanzees at Wamba. However, as both species of chimpanzee are flexible in the formation of temporary associations, this anomaly is more likely to be a reflection of differential intraspecific ecological pressures between the study sites. In the Lomako, fluctuations in group size and variations in the fruiting patterns of favored fruit trees form an interrelated complex that remains to be fully elucidated. It may be found that a different foraging strategy is appropriate for each particularly important food species or group of species. An example of this interrelationship between the production of a particular food plant and group size can be seen in the positive correlation between the frequency of small groups and the phenology of *Uapaca guineensis* (Badrian and Malenky, this volume). Kuroda (1979) has also noted the occurrence of smaller groups during the fruiting of this species at Wamba.

Larger groups were generally found to be associated with very large and widely dispersed fruit trees, like *Antiaris toxicaria*, *Dialium* spp., and *Ficus* spp. Unusually large groups were encountered between December

1981 and February 1982 (see Table I). It was during these months that several species (notably *A. toxicaria* and *Dialium* spp.) that pygmy chimpanzees favored were producing ripe fruit. In the same months (December–February) of the previous year (1980–1981), pygmy chimpanzees were heavily utilizing the fruit of *U. guineensis*. This species, as noted above, is generally associated with smaller groups. During the 1981–1982 period noted above when very large groups were encountered, *U. guineensis* was not producing fruit. Thus the very large groups may only form during those periods when the particularly large trees are fruiting, and other abundant food sources are absent.

The distinct differences seen in the frequency of group sizes between *Pan paniscus* at the two study sites is not evident in the relative composition of the groups in the two populations. As can be seen from Table II, the difference in the frequency of mixed groups between the two study areas is insignificant when compared to the interspecific differences between pygmy and forest chimpanzees (e.g., Lomako and Budongo). For common chimpanzees in woodland habitats, the frequency of mixed groups was 30% of all groups at Gombe (Goodall, 1968) and 47% at Mahale (Nishida, 1968). A higher frequency of mixed groups in common chimpanzees was reported by Suzuki (1969). At his study site in the Kasakati Basin, mixed groups accounted for the majority of observed groups. Suzuki relates this to the extremely arid nature of the terrain and the consequent necessity for the chimpanzees to range over a very large area. The high proportion of mixed groups among pygmy chimpanzees cannot be explained by relating it to such limiting environmental factors.

As can be seen in Table II, the frequency of bisexual groups (adults of both sexes but without offspring) was low at both *Pan paniscus* study sites (Lomako 7.9%, Wamba 2.5%), while that for forest-dwelling common chimpanzees in Budongo was high (30.2%). These interspecific differences in the frequencies of bisexual groups are interesting, as in common chimpanzees the bisexual groups could possibly be considered "sexual groups" (Kortland, 1962; see also Tutin, 1980).

The results for unisexual groups are also very similar within species (*P. paniscus*) and differ markedly between species (*P. paniscus–P. troglodytes*). It is interesting to note that the relatively high frequency of all-male groups found in common chimpanzees (Goodall, 1968; Reynolds and Reynolds, 1965) is not observed in either population of *Pan paniscus* in which all-male groups accounted for 4.7% (Lomako) and 2.5% (Wamba), respectively. In the all-female groups of the Lomako, there did not seem to be a correlation between the estrous condition of females and their presence in all-female groups. Among common chimpanzees, unisexual female groups are reported to consist typically of anestrous females (Goodall,

1968). The socionomic sex ratio for adults in all groups was 0.69 in the Lomako (Table IV) as compared to 0.76 at Wamba (Kuroda, 1979).

In spite of the differences between the two pygmy chimpanzee populations, it seems that in terms of broad social organization they exhibit more similarities than differences. Mixed groups are the most important type of association at both sites. All-male groups, which are so important in common chimpanzee society (Bygott, 1974), are much less so among pygmy chimpanzees. One of the consequences of the relatively constant intersexual association of pygmy chimpanzees is that females with or without young appear to be ranging over essentially the same area as do males. This contrasts with the situation found in common chimpanzee groups, in which males range further than do females with dependent offspring (Wrangham, 1975). This suggests that the dispersion of pygmy chimpanzees conforms more closely to a "classic" model in which community ranges are shared equally between the sexes (Wrangham, 1979).

Although this model of pygmy chimpanzee social organization remains to be tested, it is possible to speculate that the data on chimpanzee odontometrics may support it. *Pan paniscus* is not dentally dimorphic except in their canines, *Pan troglodytes,* on the other hand, is dimorphic in virtually every tooth (Kinzey, this volume). Post *et al.* (1978) attempted to discover if there were any relationships between dietary habits, body weight dimorphism, and dental dimorphism in primates. Using an analysis of covariance as their statistical model, they assessed the effects that body weight and diet had on dental dimorphism in 29 cercopithecoid species. Their analysis showed no significant relationship between body weight and dental dimorphism. They did note that a greater degree of sexual dimorphism was to be found in species that are omnivorous. Although recognizing certain problems with their data (definitions of dietary habits and estimates of body weights) and the impossibility of using the results to establish which of the variables is causal, they did find a significant relationship between dietary habits and dental dimorphism. Wrangham (1979) has shown that common chimpanzee males and females utilize different ranges and thus have somewhat different foraging strategies and diets (see also Galdikas and Teleki, 1981). If, as we have suggested here, male and female pygmy chimpanzees are utilizing the same ranges and foraging strategies, and if this proves to correlate with a more similar diet, then perhaps this may partially account for the virtual absence of dental dimorphism (relative to common chimpanzees) in *Pan paniscus.*

Our impression is that members of *Pan paniscus* groups may associate over quite extended periods. This cohesiveness of small groups was first seen in 1974–1975, when we observed the intermittent association of six recognizable pygmy chimpanzees over a period of 17 days (Badrian *et al.*, 1981). The "group" consisted of two adult males, an adult female with an infant, an adolescent male, and a juvenile male. Various members of the "group" fed and nested together during most of this period and were not seen to associate with any other pygmy chimpanzees during this time. Thus, although pygmy chimpanzee group dynamics are still similar to the flexible groupings of common chimpanzees, they do show some tendency toward increased stability and cohesiveness.

This cohesiveness is facilitated by the high interindividual afinity that can be seen from an examination of such social interactions as mutual grooming, food sharing, sociosexual and agonistic behaviors, as well as their grouping patterns.

Agonistic behavior among pygmy chimpanzees of the Lomako is generally confined to vocal and visual displays, with few actual attacks witnessed so far. As previously mentioned, avoidance of outright aggression was evident both in intra- and intercommunity confrontations. Avoidance of contact between pygmy chimpanzee communities (unit groups) is also reported by Kano (1982*a*) and has been noted among common chimpanzees as well (Nishida, 1979; Sugiyama, 1973). The extreme forms of violent aggression observed between chimpanzee communities (Goodall *et al.*, 1979) has not been observed to date among *Pan paniscus*.

Goodall *et al.* (1979) noted the tendency for animals taking part in patrols to remain silent while in peripheral areas of their range. Badrian and Malenky (this volume) have commented on the unusual silence accompanying the first two meat-eating episodes recorded, which contrasted with the very noisy third episode. This phenomenon could be related to the fact that the first two episodes occurred in the overlap zone, with the third taking place well within the range of one community (Bakumba, Fig. 1). While in the overlap zone, animals are perhaps more inclined to remain silent, especially when their group is small. This strategy would help to avoid confrontations between members of the two communities.

Mutual grooming reflects the degree of affinity existing between individuals in great ape societies. As can be seen from Table V, male–female grooming partnerships were the most important between adults, female–female partnerships were the next most common, and male–male dyads the least common. Our results agree with those from Wamba (Kano, 1980) regarding relative frequency for grooming dyads, but Kano and coworkers did not find it to be associated with sexual behavior as was the case in some instances in the Lomako. Pygmy chimpanzee grooming proclivities differ from those of common chimpanzees. At both Gombe and Mahale, intermale grooming partnerships are the most frequent between adults (Goodall, 1965; Nishida, 1979). The very large grooming clusters described by Goodall (1968) were not observed in the Lomako.

Only 2% of all grooming sessions recorded involved more than three individuals.

Food sharing is another behavior that would tend to cement societal bonds. At our study site, a male "shared" his captured prey with females and young. This incident was associated with copulations between the captor and the begging females. At Wamba, where plant foods are often shared between adults, females share with other females more frequently than males share with females (Kano, 1980). Both heterosexual and homosexual behavior was observed to take place between the donor and recipient of shared food (Kano, 1980; Kuroda, 1980). It is interesting to note that some of the plant foods shared between adults at Wamba are the very large fruits of *Anonidium mannii* and the seeds of the very large fruits of *Treculia africana*. These foods by their nature can be possessed by only one individual, although providing sufficient food for more than one. In this respect they can be seen as equivalent perhaps to a prey animal. The other plant foods that are shared at Wamba do not have these characteristics (i.e., are not functionally equivalent to prey animals) and indeed are not shared in the same way [i.e., mostly between females and young rather than between adults (Kano, 1980; Kuroda, 1980)]. In common chimpanzees, most meat sharing takes place between males, and less often with females and young. Females share meat and plant foods most often with infants, and rarely with any other adults (McGrew, 1975; Teleki, 1973).

As noted above, pygmy chimpanzees copulate even when the potential for reproduction is low. Mothers with very young infants were seen to be in estrus and to be sexually active. Kano (1982a) notes that pygmy chimpanzee mothers will begin to cycle again within 1 year of parturition. Common chimpanzee mothers, on the other hand, are known to resume estrous cycling about 3.5 years after parturition (Tutin and McGinnis, 1981). This is not invariable, however, as Sugiyama and Koman (1979) note some common chimpanzee mothers cycling and mating about 1 year after parturition. Nevertheless, it appears that, in general, pygmy chimpanzee females are more sexually receptive over longer periods than are common chimpanzee females. It seems, therefore, that for *Pan paniscus* copulation serves not only a reproductive function but a social one as well, possibly strengthening bonds between males and females in the community.

Homosexual behavior between *Pan paniscus* females occurred at our study site (Handler *et al.*, this volume) and also at Wamba (Kano, 1980). Kano (1980) has suggested that GG rubbing is an example of female affinitive behavior. He notes that it is frequently associated with feeding situations, including food "sharing," as females joining begging clusters often performed GG rubbing. It is interesting to note that females who were recipients of "shared" meat (see above) were involved in both heterosexual and homosexual interactions. Both Kano (1980) and Kuroda (1980) have noted that successful begging among females was often preceded by sexual behavior between the "beggar" and the possessor of the food concerned.

5. Conclusion

The preeminence of mixed groups in pygmy chimpanzee society appears to represent a different strategy from that seen in the other great apes. It combines elements of the cohesiveness of gorilla society with the flexibility of chimpanzee social organization. Wrangham (1975) suggests that the degree to which nonterritorial individuals associate may depend on the distribution of food resources. He notes that the continuous distribution of gorilla foods could be responsible for their living in permanent social groups (see also Reynolds and Reynolds, 1965). The major proportion of pygmy chimpanzee foods are not continuously distributed and available throughout the year, even though an important part of their diet consists of herbaceous foods that are both ubiquitous and nonseasonal. The proportion of these latter foods, however, does not seem great enough to account for the divergence from what Wrangham sees as the optimum foraging strategy of solitary females.

Part of the explanation for the divergence of pygmy chimpanzees from Wrangham's chimpanzee model may be found in the characteristics and phenology of major food resources. The most important fruit trees are generally widely dispersed and may fruit asynchronously and/or erratically. Thus, to ensure a constant food supply a solitary individual requires a relatively large home range, whose size would render it undefendable. In addition, individual trees are generally large enough to provide sufficient food and feeding sites to support a large group without undue competition. Feeding competition within these groups may also be reduced through utilization of the evenly distributed and continuously available herbaceous foods. Thus, pygmy chimpanzee females may associate together (and with males) without reducing their foraging efficiency. The affiliative behavior found in interfemale interactions further reduces tension between them. The sociability of female pygmy chimpanzees is in contrast to female common chimpanzees, who are "individualistic" (Nishida, 1979) and less "gregarious" (Wrangham and Smuts, 1980). A result of this increased female sociability in *Pan paniscus* is a greater tendency to aggregate, and is possibly responsible for the greater number of females in larger groups.

Nishida (1979) reports that synchrony of estrus in female common chimpanzees may result from the stimulus within the group of a single estrous female. Synchrony of estrus has also been reported among langurs, (Hrdy, 1977; Rudran, 1973), hamadryas baboons (Kummer, 1968), and humans (McClintock, 1971). The high proportion of estrous females in pygmy chimpanzee groups may thus be due to the operation of a mechanism of stimulation to estrus among them.

The almost continuous estrous condition of female pygmy chimpanzees encourages males to associate with them. As discussed above, this close association of males with females may not necessarily lead to increased food competition. The high frequencies of intersexual association, grooming, and food sharing together with the low level of male–female aggression in pygmy chimpanzees may be a factor in male reproductive strategies. Tutin (1980) has demonstrated that a high degree of reproductive success for male common chimpanzees was correlated with male–female affiliative behaviours. These included males spending more time with estrous females, grooming them, and sharing food with them. Although male affiliative behavior is relatively low, so is intermale aggression within the community. That males associate together within a group with females may be due to the fact that larger groups are more successful during competitive intercommunity interactions [see Section 3.3.2 and Kano and Mulavwa (this volume)]. The males of a community are probably related, as Kano (1982) notes that females emigrate from one community to another at Wamba. A more cohesive type of group structure may also offer a selective advantage in foraging efficiency, given the unpredictability and dispersed nature of many of the pygmy chimpanzee's major foods.

From the foregoing, the social strategy of pygmy chimpanzees is seen to be different not only from the male-bonded common chimpanzee group (Nishida, 1979), but also from the male-dominated gorilla group, in which, although females associate with one another, interfemale affinity is low (Harcourt, 1979). In contrast, the core of pygmy chimpanzee society is characterized by the presence of strongly bonded females, and males associating with them. Thus, pygmy chimpanzee social organization does not appear to be simply a compromise between that of chimpanzees and gorillas, but a system that is unique to this species.

ACKNOWLEDGMENTS. We would like to thank the government of the Republic of Zaire for permission to conduct research in the Lomako forest. We are especially indebted to the Institute de Recherche Scientifique and

Dr. Kankwenda Mbaya in particular for the tremendous support and cooperation that we have received in Zaire.

Our work was funded by grants from the National Science Foundation and the National Geographic Society. We would also like to thank Andrea Guskin, the Boise Fund, and the Poulton Fund for helping to support our earlier research in 1974–1975. We would also like to acknowledge our debt to the Department of Anatomical Sciences at Stony Brook and its Chairman, Dr. Maynard Dewey, for support prior to our leaving for the field.

We thank the Peace Corps in Boende, Habitat in Mbandaka, the British and U. S. Embassies in Kinshasa, and Cultures Zairoises. So many other people and organizations have helped to make the project possible that it would not be possible to list them separately here, but we thank them all. We also wish to thank the people of Bokoli and Lifengo and the members of Plantation Bokoli, whose friendship and assistance were invaluable. To our guides, who taught us so much about the ways of the forest and who extended to us their friendship and affection, we have no words adequate enough to express our gratitude. We wish to thank Drs. Fleagle and Gwynn for reading and commenting on earlier drafts of this manuscript.

Finally, we would like to acknowledge a special debt to the doctors, missionaries, and especially the nuns of Boende, Befale, and Baringa for the wonderful help and care they gave us when our daughter and one of us (AB) were in urgent need of medical assistance. This debt must be extended to Dr. Craig Baldwin and other members of the University Hospital at SUNY-Stony Brook, who helped us to prepare for the emergencies that we encountered.

References

Badrian, A., and Badrian, N., 1977, Pygmy chimpanzees, *Oryx* **13**:463–468.

Badrian, A., and Badrian, N., 1978, Wild bonobos of Zaire, *Wildl. News* **13**(2):12–16.

Badrian, A., and Badrian, N., 1980, The other chimpanzee, *Animal Kingdom* **83**(4):8–14.

Badrian, N., Badrian, A., and Susman, R. L., 1981, Preliminary observations on the feeding behavior of *Pan paniscus* in the Lomako Forest of central Zaire, *Primates* **22**(2):173–181.

Bygott, J. D., 1974, Agonistic Behavior and Dominance in Wild Chimpanzees, Ph.D. Thesis, University of Cambridge.

Coolidge, H. J., 1933, *Pan paniscus:* Pygmy chimpanzee from south of the Congo River, *Am. J. Phys. Anthropol.* **18**(1):1–57.

Galdikas, B., and Teleki, G., 1981, Variations in subsistence activities of female and male pongids: New perspectives on the origins of hominid labor division, *Curr. Anthropol.* **22**(3):241–256.

Goodall, J., 1965, Chimpanzees of the Gombe Stream Reserve, in: *Primate Behavior* (I. DeVore, ed.), Holt Rinehart and Winston, New York, pp. 425–473.

Goodall, J., 1968, The behavior of free-living chimpanzees in the Gombe Stream Reserve, *Anim. Behav. Monogr.* **1**:161–311.

Goodall, J., Bandora, A., Bergman, E., Busse, C., Matama, H., Mpongo, E., Pierce, A., and Riss, D., 1979, Inter-community interactions in the Gombe National Park, in: *The Great Apes* (D. Hamburg and E. McCown, eds.), Benjamin/Cummings, Menlo Park, California, pp. 13–53.

Harcourt, A. H., 1979, The social relations and group structure of wild mountain gorillas, in: *The Great Apes* (D. Hamburg and E. McCown, eds.), Benjamin/Cummings, Menlo Park, California, pp. 187–192.

Horn, A. D., 1977, A Preliminary Report on the Ecology and Behavior of the Bonobo Chimpanzee (*Pan paniscus*, Schwarz 1929) and a Reconsideration of the Evolution of the Chimpanzees, Ph.D. Thesis, Yale University, New Haven, Connecticut.

Horn, A. D., 1980, Some observations on the ecology of the bonobo chimpanzee (*Pan paniscus*, Schwarz 1929) near Lake Tumba, Zaire, *Folia Primatol.* **34**:145–169.

Hrdy, S., 1977, *The Langurs of Abu*, Harvard University Press, Cambridge.

Kano, T., 1979, A pilot study on the ecology of pygmy chimpanzees, *Pan paniscus*, in: *The Great Apes* (D. Hamburg and E. McCown, eds.), Benjamin/Cummings, Menlo Park, California, pp. 123–135.

Kano, T., 1980, Social behavior of wild pygmy chimpanzees (*Pan paniscus*) of Wamba: A preliminary report, *J. Hum. Evol.* **9**:243–260.

Kano, T., 1982a, The social group of pygmy chimpanzees (*Pan paniscus*) of Wamba, *Primates* **23**(2):171–188.

Kano, T., 1982b, The use of twigs for rain cover by the pygmy chimpanzees of Wamba, *Primates* **23**(3):453–457.

Kortland, A., 1962, Chimpanzees in the wild, *Sci. Am.* **206**:128–138.

Kummer, H., 1968, *Social Organization of Hamadryas Baboons*, University of Chicago Press.

Kuroda, S., 1979, Grouping of the pygmy chimpanzee, *Primates* **20**(2):161–183.

Kuroda, S., 1980, Social behavior of the pygmy chimpanzees, *Primates* **21**(2):181–197.

MacKinnon, J., 1971, The orang-utan in Sabah today, *Oryx* **11**:141–191.

Mackinnon, J., 1974, The behaviour and ecology of wild orangutans (*Pongo pygmaeus*), *Anim. Behav.* **22**:3–74.

McClintock, M. K., 1971, Menstrual synchrony and suppression, *Nature* **229**:244–245.

McGrew, W. C., 1975, Evolutionary implications of sex differences in chimpanzee predation and tool use, in: *The Great Apes* (D. Hamburg and E. McCown, eds.), Benjamin/Cummings, Menlo Park, California, pp. 441–463.

Nishida, T., 1968, The social group of wild chimpanzees in the Mahali Mountains, *Primates* **9**:167–224.

Nishida, T., 1979, The social structure of chimpanzees of the Mahale Mountains, in: *The Great Apes* (D. Hamburg and E. McCown, eds.), Benjamin/Cummings, Menlo Park, California, pp. 73–121.

Post, D., Goldstein, S., and Melnick, D., 1978, An analysis of cercopithecoid odontometrics: II. Relations between dental dimorphism, body size dimorphism and diet, *Am. J. Phys. Anthropol.* **49**:533–543.

Reynolds, V., and Reynolds, F., 1965, Chimpanzees in the Budongo Forest, in: *Primate Behavior* (I. DeVore, ed.), Holt Rinehart and Winston, New York, pp. 368–424.

Rijksen, H., 1978, A field study on Sumatra orang-utans (*Pongo pygmaeus abelii:* Leeson

1827). Ecology, behavior and conservation, *Meded. Landbouwhogesch. Wageningen* **78**(2):1–420.

Rudran, R., 1973, The reproductive cycles of two subspecies of purple-faced langurs (*Presbytis senex*) with relation to environmental factors, *Folia Primatol.* **19**:41–46.

Schaller, G. B., 1963, *The Mountain Gorilla*, University of Chicago Press.

Sugiyama, Y., 1973, The social structure of wild chimpanzees, in: *Comparative Ecology and Behaviour of Primates* (R. Michael and J. H. Crook, eds.), Academic Press, London, pp. 375–410.

Sugiyama, Y., and Koman, J., 1979, Social structure and dynamics of wild chimpanzees at Bossou, Guinea, *Primates* **20**(3):323–339.

Susman, R. L. S., Badrian, N., and Badrian, A., 1980, Locomotor behavior of *Pan paniscus* in Zaire, *Am. J. Phys. Anthropol.* **53**:69–80.

Susman, R. L., Badrian, N., Badrian, A., and Handler, N. T., 1981, Pygmy chimpanzees in peril, *Oryx* **16**:179–184.

Suzuki, A., 1969, An ecological study of chimpanzees in a savanna woodland, *Primates* **10**:103–148.

Teleki, G., 1973, *The Predatory Behavior of Wild Chimpanzees*, Bucknell University Press, Lewisburg.

Tutin, C., 1980, Reproductive behaviour of wild chimpanzees in Gombe National Park, Tanzania, *J. Reprod. Fertil. Suppl.* **28**:43–57.

Tutin, C., and McGinnis, P., 1981, Chimpanzee reproduction in the wild, in: *Reproductive Biology of the Great Apes* (C. E. Graham, ed.), Academic Press, New York, pp. 239–264.

Tutin, C., and McGrew, W., 1973, Chimpanzee copulatory behaviour, *Folia Primatol.* **19**:237–256.

Wrangham, R., 1974, Artificial feeding of chimpanzees and baboons in their natural habitat, *Anim. Behav.* **22**:83–93.

Wrangham, R., 1975, The Behavioral Ecology of Chimpanzees in the Gombe National Park, Tanzania, Ph.D. Dissertation, University of Cambridge.

Wrangham, R., 1979, Sex differences in chimpanzee dispersion, in: *The Great Apes* (D. Hamburg and E. McCown, eds.), Benjamin/Cummings, Menlo Park, California, pp. 481–489.

Wrangham, R., and Smuts, B., 1980, Sex differences in the behavioural ecology of chimpanzees in the Gombe National Park, Tanzania, *J. Reprod. Fertil.* **28**:13–31.

CHAPTER 14

Sexual Behavior of Pan paniscus under Natural Conditions in the Lomako Forest, Equateur, Zaire

NANCY THOMPSON-HANDLER, RICHARD K. MALENKY, AND NOEL BADRIAN

1. Introduction

Allusions to the flexible sociosexual behavior of the pygmy chimpanzee (*Pan paniscus*) have been incorporated into the "bonobo model" of human evolution (Zihlman, 1979; Zihlman and Cramer, 1978; Zihlman *et al.*, 1978), although data to substantiate pygmy chimpanzee sexuality is scant due to the limited number of subjects available for study in captivity and the recency of field studies of free-ranging animals. However, all laboratory and field studies to date that have examined aspects of sexual behavior (Jordan in Neugebauer, 1980; Kano, 1980; Kuroda, 1980; Patterson, 1979; Savage and Bakeman, 1978; Savage-Rumbaugh and Wilkerson, 1978; Savage-Rumbaugh *et al.*, 1977; Tratz and Heck, 1954) have noted that the species is characterized by a high frequency of ventroventral copulation, a prolonged period of female sexual responsivity and homosexual behavior, especially between females in the form of genitogenital rubbing. Behavioral data coupled with a number of paedomorphic characters, including small size, gracile build, low degree of sexual di-

NANCY THOMPSON-HANDLER • Department of Anthropology, Yale University, New Haven, Connecticut 06520. RICHARD K. MALENKY • Department of Ecology and Evolution, State University of New York at Stony Brook, Stony Brook, New York 11794. NOEL BADRIAN • Department of Anthropology, State University of New York at Stony Brook, Stony Brook, New York 11794.

morphism, reduced dentition and facial prognathism, have led some (Coolidge, 1933; Lowenstein and Zihlman, 1980; Zihlman and Cramer, 1978; Zihlman, 1979; Zihlman *et al.*, 1978) to suggest that this "generalized" species might provide the best model for the last common ancestor of *Pan, Gorilla,* and *Homo.* A number of workers, however, have challenged this assertion (Latimer *et al.*, 1981; Johnson, 1981). Although this model has stimulated considerable interest in *Pan paniscus* among students of human evolution, a wider data base is needed before the efficacy of such a model for human origins and all of its socioecological implications can be judged.

This study considers the sociosexual behavior of *Pan paniscus* under natural conditions in the Lomako Forest of Equateur, Zaire. Data from the study populations at Wamba (Kano, 1980, 1982, and this volume; Kitamura, 1983; Kuroda, 1979, 1980, and this volume) and the Lomako Forest (Badrian and Badrian 1977, 1978, 1980; Badrian and Malenky, this volume; Badrian *et al.*, 1981; Susman, 1980, and this volume; Susman *et al.*, 1980, 1981) will be compared and contrasted with the laboratory evidence (Dahl, 1984 and personal communication; Patterson, 1979; Savage and Bakeman, 1978; Savage-Rumbaugh and Wilkerson, 1978; Savage-Rumbaugh *et al.*, 1977). Literature on the sexual behavior of the other African apes accumulated from long-term field (Fossey, 1982; Galdikas, 1981; Goodall, 1968, 1983; Harcourt, 1981; Harcourt *et al.*, 1981; McGinnis, 1973; Tutin, 1975; Tutin and McGinnis, 1981) and laboratory studies (Graham, 1981; Nadler, 1981; Nadler *et al.*, 1981; Yerkes and Elder, 1936*a,b;* Yerkes, 1939; Young and Yerkes, 1943) will be discussed in relation to the sociosexual behavior of the pygmy chimpanzee.

2. Methodology

The original data presented in this paper were collected over an 18-month period (December 1980–June 1982) during the initial phase of the Lomako Forest Pygmy Chimpanzee Project. A description of the study site is given by Badrian and Malenky (this volume). During the course of the study, the Lomako pygmy chimpanzee population was unhabituated and no attempt was made to artificially provision the animals. The study population consisted of at least two geographically distinct communities (the Bakumba and Eyengo communities). Animals were most frequently encountered in small groups of 2–10 animals (the modal group being 2–5 pygmy chimpanzees). In the smaller groups, males and females were about equally represented; a bias toward females in larger groups brought the estimated socionomic sex ratio to 0.69. Data from the group counts indicates that adult females were more or less equally divided into mothers (females with dependent young) and females that were not obviously accompanied by infants. Further detail on the social dynamics of the Lomako study population are given in Badrian and Badrian (this volume).

Whenever sexual behavior (herein defined as genitogenital contact between animals) was observed, the degree of female sexual swelling, age and reproductive status of participants, copulatory posture, duration of copulation, substrate and above ground height, time of day, and preceding activity were recorded. Data have been aggregated from four observers covering a total of 680 hr of direct visual contact with the animals.

It is difficult to ascertain the phases of the menstrual cycle of great apes even under laboratory conditions (Graham, 1981; Nadler *et al.*, 1981) and it was nearly always impossible to verify menses in the field. Therefore, the female's cycle phase was classified according to the size and turgescence of the perineal swelling. A simple four-grade classification which rated the size of the swelling on a scale from 0 to 3 (0 minimum to 3 maximum) was used to increase interobserver reliability under the frequently poor observational conditions of the dense rain forest. Females with totally flat sexual skins were rarely observed in the Lomako, consisting primarily of young adolescent females who had yet to begin cycling and mothers of very young infants. We assumed that females of maximum swelling (E3) were approaching ovulation, as has been indicated in clinical studies of common chimpanzees (Yerkes and Elder, 1936b; Graham, 1981), and that E0 and E1 swellings were indicative of lower fertility. Figure 1 illustrates sexual swellings classes 1–3.

The age and sex classes referred to here are based on the following characteristics:

1. Adult male: full size, scrotum fully pendant.
2. Adult female: full size (slightly smaller than large males), genital swellings developed.
3. Adolescent male: smaller and more gracile than adult (approximately three-quarters size), scrotum not fully pendant.
4. Adolescent female: smaller than adult animals, genital swellings small at all times observed.
5. Juvenile: no more than half size of adult; generally independent but closely associated with mother and occasionally seen to cling; development of secondary sexual characteristics slight.
6. Infant: keeps in frequent proximity to mother; commonly rides ventrally but also may ride dorsally during progression; nurses frequently.

Figure 1. Sexual swellings of adult females. Swellings representing grades 1–3 are pictured (left to right). Grade 0 (minimum) is not shown.

We rarely saw evidence of ejaculation in the field and so the assumption was made in this analysis that mounting, intromission, and thrusting resulted in ejaculation. Although a dominance hierarchy was not determined during this study, data from the Lomako are consistent with those from Wamba in suggesting that *Pan paniscus* society is less rigidly structured than that of *Pan troglodytes*. Although the type of mounting behavior so characteristic of dominant–subordinate relationships among common chimpanzees and baboons was occasionally observed, it was relatively rare among pygmy chimpanzees, indirectly strengthening the assumption made above that mounting between males and females constituted sexual behavior rather than a dominance interaction.

When the study animals were on the ground, they generally did not tolerate the observers' presence and, especially during the first year, fled when attempts were made to follow them. Thus, the data presented here are biased toward those matings that occurred in the trees. Observations were made through binoculars (10 × 30, 10 × 40) and data were recorded in field notebooks and later transcribed in base camp.

3. Results

A total of 105 instances of sexual behavior were noted during 414 hr (corrected for sightings of multiple observers) of direct visual contact over the 18-month period. These observations break down into 75 cases of male–female copulation, 25 cases of female–female genital rubbing, and five cases of male–male genital contact. The heterosexual copulatory rate was thus 0.18/hr.

3.1. Male–Female Sexual Behavior

Data on the 75 heterosexual matings are presented in Table I. Adult males were involved as partners in 84% of all couplings. Adolescent and infant males were also sexually active (accounting for 7 and 8% of copulations, respectively). Juvenile males accounted for only 1% of the total heterosexual matings observed.

The mating data between adults where the genital swellings of females were classified (columns 3–10, rows 1–3 of Table I) illustrates the tendency of female pygmy chimpanzees to copulate throughout the major part of their menstrual cycle. There were 42 cases in which the perineal swelling of the female was rated. The E0 class of female was not observed to copulate (although this was also the least frequently observed class of female). The E2 class copulated most frequently (57%). The E1 class

Table I. Frequency Distribution of Heterosexual Matings by Age and Reproductive Status of Partners[a]

Males		E? F	E? Mo	E0 F	E0 Mo	E1 F	E1 Mo	E2 F	E2 Mo	E3 F	E3 Mo	Ad F	J F	I F	Row total
Adult	dv	9	4	—	—	8	3	7	13	4	1	1	—	—	50
	vv	1	3	—	—	—	—	—	3	—	1	—	—	—	8
	?	2	1	—	—	1	—	1	—	—	—	—	—	—	5
Ad	dv	—	—	—	—	—	—	—	—	1	—	—	—	—	1
	vv	—	—	—	—	—	—	2	3	—	—	—	—	—	5
J	dv	—	—	—	—	—	—	—	—	—	—	—	—	—	0
	vv	—	—	—	—	—	—	1	—	—	—	—	—	—	1
I	dv	—	—	—	—	—	—	—	—	—	—	—	—	1	1
	vv	—	—	1	—	2	—	1	—	—	—	—	—	—	4
Column total		12	8	1	0	11	3	12	19	5	2	1	0	1	75

[a]E, swelling state; Ad, adolescent; J, juvenile; I, infant; dv, dorsoventral; vv, ventroventral; F, female; Mo, mother of dependent infant; ?, unknown.

(mothers and females combined) contributed 28% of total copulations, and the maximally swollen E3 class only 14%.

Mothers and females, defined by the presence or absence of dependent offspring, were about equally represented in our study population based on the composition of our group counts. Of 1680 animals counted and classified over the course of 268 sightings, mothers represented 19% of the total and females 18%. Again referring to Table I (columns 1–10, rows 1–9), a comparison of copulations by mothers and by adult females shows that the mother class was clearly sexually active, accounting for 44% of 73 copulations. This was slightly less than the grouped female class (56%).

3.1.1. Copulatory Position

Table I also indicates the distribution of copulatory position across age and reproductive classes. Of the 70 heterosexual couplings in which position was clearly observed, 74% were in the dorsoventral and 26% in the ventroventral position. Selection of the ventroventral position was not evenly distributed across male age classes. Adult males rarely used this position, mating face to face in only eight of 58 matings between adults. Combining the three subadult male age classes shows that ventroventral positioning was observed in ten of 12 cases. Of the total 18 ventroventral matings, ten took place with known mothers. Further, seven of the eight adult face-to-face matings involved mothers with dependent young, compared to 21 of 50 dorsoventral matings between adults.

3.1.2. Duration of Copulation

An estimate of duration of copulation was obtained by counting thrusts (delivered at approximately 2 per sec) or by counting (1 and 2 and 3 and . . .). Converting our estimates to seconds gave a mean duration of 12.2 sec (range 1.5–45.0 sec, $N = 51$).

3.1.3. Temporal Pattern of Mating

Figure 2 correlates diurnal patterns of mating and feeding. The highest frequency of matings was observed between 7:00 and 9:00, which coincided with the morning feeding peak. Copulations increased again during the evening hours as the pygmy chimpanzees began to feed again, prior to nesting for the night.

3.1.4. Courtship

Courtship did not appear particularly elaborate or aggressive. Most frequently, a male approached within 15 feet of a female and sat, leaning back and displaying his erect pink penis while gazing intently in the female's direction. This period of penile erection and gazing was sometimes prolonged: one male was recorded in courtship posture for 18 min. If the female was not responsive, the courting male occasionally shook a branch to gain her attention. When the female was receptive, she usually backed up to him, presenting her genitals while standing quadrupedally with limbs

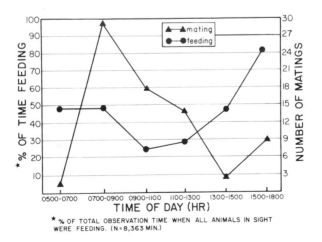

Figure 2. Diurnal patterns of mating and feeding. The percent of time feeding is the percent of total observation time when all animals in sight were feeding ($N = 8363$ min).

slightly flexed. Mounting, intromission, and thrusting followed immediately, the male crouching or standing bipedally behind the female. In other variations of dorsoventral copulation, the female sat on the lap of the male facing away from him or lay prone on her belly while the male thrusted from a sitting position. When the female presented ventrally, she generally leaned back, supporting herself in the branch network, or lay on her back with her legs wrapped around the waist of the male. On two occasions, the male was observed in the inferior position during ventroventral copulation.

3.1.5. Vocalizations

Matings were generally silent, but twice during observed matings, the female emitted drawn-out, nasalized squeals. Similar vocalizations were frequently heard while searching for the animals. Once an old male was observed to lip smack during a mating.

3.1.6. Other Behaviors Associated with Copulation

Eye contact was always maintained during ventroventral copulation, and occasionally during dorsoventral mating the female turned her head to gaze at her partner. Females were twice noted to grasp their partners' testicles. Copulation was completed when, after a generally brief period of thrusting, the couple suddenly broke apart and one or both moved away, although on several occasions the partners stayed to groom each other briefly. Males often maintained their erections following mating.

As mentioned in Section 2, it was difficult to determine whether ejaculation had occurred from the presence of semen on either partner's sexual organs. Twice, however, *coitus interruptus* with a visible stream of ejaculate was observed; one instance involved a male (Ado) with only one, partially descended testicle.

Other members of the group appeared to be drawn to mating (and homosexual) activity. Infants were frequently involved. On nine occasions, mothers were observed to mate while holding their infants ventrally. The infants either clung to their mother's back or hung above or below the mating couple. Subadults and adults also frequently moved toward copulating pairs and sat in close proximity, watching the activity. During one observation, two adolescents, one juvenile, and one adult male mated in succession with a maximally swollen female within a period of 5 min. Female infants were twice seen to approach adult males, handle the male's genitals, and rub their clitorides against them. The males, while tolerant, did not respond sexually and showed little interest. Among immature pygmy chimpanzees, interference during mating appears to be motivated more by a desire to participate than to disrupt as is seen in immature common chimpanzees (Tutin, 1979a).

3.2. Female–Female Genital Contact

Pygmy chimpanzee females show a unique form of homosexual coupling that has been termed genitogenital (GG) rubbing by Kuroda (1980). Two females embrace face to face, stare into each other's eyes, and rub their genitals together in rapid, lateral movements. In general, one of the partners wraps her legs around the other's waist as they rub. The behavior has been described previously for both captive (Jordan, in Neugebauer, 1980; Savage and Bakeman, 1978; Savage-Rumbaugh and Wilkerson, 1978) and free-ranging (Kano, 1980; Kuroda, 1980, and this volume) pygmy chimpanzees. In a Yerkes study Savage-Rumbaugh and Wilkerson (1978) noted that occasionally intromission of the erect clitoris is achieved, upon which the partners shift to thrusting behavior more typical of male–female copulations: this variation of GG rubbing has not been reported in the wild.

In the Lomako Forest, GG rubbing between females was observed 25 times. On six of these occasions, the participants repeated the act two to three times either in direct succession or within a 10-min period. Average duration of GG rubbing was 14.82 sec (range 5–34 sec, $N = 14$), slightly longer than the mean for heterosexual copulation. During the Lomako study, this behavior was observed only in the trees, although Kuroda (1980; Fig. 2) illustrates the behavior with photographs taken on the ground. In general, GG rubbing was performed silently; however, twice the partner in the inferior position was noted to "grin" with full lip retraction and to vocalize (a nasal "eee eee eee").

Genitogenital contact was seen in a variety of contexts. Most commonly, GG rubbing was observed during feeding sessions (16 times). It was also seen during travel between feeding trees (four times), during reunion of parties (twice), and during play between infants (three times). The behavior has been strongly associated with feeding by both the Yerkes researchers and those of Wamba.

From our observations, it appears that GG rubbing begins in infancy. Infants were commonly observed on the ventrum of one of the participants during female homosexual contact. The behavior has also been observed between mothers and daughters (two pairs). Infant females also employ genitogenital rubbing during play sessions. Savage-Rumbaugh and Wilkerson (1978) have reported as many as five females simultaneously rubbing genitals in a zoo colony; we observed only one sexual triad, consisting of a maximally swollen and two subadult females, during our study.

At Wamba, GG rubbing is most frequently observed between swollen females, although the behavior is not restricted to this class (Kuroda, 1980). Similarly, Savage-Rumbaugh and Wilkerson (1978) reported that the frequency of GG rubbing increased between two captive females as the adolescent began to show more regular swellings. Table II illustrates the frequency of female homosexual behavior in various age and cycle phases, showing that E2 mothers and females were the most active participants, as they also were in heterosexual coupling.

The E2 class participated as one or both partners in 38% of observed combinations. Adult females of classes E1 and E3 participated equally (12% each). Subadults account for another 26% of female–female sexual behavior, infants being the most frequently selected immature partner. The only females that were never observed to participate in GG rubbing were E0 females, and one E0 mother was seen to participate once. Mothers and females GG rubbed in about equal proportions.

Although the sample size is small, the data suggest that females in swelling phases E1–E3 are attractive and responsive, at least to other females. Other observations suggest that GG rubbing may also function to stimulate male interest. In several instances, GG rubbing was closely associated in time with heterosexual copulation. On one occasion, a mother

Table II. Frequency Distribution of Genitogenital Rubbings between Female Age–Sex Classes[a]

	E? F	E? Mo	E0 F	E0 Mo	E1 F	E1 Mo	E2 F	E2 Mo	E3 F	E3 Mo	Ad F	J F	I F	Row total
E? F	1	—	—	—	—	—	—	—	—	—	1	—	—	2
E? Mo	—	—	—	—	—	—	—	—	—	—	—	—	2	2
E0 F	—	—	—	—	—	—	—	—	—	—	—	—	—	0
E0 Mo	—	—	—	—	—	—	1	—	—	—	—	—	—	1
E1 F	—	—	—	—	—	—	4	—	—	—	—	—	—	4
E1 Mo	—	—	—	—	—	—	—	2	—	—	—	—	—	2
E2 F	—	—	—	—	—	—	—	2	—	—	1	—	—	3
E2 Mo	—	—	—	—	—	—	1	1	—	3	—	—	—	5
E3 F	—	—	—	—	—	—	—	—	—	1	1	—	—	2
E3 Mo	—	—	—	—	—	—	—	—	—	—	—	—	—	0
Ad F	—	—	—	—	—	—	—	—	—	—	—	1	—	1
J F	—	—	—	—	—	—	—	—	—	—	—	—	—	0
I F	—	—	—	—	—	—	—	—	—	—	—	—	3	3
Column total	1	0	0	0	0	0	6	5	0	4	3	1	5	25

[a]See Table I for key.

(STI) with an E2 swelling was mated by two different males in the space of 3 min. Ten minutes later, another E2 female joined her and they rubbed for 5 sec. The first male that had mated with STI approached and the E2 female turned, presented and they mated briefly dorsoventrally. The two females again rubbed genitals for 15 sec, during which the E2 female "grinned" and vocalized. Following the GG rubbing, the E2 female arm-swung away, followed by the same male, and both were lost to sight.

As in heterosexual copulation, other group members appeared to be stimulated by female–female sexual activity. Late one afternoon, two females, a male, and a juvenile male were spotted. The E1 and E2 females began to GG rub while the male and juvenile sat above them, watching. The females briefly parted and rejoined three times. As the E2 female began to groom the E1's back, the male approached them, carrying a sprig of *Dialium* fruit. Suddenly, a third female (E2) emerged from the foliage, backed up to the male and the two mated dorsoventrally. The four adults then moved off together, leaving the juvenile to feed in the *Dialium*.

Both Kano (1980) and Kuroda (1980) have suggested that a possible function of GG rubbing is the easing of tension during group excitement. They have also noted a high frequency of female–female sexual behavior in the context of begging for food (Kuroda, this volume). A third instance of female homosexual behavior associated in time with heterosexual mating has elements of both situations. While watching a group of animals that were slowly dispersing after a 2½ hr period of mixed feeding and resting, we heard the distress cries of an infant duiker intermixed with the vocalizations of an excited group of pygmy chimpanzees. We soon discovered a small adult male sitting in a tree, grasping an infant black-fronted duiker (*Cephalophus nigrifrons*) by the throat. The rest of the group was scrambling about and vocalizing intensely. The duiker ceased to struggle and bleat shortly afterward and other members of the group began to move toward the male holding it. Among the first to arrive were two swollen (E2 and E3) mothers carrying their young. Although the male apparently tried to avoid contact by moving frequently, these two mothers persistently followed him for the next hour, whereupon we lost contact with the male altogether. During the period of observation, the male mated four times with one or the other of these mothers. When the mothers were able to retain proximity to the male, they and their infants encircled him, staring into his eyes and extending their hands palms up in a begging gesture. At least one of the mothers was rewarded for her efforts by receiving a small piece of the carcass.

During one period of intense group excitement, the male carrying the duiker was chased by another male. Following the chase, the two mothers GG rubbed and the male with the duiker approached and mated with one of them, then sat with an erection clutching the duiker to his chest. He then transferred to another tree, followed once again by one of the mothers and was soon after lost to sight.

These observations demonstrate the wide variety of contexts under which genitogenital rubbing and sexual behavior in general occur among pygmy chimpanzees and underline the difficulty of ascribing any one function to the observed behavior.

3.3. Male–Male Genital Contact

Sexual behavior in the form of genital contact was seen much less frequently among males. Five cases were recorded in 18 months, including dorsoventral mounting, ventroventral mounting with pseudocopulation, and what Kano (1980) has termed rump contact, in which two males, standing quadrupedally, face in opposite directions and rub their perianal regions together.

Dorsoventral mounting was observed once between two adult males and resembled the dominance mounting behavior seen more typically in common chimpanzees and baboons. The male in the superior position thrusted for approximately 30 secs.

Male–male mounting in the ventroventral position was observed twice; once between two juveniles and once between an adult and an adolescent. On both occasions, the animals involved had been quietly feeding prior to contact and neither incident appeared aggressive in nature. All males observed had erect penises.

We observed only one incidence of rump–rump contact, the day following the incident of adult–adolescent ventroventral mating mentioned briefly above. It occurred in the same fig tree. On this occasion, an adult male (RAN) had been feeding and resting alone for 1½ hr when a smaller, past-prime male (ALB) entered the tree. RAN promptly came scrambling across the breadth of the tree toward ALB, who fled, squealing. Suddenly, the the older male stopped and presented his posterior. RAN responded by turning and briefly rubbing his anus against ALB's. The two males then moved out of sight as another influx of pygmy chimpanzees entered the tree. Ten minutes later, an E1 mother with ventral infant and STU, a past-prime female, GG rubbed shortly after their arrival in the fig tree and then sat feeding in close proximity as the final members of the party moved into the tree for the morning.

As in female–female sexual behavior, it is difficult to ascribe any one function to male–male genital contact, but incidences between males are far less frequently observed. Dominance, however, appears to play a greater role in male–male sexual contacts than in those between females.

Use of the genitals in a nonreproductive context may very well serve as affiliative behaviors in both sexes of pygmy chimpanzees.

4. Discussion

The data presented here suggest that the social and sexual behavior of *Pan paniscus* differs from that of other apes, particularly the common chimpanzee. Published material from the Wamba study site and studies of captive pygmy chimpanzees support this conclusion. Ventroventral and other copulatory positions rarely assumed by mature common chimpanzees and gorillas, GG rubbing between females, and a high incidence of copulation outside the period of maximum swelling are all peculiarities of *Pan paniscus* that have been noted in captive and free-ranging animals.

Although it is commonly believed (e.g., Morris, 1967) that face-to-face mating is unique to humans, other mammals, such as the whale and porpoise (Harrison, 1969), commonly employ this copulatory position. Among adult anthropoids, the highly arboreal orangutan (Galdikas, 1981; MacKinnon, 1971, 1974; Nadler, 1977) adopts this mating posture in both arboreal and terrestrial mating, and Chivers (1978) reports that siamangs have also been observed to copulate face to face, although dorsoventral positioning is the hylobatid norm. Among common chimpanzees and gorillas ventroventral copulation is observed only in age- or size-discrepant mounts (Nadler, 1975b; Tutin, 1975; Tutin and McGrew, 1973; Goodall, 1968), and wild gorillas have never been observed to mate face to face (Fossey, 1982; Harcourt *et al.,* 1981).

The two species of chimpanzee offer an interesting contrast. Laboratory (Yerkes, 1939; Savage-Rumbaugh and Wilkerson, 1978; Nadler *et al.,* 1981) and field (Goodall, 1968; McGinnis, 1973; Tutin, 1975) studies report that adult *Pan troglodytes* copulate only in the dorsoventral position. On the other hand, Savage and Bakeman (1978) report a frequency of 53% ventroventral copulations in their three captive pygmy chimpanzee subjects, and Patterson's (1979) adult pygmy chimpanzee subjects copulated only in the ventroventral position, save on one occasion. Free-ranging pygmy chimpanzees apparently adopt this mating posture less frequently. Kano (1980) reports that 40 of 106 observed copulations at Wamba were completed *en face*. This figure is close to the proportion of ventroventral matings reported earlier in this chapter for the Lomako study population. Thus, while it is characteristic of *Pan paniscus,* ventroventral mating appears to be less frequent in the wild than in captivity.

Savage-Rumbaugh and Wilkerson (1978, p. 335) note that the ventroventral position is preferred by the female when she is less than maximally swollen and suggest that "increased flexibility in positioning is linked to the tendency to expand times of copulatory activity to other than the female's maximum tumescent phase." Patterson (1979) postulates that face-to-face mating may result in increased stability during arboreal copulation and/or is linked to the more ventral (relative to the common chimpanzee) positioning of the female's genitals. The data from the Lomako suggest that perhaps the high frequency of face-to-face copulations observed in Savage-Rumbaugh and Wilkerson's (1978) study was more strongly biased by the age of the male, an adolescent male who had only recently matured. Bosondjo is notably smaller in their photographs than the mature female. The choice of copulatory position may be influenced by a number of factors, including substrate and support available, individual preference, and degree of swelling of the female. However, observations of free-ranging pygmy chimpanzees suggest that age (and size) of the partners plays the greater role and that retention of the position by adult partners might possibly represent a behavioral paedomorphism.

4.1. Menstrual Cycles and Sexual Swellings

The reproductive biology of the common chimpanzee has been a frequent subject of research for its own sake and also for possible analogies to human reproductive physiology and behavior. The sexual skin of the female provides an external indicator of endocrinological processes, which has proven extremely useful in behavioral studies. Although the sexual swellings (the bare area of skin of and around the external genitalia that undergoes gradual tumescence and detumescence, respectively, during the follicular and luteal phases of the menstrual cycle) of common chimpanzees vary both between and within individuals and with age and season, there is a degree of uniformity in the pattern of swelling shown within a nonpregnant cycle (measured from the day following the cessation of menstrual flow to the last day of menses). Young and Yerkes (1943) presented data on 653 cycles of 22 mature female common chimpanzees of mixed age and found a mean menstrual cycle length of 37.3 ± 0.14 days. Following Young and Yerkes' classification of swelling into four phases, many modern researchers divide the cycle into *preswelling* (period of quiescence following cessation of menses), *swelling* (tumescence, including maximum tumescence), *postswelling* (rapid detumescence followed by a period of quiescence), and *menses*.

Adolescent common chimpanzees typically show longer menstrual cycles than adults, with longer preswelling and swelling phases during the half of the cycle dominated by estrogen (the follicular phase) and shortened postswelling phases, when in the course of a normal adult, nonpregnant cycle, progesterone apparently inhibits the effect of estrogen on the sexual swelling (the luteal phase). Young and Yerkes (1943) also indicated that their subjects tended to show prolonged preswelling phases during the winter months in Orange Park (November–February) which in some instances lengthened the intermenstrual interval into what Graham (1981) has described as "winter amenorrhea."

Dahl (personal communication) has begun preliminary data collection on the menstrual cycles of three parous pygmy chimpanzees. Data collected from the first 10 months of research (December 1981–October 1982) are intriguing in that they suggest that the pygmy chimpanzee exhibits a menstrual cycle similar to that of adolescent common chimpanzees. However, the small sample size and the different reproductive states of the three females in this study make it difficult to generalize from these data. Preliminary findings indicate a mean cycle length of 46.0 days with a long and invariable swelling phase of 22.4 days for what Dahl considers the normal cycle of adult pygmy chimpanzees. Young and Yerkes (1943) in a separate sample of adolescent and adult common chimpanzees noted a mean adult cycle length of 35.2 days (swelling phase, 18.2 days) and adolescent cycle length of 42.6 days (swelling phase, 21.7 days). Savage-Rumbaugh and Wilkerson (1978) have further noted that pygmy chimpanzees do not show the same degree of maximal detumescence as do common chimpanzees. Further observations of the perineum of four females at the Yerkes Regional Primate Center by Dahl (1984) indicate that unlike *Pan troglodytes*, the labia majora are retained into adulthood and that the frenulum and clitoris of *Pan paniscus* are relocated during tumescence and elongation of the labia minora to a more anterior position between the thighs. Dahl (1984) suggests that the configuration of the labia minora resembles the immature condition in *Pan troglodytes*.

Comparable field data are not available due to the difficulty in maintaining day-to-day contact with recognizable cycling females. Kano (1980), however, mentions that one young female maintained *maximum* tumescence for 11 days of contact and another older female for at least 12 days. As mentioned previously in this chapter, the only females in the Lomako study to show what could be termed truly flat sexual skins were juvenile/adolescent females who had not begun to cycle and mothers of very young infants.

Although a larger sample is necessary, it appears that adult female pygmy chimpanzees show a different pattern of sexual swelling than do common chimpanzees and spend a relatively longer period in the swelling phase of the menstrual cycle. To our knowledge, the endocrinological basis of pygmy chimpanzee menstrual cycles remains unknown.

4.2. Distribution of Copulations across Swelling Phases

For over half a century, researchers have been debating whether estrus, a circumscribed period of time in mammals during which the female is responsive to courtship and mating, generally coincident with peak fertility at ovulation, is characteristic of higher primates. Although noting a peak in frequency of copulation during midcycle (Rowell, 1972; Saayman, 1975; Yerkes and Elder, 1936a,b; Zuckerman, 1932). Zuckerman (1932) argued that a loss of periodic estrus and permanent sexual receptivity on the part of females was correlated with living in permanent social groups in monkeys, apes, and humans. Yerkes and co-workers (Yerkes and Elder 1936a,b; Yerkes, 1939) maintained that despite occasional observation of copulation outside the midpoint of the menstrual cycle, common chimpanzees as well as other primates exhibit estrus. Yerkes and Elder (1936a,b) observed between 500 and 600 controlled matings between 13 mature females and four mature male common chimpanzees and speculated that in a state of nature, experienced and congenial consorts would copulate only during a very limited period of the cycle when the genital swelling was maximal or nearly so.

Yerkes and Elder's insight into common chimpanzee sociosexual behavior was vindicated by the Gombe Stream studies. Tutin (1975) provides data on distribution of copulations with respect to swelling phase of the menstrual cycle for seven common chimpanzees. She rated swellings as a fraction of full size: 0, 1/4–1/2, 3/4, and 1. Copulations (expressed as a percentage of total copulations) were distributed as follows across the four swelling phases: 0.01, 0.02, 0.06, and 0.91 ($N = 1101$). Tutin thus refers to the stage of maximal swelling as the estrus period and found the mean length of estrus to be 9.8 days ($N = 37$ cycles of seven females, range 7–17 days). Tutin's mean is longer than that given by Goodall (1968) of 6.5 days and shorter than the mean determined by McGinnis (1973) of 16.3 days. These discrepancies might be attributed to differences in the ages of the females observed in the different studies, as Tutin (1975, p. 60) notes that young females show longer periods of estrus than mature females.

It is more difficult to postulate a discrete period of estrus for the pygmy chimpanzee on the basis of a copulatory peak at maximum tumescence. Data from the Lomako Forest, while not strictly comparable to those of Tutin and based on a small number of copulations, indicate a less restricted period of copulation. On a four-point scale rating sexual swellings from 0 to 3 (none, small, medium, large), the frequency of copulations observed between adults were distributed 0.00, 0.28, 0.57, and 0.14, respectively. Since copulations were observed on an opportunistic basis and females were not followed each day, classification of swelling may have been confounded by individual idiosyncracies. Still, it is difficult to explain the low frequency of copulations during the E3 phase of maximal swelling. Savage-Rumbaugh and Wilkerson (1978), however, have also noted the tendency of their two female pygmy chimpanzees to copulate throughout the menstrual cycle.

As noted previously, the swelling phase of the menstrual cycle of the pygmy chimpanzee may be longer than that of the adult common chimpanzee and endocrinological factors may affect female attractiveness, receptivity, and proceptivity (*sensu* Beach, 1978). Clearly, research is necessary on the endocrinology of *Pan paniscus* and its role in sexual behavior. Badrian and Badrian (this volume) discuss the possible effects of increased female responsiveness to both males and other females on the social grouping of the pygmy chimpanzee in the Lomako Forest. The physiological and social factors affecting the occurrence and distribution of sexual behavior in this species remain to be elucidated by both field and laboratory studies before the loss of estrus can be postulated for the pygmy chimpanzee.

4.3. Lactational Amenorrhea

Another apparent difference between the two species of chimpanzees is the degree of sexual activity exhibited by mothers of young infants. Following birth many mammals, including the great apes and some monkeys, show an inhibition of the menstrual cycle (and sexual activity) referred to as lactational amenorrhea. Unfortunately, the need to maintain high breeding rates in captive colonies has prevented detailed studies of the phenomenon. Nadler *et al.* (1981) found significant differences in the length of amenorrhea between common chimpanzee mothers who had their infants removed within 48 hr after birth and those who were allowed to keep their offspring with them 3–8 months or longer than 1 year. In this study, 27 mothers who retained their infants for less than 2 days had a mean duration of 118 ± 22.7 days amenorrhea. Fourteen mothers who nursed their infants for longer than 1 year had a mean amenorrhea of 370 ± 48.0 days; five of them resumed cyclicity before infant removal.

In the wild when the infant survives, duration of acyclicity is considerably prolonged. Tutin and McGinnis (1981) report a median latency from parturition to maximum tumescence of 43 months (range 11–81 months) for nine females at Gombe Stream. Even after a return to cycling, seven females went through a median of 3.6 cycles (range 1–11) before the next conception. Two of the females in this study who resumed cycling while still suckling were apparently less attractive to males and showed low copulatory frequencies. Goodall (1983) reports that primiparous common chimpanzees at Gombe show a shorter mean period of lactational amenorrhea (30 months) than do multiparous mothers (48 months). However, the primiparous mothers cycle and mate regularly on average for a longer period of time (29 months) than do the multiparae (5 months) before conception.

The Wamba and Lomako study populations lack long-term demographic data on known individuals, but researchers from both sites have noted that mothers of young infants appear to be cycling and sexually active, suggesting that pygmy chimpanzees may have a relatively shorter period of postpartum amenorrhea than do wild common chimpanzees. As more data are accumulated, perhaps it will be demonstrated that pygmy chimpanzees follow a pattern similar to the primiparae at Gombe.

5. Conclusion

Throughout 50 years of research on the pygmy chimpanzee, the suggestion has been made that the species evolved through neoteny. The incomplete evidence that features of the external genitalia and the menstrual cycles of adult pygmy chimpanzees resemble those of adolescent common chimpanzees, the pattern of lactational amenorrhea that more closely approaches that of primiparous common chimpanzees, coupled with a frequency of copulation throughout the menstrual cycle and ventroventral copulatory postures that are employed more commonly by subadult common chimpanzees provide a very intriguing set of observations for the study of the evolution of both hominoids and hominids. But the need for supporting data is very evident. The reproductive parameters of *Pan paniscus* under natural conditions can only be determined by the accumulation of further long-term field data, and collateral studies on the reproductive physiology of this species are urgently needed. Detailed information on the sociosexual behavior of the pygmy chimpanzee in a wider social, ecological, and demographic context must be gathered before any assertion can be made that *Pan paniscus* is a better model than any other extant hominoid for the study of the evolution of hominid morphology and behavior.

We ardently hope that efforts to protect the natural habitat of the pygmy chimpanzee will succeed and that studies focusing on all aspects of the behavioral ecology of this least known and smallest of the great apes will be able to proceed in an undisturbed environment.

ACKNOWLEDGMENTS. This study was supported by grants from the National Science Foundation, the National Geographic Society, and the Institute for Intercultural Studies. Permission to undertake research in the Lomako Forest was given by the Institut de Recherche Scientifique (I.R.S.) and we thank Professors Lleke Bochoa and Kankwenda Mbaya and their staffs for the support they gave us while in Zaire.

Unlimited appreciation is extended to our friends and guides Ikwa, Lofinda, Bofaso, and Lokuli. Camp Ndele could not have existed without the help of the many residents of the village of Bokoli who acted as guides, porters, and couriers and we remain grateful for their help and friendship.

Cultures Zairoises served as our contact point with the outside world and we acknowledge our gratitude to Cits. Monapono and Entombo, Chefs du Poste, Plantation Bokoli; Monsieur DeRon of Plantation Watsi; and Monsieur Hejja of the Kinshasa office. Madam and Monsieur Christiaan have remained friends of this project since the beginning and our appreciation is hereby acknowledged. We could not have functioned without the help of all these good people.

Many new friends were made during our 18-month sojourn and we especially wish to convey our appreciation for food, shelter, and great company to Janice Wescott, U. S. AID; Ken Lizzio, Warren Littrell, David DeCrane, and James Zumwalt, U. S. Embassy; Brenda Finucane, Cheryl Hayes, Paula Loscoco, and David Henry, Peace Corps; Father John and Pere Yope of the Catholic Mission in Befale; Pere Jos and all the fathers of the Catholic Mission in Boende; and Harry and LuAnn Goodall, Habitat for Humanity-Mbandaka. They all helped to make life inside and outside the Lomako Forest both possible and pleasurable.

For support and advice received both before and after the Zairian interlude, the senior author wishes to thank Laurie Godfrey and George Armelagos, University of Massachusetts, Amherst, and Alison Richard, Richard Potts, and John Rhoads, Yale University. Kathy Wolf, Barbara Ruth, David Sprague, Todd Preuss, Jeff Rogers, Rebecca French, Art Mitchell, and Eleanor Sterling offered further advice and support during the preparation of this manuscript. Jeremy Dahl and Sue Savage-Rumbaugh were exceptionally generous with their time during the conference from which this chapter arose. We also thank them for giving us the opportunity to meet Matata, Bosondjo, Laura, Lorel, Linda, Lisa, and Kanzi. Special thanks to Jeremy for keeping me up to date on his ongoing research. The venture would not have been undertaken without the lifelong love and support of my parents, Ralph and Annette Thompson. Randall Susman deserves the Order of Merit for the patience he has shown as project director, editor, and good friend.

Drs. Mary Leakey, Harold Coolidge, and Russell Mittermier were all instrumental in helping me get started in my career with pygmy chimpanzees and their support is gratefully acknowledged.

Finally, we extend our heartfelt thanks to Stubbles, Phantom, Stitches, Lolita, Randall, Bulldog, Adolf, Hamburg, and other Hedon group members of the Bakumba Community for allowing a new group of nonchimpanzee primates to share their forest and observe their daily lives.

References

Badrian, A., and Badrian, N., 1977, Pygmy chimpanzees, *Oryx* **14**:463–472.

Badrian, A., and Badrian, N., 1978, Wild bonobos of Zaire, *Wildl. News* **13**:12–16.

Badrian, A. and Badrian, N., 1980, The other chimpanzee, *Animal Kingdom* **83**:8–14.

Badrian, N., Badrian, A., and Susman, R., 1981, Preliminary observations on the feeding behavior of *Pan paniscus* in the Lomako Forest of Central Zaire, *Primates* **22**(2):173–181.

Beach, F. A., 1976, Sexual attractivity, proceptivity and receptivity in female mammals, *Horm. Behav.* **7**:105–138.

Chivers, D. J., 1978, Sexual behavior of wild siamang, in: *Recent Advances in Primatology*, Vol. 1—*Behaviour* (D. J. Chivers and J. Herbert, eds.), Academic Press, New York, pp. 609–610.

Coolidge, H. J., 1933, *Pan paniscus*. Pygmy chimpanzee from south of the Congo River, *Am. J. Phys. Anthropol.* **18**:1–57.

Dahl, J. F., 1984, The external genitalia of female pygmy chimpanzees (submitted).

Fossey, D., 1982, Reproduction among free-living mountain gorillas, *Am. J. Primatol. Suppl.* **1**:97–104.

Galdikas, B. M. F., 1981, Orangutan reproduction in the wild, in: *Reproductive Biology of the Great Apes* (C. E. Graham, ed.), Academic Press, New York, pp. 281–300.

Goodall, J., 1968, The behavior of free-living chimpanzees in the Gombe Stream Reserve, *Anim. Behav. Monogr.* **1**:161–311.

Goodall, J., 1983, Population dynamics during a 15 year period in one community of free-living chimpanzees in the Gombe National Park, Tanzania, *Z. Tierpsychol.* **61**:1–60.

Graham, C. E., 1981, Menstrual cycle of the great apes, in: *Reproductive Biology of the Great Apes* (C. E. Graham, ed.), Academic Press, New York, pp. 1–44.

Harcourt, A. H., 1981, Intermale competition and the reproductive behavior of the great apes, in: *Reproductive Biology of the Great Apes* (C. E. Graham, ed.), Academic Press, New York, pp. 301–318.

Harcourt, A. H., Stewart, K. J. and Fossey, D., 1981, Gorilla reproduction in the wild, in: *Reproductive Biology of the Great Apes* (C. E. Graham, ed.), Academic Press, New York, pp. 265–280.

Harrison, R. J., 1969, Reproduction and reproductive organs, in: *The Biology of Marine Mammals*, (H. T. Andersen, ed.), Academic Press, New York, pp. 145–169.

Johnson, S. C., 1981, Bonobos: Generalized hominid prototypes or specialized insular dwarfs? *Curr. Anthropol.* **22**:363–374.

Kano, T., 1979, A pilot study on the ecology of the pygmy chimpanzee, *Pan paniscus*, in: *The Great Apes* (D. Hamburg and E. McCowan, eds.), Academic Press, London, pp. 123–135.

Kano, T., 1980, Social behavior of wild pygmy chimpanzees (*Pan paniscus*) of Wamba: A preliminary report, *J. Hum. Evol.* **9**:243–260.

Kano, T., 1982, The social group of pygmy chimpanzees (*Pan paniscus*) of Wamba, *Primates* 23(2):171–188.

Kano, T., 1984a, Observations of physical abnormalities among the wild bonobos (*Pan paniscus*) of Wamba, Zaire, *Amer. J. Phys. Anthropol.* 63:1–11.

Kitamura, K., 1983, Pygmy chimpanzee association patterns in ranging, *Primates* 24:1–12.

Kuroda, S., 1979, Grouping of the pygmy chimpanzees, *Primates* 20:161–183.

Kuroda, S., 1980, Social behavior of the pygmy chimpanzee, *Primates* 21(2):181–197.

Latimer, B. M., White, T. D., Kimbel, W. H., Johanson, D. C., and Lovejoy, C. O., 1981, The pygmy chimpanzee is not a living missing link in human evolution, *J. Hum. Evol.* 10:475–488.

Lowenstein, J. M., and Zihlman, A. L., 1980, A watered-down version of human evolution, *Oceans* 13(3):3–6.

MacKinnon, J. R., 1971, The orangu-tan in Sabah today, *Oryx* 11:141–191.

MacKinnon, J. R., 1974, The behavior and ecology of wild orangu-tans (*Pongo pygmaeus*), *Anim. Behav.* 22:3–74.

McGinnis, P. R., 1973, Patterns of Sexual Behavior in a Community of Free Living Chimpanzees, Ph.D. Dissertation, University of Cambridge, Cambridge, England.

Morris, D., 1967, *The Naked Ape*, McGraw-Hill, New York.

Nadler, R. D., 1975, Face to face copulation in nonhuman mammals, *Med. Aspects Hum. Sexuality* 1975(May):173–174.

Nadler, R. D., 1977, Sexual behavior of captive orangutans, *Arch. Sex. Behav.* 6:457–475.

Nadler, R. D., 1981, Laboratory research on sexual behavior of the great apes, in: *Reproductive Biology of the Great Apes* (C. E. Graham, ed.), Academic Press, New York, pp. 192–238.

Nadler, R. D., Graham, C. E., Collins, D. C., and Kling, C. R., 1981, Postpartum amenorrhea and behavior of apes, in: *Reproductive Biology of the Great Apes* (C. E. Graham, ed.), Academic Press, New York, pp. 69–82.

Neugebauer, W., 1980, The status and management of the pygmy chimpanzee, *Int. Zoo Yearb.*, 20:64–70.

Patterson, T., 1979, The behavior of a group of captive pygmy chimpanzees (*Pan paniscus*), *Primates* 20(3):341–354.

Rowell, T. E., 1972, Female reproductive cycles and social behavior in primates, in: *Advances in the Study of Behavior* (D. S. Lehrman, R. A. Hinde, and E. Shaw, eds.), Academic Press, London, pp. 69–105.

Savage, S., and Bakeman, R., 1978, Sexual morphology and behavior in *Pan paniscus*, in: *Recent Advances in Primatology*, Vol. 1, *Behavior* (D. J. Chivers and J. Herbert, eds.), Academic Press, London, pp. 613–616.

Savage-Rumbaugh, E. S., and Wilkerson, B. J., 1978, Socio-sexual behavior in *Pan paniscus* and *Pan troglodytes:* A comparative study, *J. Hum. Evol.* 7:327–344.

Savage-Rumbaugh, E. S., Wilkerson, B. J., and Bakeman, R., 1977, Spontaneous gestural communication among conspecifics in the pygmy chimpanzee (*Pan paniscus*) in: *Progress in Ape Research* (G. H. Bourne, ed.), Academic Press, New York, pp. 97–116.

Saayman, G. S., 1975, The influence of hormonal and ecological factors upon sexual behavior and social organization in Old World Primates, in: *Socioecology and Psychology of Primates* (R. H. Tuttle, ed.), Mouton, The Haque, pp. 181–204.

Susman, R. L., 1980, Acrobatic pygmy chimpanzees, *Natural History* 89(9):32–39.

Susman, R. L., Badrian, N., and Badrian, A., 1980, Locomotor behavior of *Pan paniscus* in Zaire, *Am. J. Phys. Anthropol.* 53:69–80.

Susman, R., Badrian, N., Badrian, A., and Handler, N. T., 1981, Pygmy chimpanzees in peril, *Oryx* 16:180–183.

Tratz, E., and Heck, H., 1954, Der Afrikanische anthropoide "bonobo", eine neue Menscheauaffengattung, Saugetier, *Mitteilungen* (Stuttgart) **2**:97–101.

Tutin, C. E. G., 1975, Sexual Behavior and Mating Patterns in a Community of Wild Chimpanzees (*Pan troglodytes schweinfurthii*), Ph.D. Dissertation, University of Edinburgh, Edinburgh, Scotland.

Tutin, C. E. G., 1979a, Responses of chimpanzees to copulation, with special reference to interference by immature individuals, *Anim. Behav.* **27**:845–854.

Tutin, C. E. G., 1979b, Mating patterns and reproductive strategies in a community of wild chimpanzees (*Pan troglodytes schweinfurthii*), *Behav. Ecol. Sociobiol.* **6**:29–38.

Tutin, C. E. G., 1980, Reproductive behaviour of wild chimpanzees in the Gombe National Park, Tanzania, *J. Reprod. Fert., Suppl.* **28**:43–57.

Tutin, C. E. G., and McGinnis, P. R., 1981, Chimpanzee reproduction in the wild, in: *Reproductive Biology of the Great Apes* (C. E. Graham, ed.), Academic Press, New York, pp. 239–264.

Tutin, C. E. G., and McGrew, W. C., 1973, Chimpanzee copulatory behavior, *Folia primatol.* **19**:237–256.

Yerkes, R. M., 1939, Sexual behavior in the chimpanzee, *Hum. Biol.* **11**:78–111.

Yerkes, R. M., and Elder, J. H., 1936a, Oestrus, receptivity and mating in chimpanzees, *Comp. Psychol. Mongr.* **13**:1–39.

Yerkes, R. M., and Elder, J. H., 1936b, The sexual and reproductive cycles of chimpanzee, *Proc. Nat. Acad. Sci. U. S. A.* **22**:276–283.

Young, W. C., and Yerkes, R. M., 1943, Factors influencing the reproductive cycle in the chimpanzee: The period of adolescent sterility and related problems, *Endocrinology* **33**:131–154.

Zihlman, A. L., 1979, Pygmy chimpanzee morphology and the interpretation of early hominids, *So. Afr. J. Sci.* **75**:165–168.

Zihlman, A. L., and Cramer, D. L., 1978, Skeletal differences between pygmy (*Pan paniscus*) and common chimpanzees (*Pan troglodytes*), *Folia Primatol.* **29**:86–94.

Zihlman, A. L., Cronin, J. E., Cramer, D. L., and Sarich, V. E., 1978, Pygmy chimpanzee as a possible prototype for the common ancestor of humans, chimpanzees, and gorillas, *Nature* **275**:744–746.

Zuckerman, S., 1932, *The Social Life of Monkeys and Apes.* Revision of 1932 Edition, Rutledge & Kegan Paul, London.

CHAPTER 15

The Locomotor Behavior of
Pan paniscus in the Lomako Forest

RANDALL L. SUSMAN

1. Introduction

Studies of free-ranging apes date to the 1930s and 1940s [Nissen (1931) on the common chimpanzee; Bingham (1932) on the gorilla; Carpenter (1940) on the white handed gibbon], and long-term field studies on great apes have been in progress since the 1960s (Fossey, 1972; Nishida, 1968; Rodman, 1973; Goodall, 1968). We have come to understand much about the ecology, social behavior, diet, and life histories of our closest living relatives, the Pongidae. As we have learned more about the great apes, our definition of both ape and human has changed. Whereas it was once thought that among primates only humans (and our fossil forebears) made and used tools, hunted and ate meat, and possessed the capacity for symbolic communication, studies of the great apes have revealed the subtlety of these definitions of humankind. With the advance in our understanding of the fossil record, the subtle transition in brain size and dental reduction from ape to human has been revealed. It is now widely agreed that the initial morphological and behavioral change from ape to human came in the locomotor apparatus and bipedalism. The shift from four- to two-legged progression may well have been the one that initiated the hominid trajectory. In spite of the importance of locomotion in the hominid career, relatively little is known of the locomotion of free-ranging primates, particularly the great apes. We will not be able to reconstruct

RANDALL L. SUSMAN ● Department of Anatomical Sciences, School of Medicine, State University of New York at Stony Brook, Stony Brook, New York 11794.

the behavior of ape and human ancestors until we have a fuller understanding of locomotor behavior (and its relationship to morphology) in living hominoids.

Studies of ape locomotion thus far have presented only preliminary accounts of ape locomotion (Schaller, 1963; Reynolds and Reynolds, 1965; Goodall, 1968; MacKinnon, 1971, 1974; Kortlandt, 1972; Kano, 1979, 1983; Sugardjito, 1982; Susman et al., 1980). The literature does not contain a single systematic, in-depth study of locomotor behavior in any of the great apes. Only two great ape studies have focused on great ape locomotion (Susman et al., 1980; Sugardjito, 1982), and Sugardjito is the only investigator who has considered the critical mode of vertical climbing (Fleagle et al., 1981; Stern and Susman, 1981) as a separate category of locomotion.

Captive studies have revealed that the African apes share the locomotor mode of knuckle-walking. Since, however, the three apes vary in body size, limb proportions, and habitat, aspects of the overall locomotor profile of the three African apes should vary in some manner. In fact, preliminary field reports of pygmy chimpanzees, common chimpanzees, and mountain gorillas thus far suggest that the three species exhibit differences in their locomotor profiles. One such difference in the locomotion of pygmy chimpanzees, common chimpanzees, and gorillas is in the amount and type of arboreal behavior they exhibit. Recording the amount of time members of each species spend on the ground and in the trees is the ideal way to document this behavioral difference. However, since local environmental conditions and availability of preferred foods, which determines locomotor behavior and the frequency of different modes, varies from place to place, "standardized" locomotor data are necessary. Some control can be achieved by comparisons of animals from broadly similar habitats. Pygmy chimpanzees from the low, humid Zairian basin can best be compared with common chimpanzees from lowland rain forest (e.g., Reynolds and Reynolds, 1965). Less comparable common chimpanzee habitats, such as the very dry savanna at Mt. Assirik or the more open woodland at Gombe (Goodall, 1968) are less desirable for these comparisons. Further, in broadly (structurally) similar forests the methods of recording data on substrate usage circumvent the problem of *absolute or relative time* spent or *distance moved* in various locomotor modes. Instead of absolute time, we view the question as: How do pygmy chimpanzees, common chimpanzees, or gorillas each locomote on various supports? By recording the frequencies of various locomotor modes on different supports (foliage, branches, boughs, and trunks) data can be directly compared from one species or population to another. Locomotor differences viewed in terms of how animals handle different substrates can, in turn, can be related to differences in body proportions and musculoskeletal morphology, and to theoretical expectations from biomechanical modeling and laboratory studies of locomotion and gait.

2. Methods

In this study I have extended the preliminary account of the loco-motor behavior of the pygmy chimpanzee (*Pan paniscus*) of the Lomako Forest [for a full description of the study area see Badrian and Malenky (this volume)]. I have attempted here to (1) describe locomotion in terms of the frequency of important types of behavior and, (2) relate each category of behavior to the substrate on which it occurs. These categories were described earlier (Susman *et al.*, 1980). The same definitions apply here. As in the earlier report, most of the locomotion reported here occurs during feeding. Because the study population is not yet habituated (see below), it is difficult for the observer to follow animals during travel. Also, because animals are more difficult to observe on the ground, the data on arboreal locomotion are more extensive. The relative attention paid here to arboreal versus terrestrial locomotion is not meant to imply that ar-boreality is more "important" or more "frequent" (in terms of absolute time) than terrestriality in pygmy chimpanzees. I have not attempted to depict the absolute time spent in each of the various locomotor modes. This would be most difficult, as locomotion is affected by many factors, including the context in which the animals are encountered (feeding, trav-eling, social interaction), distribution of food resources, weather, and the degree to which the individuals are habituated to the observers. Animals just beginning to feed in the early morning may be more willing to suffer the annoyance of humans nearby. But even on these occasions animals will be wary while feeding. At other times animals may flee upon contact and the brief bouts of locomotion are not representative of those that might have been exhibited in an "unexcited" state. We have evidence (see Section 4) that as time passes and animals become more habituated to the observers, behaviors such as armswinging and leaping/diving used for escape by excited individuals become less frequent. Interesting also is the decline over time of bipedalism—a favored mode of movement during display and agonistic behavior. Thus the frequency of each of the locomotor modes (as a percent of the total number of bouts) is only an approximation of the actual profile. As our studies continue further "strat-ification" of our sample will be made. Such conditions as locomotion during travel versus feeding; locomotion during agonistic behavioral en-counters (inter- and intraspecific); locomotion during sociosexual activity and play; and (with a focal animal approach to various age and sex classes), locomotor frequency over a complete daily activity cycle must eventually be considered.

The second point, the association of locomotor behavior with substrate, is a firmer and more meaningful one here. It is also a valid basis for comparision if what one is ultimately interested in is the comparative association of morphology and behavior in different species and ultimately in fossils. With the data presently at hand we can ask what forms of locomotion are executed by pygmy chimpanzees when they are on various substrates [vertical and oblique trunks, boughs, branches, foliage, and lianes; see Susman et al. (1980) for definitions]. Here we assume that the use of a particular locomotor mode is dictated by the physical properties of the support. An animal fleeing or moving calmly would meet the same *general* physical problems of descending a large trunk or traversing a firm horizontal bough.

Our methods have been described previously (Susman et al., 1980). Following Fleagle (1976), we used the "bout" as our unit of measure. A bout of locomotion is defined as a movement from one stationary posture to another or as a definable segment of a continuous sequence of progression. This method is exclusive of both time and distance. Timing short bouts of locomotion is prone to inaccuracy and estimating distances moved is also difficult when distance of the observer from the subjects varies. The recording of time spent in different locomotor modes might be better suited to future situations in which focal animals can be observed over a daily activity cycle. Recording of locomotor bouts and their breakdown into major categories of quadrupedal, quadrumanous climbing and scrambling, armswinging, leaping and diving, and bipedal behavior also has some inherent problems such as occur whenever continuous patterns of behavior are broken into categories. *Ad libitum* sampling with recognition of major types of locomotor modes provides a basis for a maximum of observations per unit time of contact. As the sample size of locomotor bouts increases, the effects of nonrandomness of this sampling technique and the consistency of observations become less problematic.

The following includes data collected from November 1980 through June 1982. I have not included data from our earlier study (Susman et al., 1980) here. The earlier data were excluded so that I could compare them with the present, more comprehensive sample. The comparison of the two sets is presented below (see Section 4). Here I report data from 1722 bouts of locomotion from unprovisioned, and as yet unhabituated subjects. Data were recorded by four different investigators. Analysis of data from individual observers indicates that the main interobserver differences were in the distinction between (terminal) boughs and (central) branches [for definitions of boughs, branches, etc., see Susman *et al.* (1980)]. Data from three of the observers showed highly consistent percentages of bough and branch activity (approximately 13% and approximately 65%, respectively). Data from the fourth investigator revealed an overall percentage of bough use at 26%, while branch behaviors accounted for only 50% of the total. When, however, both bough and branch percentages caculated from all four observers are combined it can be seen that the totals are roughly equal (78% and 76%, respectively). Since the predominant modes of locomotion on boughs and branches are quadrupedalism and quadrumanous climbing and scrambling, the more conservative approach, which would obviate the interobserver effects, would be to combine boughs and branches into one substrate category. We must be cautious of overinterpreting the small percentage differences relating to the bough–branch distinction (Fleagle, 1976; Susman *et al.*, 1980), since the accurate identification of thin boughs versus thick branches is subject to error.

3. Results

3.1. Terrestrial Locomotion

First sightings of animals on the ground were rare in the beginning but have become more frequent as the study has progressed. There have been 17 recorded episodes of terrestrial first sightings thus far (Fig. 1). Figure 1 represents only a fraction of our total sightings, but these are ones for which we have accurate estimates of the location of the animals upon contact. For the most part animals move on the ground by quadrupedal knuckle walking. Pygmy chimpanzees also engage in tripedal and bipedal gaits, particularly when carrying food. The most commonly observed food item that is transported for short distances is the canelike plant, *Haumania liebrechtsiana* (Badrian and Malenky, this volume). Mammalian prey is also carried quadrupedally or tripedally. In a well documented meat-eating episode, an adult male ran tripedally while carrying a young duiker. During displays animals assume bipedal postures and walk for short distances dragging branches or food.

We continue to observe pygmy chimpanzee tracks along stream beds and in channel sands. The animals also wade in shallow streams while feeding and traveling. While we have observed numerous footprints along the banks of streams, we have not observed knuckle prints, suggesting that pygmy chimpanzees, like common chimpanzees, avoid getting their hands wet by assuming bipedal postures when crossing streams.

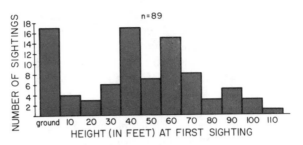

Figure 1. First sightings.

It is important to emphasize that terrestrial quadrupedalism occupies a large part of the daily activity of pygmy chimpanzees of the Lomako Forest. As yet, and until we can undertake protracted observation periods, we cannot state what percentage of a day's activity is carried out on the ground. It is, however, the more diverse arboreal activities that are the most easily observed and it is arboreal activities that place the animals at the highest risk of injury and impose rigorous physical and energetic demands on large-bodied primates such as pygmy chimpanzees.

3.2. Arboreal Locomotion

Arboreal locomotion, and vertical climbing in particular, place great morphological and physiological demands on relatively large-bodied primates [for experimental studies of climbing see Vangor (1979), Fleagle *et al.* (1981), Stern and Susman (1981)]. Morphological correlates of arboreality have been discussed elsewhere (Keith, 1923; Gregory, 1928; Washburn, 1968; Tuttle, 1974), and, like the other apes, *Pan paniscus* reveals a number of features of its anatomy that attest to its arboreal proclivity (Coolidge, 1933; Susman, 1979; Jungers and Susman, this volume; Zihlman, this volume).

The five basic modes of arboreal locomotion are (1) quadrupedalism, (2) quadrumanous climbing and scrambling, (3) bimanual suspension (armswinging), (4) leaping and diving, and (5) bipedalism [see Susman *et al.* (1980) for details].

3.2.1. Quadrupedalism

This mode of arboreal locomotion employs a definable (usually diagonal sequence) gait. The majority of quadrupedalism (64% of quadrupedal bouts) occurred on branches. Thirty-five percent of quadrupedal bouts occurred on boughs, while 1% (three bouts) were recorded in animals moving over horizontal lianes (one bout) or juveniles in terminal branches and foliage (two bouts). Out of 532 bouts of quadrupedalism, 512 were carried out with the hands in palmigrade position (Fig. 2). The data reveal that (above-bough) palmigrade quadrupedalism is by far the most common means of locomotion on boughs. Only 20 instances of knuckle-walking were recorded; of these, 18 (90%) occurred on boughs.

During bough and branch quadrupedalism the feet were normally utilized with an abducted hallux, whether or not it was actually grasping. In a number of cases individuals were encountered who were missing fingers or toes. One such individual was an adult male member of the Bakumba community (Badrian and Badrian, this volume), who lacked all but the thumb of his left hand (Fig. 3). This individual displayed locomotor hand postures reminiscent of the fist-walking of orangutans in combination with a modified quadrupedal gait in which most of the animal's weight was borne by the hindlimbs.

3.2.2. Quadrumanous Climbing and Scrambling

Quadrumanous locomotion includes use of all four hands and feet in varying combinations during unpatterned, sometimes suspended, diverse gaits on all substrates. This category comprises 31% of the total locomotor bouts observed, and, with quadrupedalism (also 31% of the total), comprises 63% of all observed instances of locomotion. Quadrumanous vertical climbing and opportunistic scrambling is the most generalized locomotor category. It takes place on all supports (trunks, bough, branches, foliage, and lianes) and is the predominant behavior on vertical trunks and in transfers between adjacent trees (Fig. 4). In 160 recorded episodes of locomotion on trunks, 147 (92%) were in a quadrumanous mode. Thirteen bouts (80%) of locomotion on trunks involved leaping from trunks to other supports or to the ground. The most frequent means of locomotion in foliage and terminal branch networks was quadrumanous scrambling. Of 120 episodes of locomotion in foliage, 74 (62%) were quadrumanous. The most common support used during quadrumanous climbing and scrambling, however, was inclined branches. In a total of 540 bouts of quadrumanous locomotion, 39% were on branches. Here quadrumanous locomotion differed from quadrupedal behavior by the lack of patterned gait and the frequent suspension below the substratum in the case of quadrumanous behavior (Fig. 5).

Figure 2. Palmigrade quadrupedalism high in the canopy (60 ft) in an adult female. Animal rises from sitting position on a bough and walks quadrupedally to a quadrupedal, standing posture on a bough-branch (sequence taken with 35 mm camera at approximately 2 frames/sec). (Facing page) Same individual as above. (f–j) Note the abnormality of the second and fourth fingers of her left hand.

3.2.3. Bimanual Suspension

Armswinging is the mode of bimanual suspension characterized by alternating hand-to-hand progression beneath branches (83% of the time), boughs (7%), in foliage (6%), and from lianes (4%). Three hundred and seventy-five bouts of armswinging were recorded, comprising 21% of the total number of bouts. As with the other locomotor categories, branches were the most frequently utilized support for armswinging (Fig. 6). The only method of movement that outstripped armswinging from branches in frequency of occurrence was branch quadrupedalism (comprising 18%

Figure 3. Adult male (lacking second and fifth fingers of his left hand) moving quadrumanously along a bough. Pattern of movement involved an awkward bipedal step (frame 1), followed by a brief period of support with the right fist (tripedal, frame 2), then another "bipedal" step. The unusual gait seen here is uncharacteristic of normal animals. Locomotion such as this was not included in main sample. (Taken at approximately 2 frames/sec.)

Figure 4. Adult male moving quadrumanously between branches in the middle canopy.

and 20%, respectively). Dropping from branches by stationary or moving animals to supports below was an infrequent form of bimanual locomotion. In this mode animals atop or hanging from one branch would drop to one below.

Thirty-four percent of the episodes of armswinging involved immature animals, including juveniles, adolescents, and (what were judged by relative size to be) subadults. Immature individuals exhibited 80% of the armswinging that was recorded on lianes, almost 50% of that seen in foliage. The majority of armswinging on boughs and branches, however, was executed by adults. Armswinging in both adults and young often comprised the end of a locomotor sequence involving a series of different modes. Such a sequence is illustrated by the second female in Fig. 7 (a_2-d_2). In this sequence the second female sidestepped her way quadrumanously down an inclined bough (a_2), pivoted on her right foot (b_2-c_2), then dove for a handhold on a branch below (d_2). Finally she armswung out of view (not pictured). The first female chose a more cautious descent down a vertical trunk (a_1-f_1). A surprising finding in this category was the great number of adult females with clinging infants that engaged in rapid armswinging. Roughly 30% of armswinging was by females with

Figure 5. Adult male climbing quadrumanously up an inclined branch. Note the lack of prehension in the left foot.

infants. In one instance a female grabbed a dead branch, which broke under the combined weight of mother and infant, and the two fell 3–4 feet into the foliage. One instance was noted of an adult male armswinging with an infant riding ventrally. Often during armswinging the mother would help support the infant by flexing one thigh. (Food items were carried in the groin in a similar manner during suspensory bouts.)

3.2.4. Leaping and Diving

Dramatic leaps, feet-first with the trunk more or less vertical, and head-first dives with the trunk horizontal (Susman *et al.,* 1980) are common in pygmy chimpanzees (Fig. 8). Ten percent of the bouts recorded were of this type (N = 159). The substrate was scored as that *from which* the animal leapt or dove. As such, 60% of leaping and diving was from branches, 14% was from boughs, and 13% was from foliage. Because of the difficulty of observation, we did not record the landing substrate, but

Figure 6. Adolescent male armswinging through the canopy (on branches).

animals most often landed on branches, in foliage, or, occasionally, on trunks. Head-first dives of 15 feet or more were not uncommon and animals often covered vertical distances greater than 15 feet. Three incidents (two involving adolescents and one involving an adult) were noted wherein the individuals landed on dead branches, which then snapped under the animal's weight. In all cases the subjects righted themselves after a short fall and no visible injuries were observed. As with armswinging, we were again surprised to find that mothers with infants often leapt or dove high in the canopy. Mothers with clinging infants accounted for 18% of the leaping and diving we observed. In one case a mother and her infant leapt downward from a branch of an emergent tree and landed by a single hand-hold on a branch some 20 feet below. Leaping and diving was the only mode save for quadrumanous climbing and scrambling that occurred on all substrates. Leaping was the most frequent method of descending to the ground. In feet-first descent of vertical trunks the final distance was most often accomplished with a leap or drop. The distance varied from a few to as many as 20–25 feet.

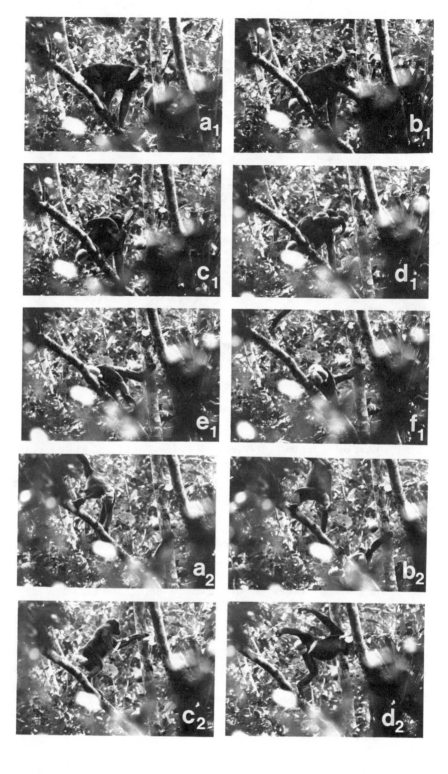

3.2.5. Bipedalism

Bipedal locomotion in the trees occurred in 6% of the observed bouts. Bipedalism was recognized as locomotion with weight borne by the hindlimbs, with the trunk vertical, even though in one-fourth of the observed bouts ($N = 103$) the hands provided some ancillary lateral balance. Sixty-seven of the bipedal episodes took place on branches. Of these, one third were by immature individuals (Fig. 9). On boughs (which accounted for 31% of bipedal activity) 23% of bipedalism was by immatures (juveniles, adolescents, and subadults).

Bipedalism in the trees was often employed when subjects were carrying food [especially in both hands (Susman *et al.*, 1980)] and when animals were threatening or displaying (usually when the latter occurred it was directed at the observers). In the trees the foods that were carried were normally broken sprigs of fruits (e.g., *Dialium pachyphyllum*). The bent hip, bent knee gait with marked pelvic rotation that characterizes the bipedalism of common chimpanzees (Elftman, 1944; Jenkins, 1972; Bauer, 1977) also characterizes the arboreal bipedalism of pygmy chimpanzees (Susman *et al.*, 1980).

4. Discussion

Given the arboreal bias in the data presented it is important to underscore a number of qualifications. First, we cannot interpret our observations of locomotion thus far as an accurate portrayal of the amount of time a pygmy chimpanzee spends in each of the various locomotor modes or the amount of time individuals spend in the trees or on the ground. Such a profile will depend on the age, perhaps sex (whether females are alone or with infants; whether adults are robust males or somewhat smaller females, etcetera) and numerous other factors (such as location and dispersion of foods). Without all-day tracking of a sizeable sample of subjects of different ages and of each sex, we refrain from statements about the absolute amount of time spent in different locomotor modes.

What we can say with confidence, and with the hope that it will serve as a basis of comparison between *Pan paniscus* and other species, is *how*

Figure 7. Two females traversing an inclined bough. The first female (a_1–f_1) reached a vertical trunk and spiraled down quadrumanously. The second female (a_2–d_2) dove from the bough into the branches of an adjacent tree and moved out of view. (Taken at approximately 1.5 frames/sec.)

Figure 8. Adults diving head first across gaps in the high canopy. The male (above) crossed a 6-m gap after a slow quadrumanous scramble through the foliage of the tree on the upper right. He landed on the foliage of the tree in the lower left. Male below dives from a thin, vertical trunk into foliage branch network of tree to the left (out of view).

(what modes) and with what relative frequency (as a percentage of locomotion observed) animals move on various substrates. It thus becomes possible to compare how different species negotiate the small branch/foliage setting, how they utilize trunks, or how they move between the trees. For example, orangutans cross wide gaps in the canopy by "tree swinging" (Sugardjito, 1982), Budongo forest chimpanzees cautiously armswing across (Reynolds, 1965; Reynolds and Reynolds, 1965), and pygmy chimpanzees in the Lomako Forest often leap or dive [as noted above; also see Susman *et al.* (1980)]. Thus when the common problem of tree transferring is considered, the frequency of locomotor mode and behavior associated with it provide an objective basis for interspecific comparison.

Perhaps the greatest problem for locomotor studies, however, is the factor of *observer effect* (and its corollary *habituation*). Our data, when viewed in temporal perspective, reveal the problem. Earlier work on an admittedly small number of observations (Susman *et al.*, 1980) revealed a greater component of bipedalism and leaping and diving than reported here. When data from Susman *et al.* (1980) are compared to the enlarged sample here, this fact is revealed (Fig. 10). No doubt one reason for the

Figure 9. Juvenile walking bipedally along a branch high in the canopy.

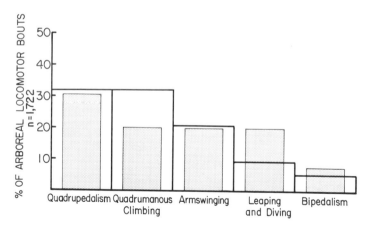

Figure 10. Shifts in proportions of various locomotor mode frequencies from the preliminary observation period (stippled histogram) to the present. Note the drop in leaping and diving and decline in bipedalism and rise in quadrumanous climbing and scrambling and in quadrupedalism. This marks a shift from locomotion often used in escape and display behaviors (leaping and diving, and bipedalism, respectively), and a rise in the normative, cautious modes of locomotion (quadrumanous climbing and scrambling, and quadrupedalism).

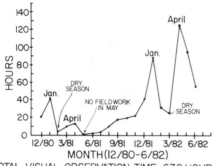

TOTAL VISUAL OBSERVATION TIME = 632 HOURS

Figure 11. *Pan paniscus* observation time over the 18-month study period. Curve represents the increase through time of minutes of observation per day. Drop in February 1981 and February–March 1982 represents the brief "dry" season during which animals spend more time on the ground and are more difficult to locate and observe. Decline in June 1982 represents termination of the initial phase of the study. Total visual observation time was 632 hr.

decrease in leaping and diving and in bipedalism is that over time there has been a lessening of observer effect along with an increased habituation of the study population. The decrease in bipedalism (in bipedal displays and agonistic behavior) is due primarily to the fact that subjects more frequently remain calm and do not move from preferred feeding spots while carrying food. The decrease in rapid flight by leaping and diving also suggests that the animals are slowly becoming habituated as the study progresses (Fig. 11). It is clear, in any case, that the above factors must be considered when locomotor profiles are constructed.

5. Comparisions of Pygmy Chimpanzees in the Lomako and at Other Sites

Our characterization of the locomotion of *Pan paniscus* in the Lomako Forest is one of versatility. Animals engage in a considerable amount of arboreality punctuated by a high frequency of armswinging (21% of recorded bouts) and leaping and diving (10%). We know that animals also spend a good part of each day foraging on the ground for herbaceous plants such as *Haumania liebrechtsiana*, *Palisota ambigua*, and *Sarcophyrynium macrostachyum*. They also prey on mammals, reptiles, and insects and perhaps eat fish and crustaceans (Badrian and Malenky, this volume). Traveling between dispersed arboreal foods also necessitates terrestriality. During the heat of the day, groups spend considerable time on the ground, where it is a great deal cooler than in the high canopy.

Kano (1979) was impressed that pygmy chimpanzees seemed to be "much quicker in their movements in the trees" than the common chimpanzees that he observed in Western Tanzania. Kano noted from his own observations that common chimpanzees appeared to regard the ground as a safe place; on the other hand, pygmy chimpanzees he observed during his early surveys throughout Equateur did not escape on the ground and sometimes hid within the thick canopy. But Kano likewise noted that pygmy chimpanzees were terrestrial. He left open the question of whether *Pan paniscus* was more or less arboreal than *Pan troglodytes* and concluded, as did Horn (1976) earlier, that pygmy chimpanzees were primarily terrestrial.

Recently Kano has reported on the locomotion of *Pan paniscus* at Yalosidi (Kano, 1983). Kano (1983) notes that vertical climbing and arboreal quadrupedalism are common but that at Yalosidi it is primarily the adult males that engage in armswinging, running, leaping, and what Kano refers to as "high-risk" types of locomotion. Kano recorded only a few incidents of armswinging and leaping by females with infants, while we have observed a large number of armswinging (30%) and leaping/diving (18%) episodes by mothers with ventrally clinging infants. Bipedalism on the ground for short distances (six steps or less) was observed on five occasions over 68 h in a 3½ month period by Kano (1983). More common at Yalosidi was bipedal standing, which was observed in 43 instances.

There appear to be significant habitat differences in the Lomako, where primary forest predominates and no settlement is found, and at the pygmy chimpanzee study sites of Wamba, Yalosidi, and Lake Tumba. At the latter localities a large portion of the study areas consists of secondary forest (areas of former cultivation) or areas where present-day villages are found. Where the forest has been depleted and replaced by secondary growth or where areas have been cleared (Kano, 1983) the effects, no doubt, seriously influence locomotor differences. This is seen not only in the increased terrestrial locomotion of pygmy chimpanzees at these sites, but in observation that the monkeys (including *Cerocebus atterimus, Cercopithecus ascanius, C. mona,* and *Colobus angolensis*) are also more frequent inhabitants of the ground as well (Kano, 1983). A secondary contributing factor to locomotor differences at these sites may be the virtual elimination of carnivores (and other large mammals) at Wamba by hunting (Kuroda, 1980). The lower nest heights recorded by Kano (1983), including a greater number of ground nests (8% at Yalosidi), may reflect a relaxed predator pressure. At Yalosidi and Wamba, pygmy chimpanzees are frequently maimed in traps and snares and an astonishing number of individuals are missing entire hands, feet, toes, or fingers (Kano, 1984). These effects obviously also directly influence the movement potential of pygmy chimpanzees at these localities.

6. Comparisons of the Locomotion of Pygmy and Common Chimpanzees

Data presently at hand allow for only the most general comparative statements about the nature of locomotor differences in *Pan paniscus* and *P. troglodytes*. Furthermore, some data suggest that one must be careful about overgeneralizing data on western and eastern common chimpanzees (Jungers and Susman, this volume). Perhaps the most meaningful comparisons of locomotion between common and pygmy chimpanzees can be made at present is between the Lomako forest population (this study) and Budongo forest chimpanzees (Reynolds, 1965; Reynolds and Reynolds, 1965). The reasons that the Budongo study is most appropriate are (1) the study area consists of dense forest, (2) the eastern long-haired chimpanzee *Pan troglodytes schweinfurthii* is the closest in size to *Pan paniscus* (Jungers and Susman, this volume), and (3) the study by Reynolds and Reynolds (1965) is still the most explicit account of the locomotor behavior of the common chimpanzee.

Reynolds and Reynolds (1965) note that chimpanzees in the Budongo Forest spent 50–75% of the daylight hours in the trees. The amount of time they spent in the trees or on the ground, however, was dependent on food resources and time of the day. For travel (50 yards or more), animals came to the ground; to flee they also came to the ground. To climb or descend large trunks animals utilized small adjacent saplings or climbers and to descend they sometimes rode saplings to the ground under their great body weight (Reynolds and Reynolds, 1965).

The Reynolds' overall characterization of chimpanzee locomotion on the ground and in the trees is similar to our profile of *Pan paniscus*. Bipedalism was more common in the trees than on the ground. In the trees bipedalism took place over distances of 15 feet and lateral support by the forelimbs was employed. Armswinging among the trees was common in Budongo Forest chimpanzees. While leaping was a rare occurrence, vertical leaps downwards of 30 feet were recorded.

Other accounts of *Pan troglodytes* [including those of Kortlandt (1962, 1975) and Goodall (1968)] suggest a more cautious form of arboreality in common chimpanzees versus pygmy chimpanzees and support the impressions of Kano (1979). Kortlandt (1962, 1975) noted that adult common chimpanzees in eastern Zaire seldom "brachiated."

We believe that pygmy chimpanzees in the Lomako Forest are accomplished arborealists with a relatively high component of armswinging and leaping and diving in their locomotor repertoires. Armswinging is a particularly good method of movement on branches and smaller supports and especially when involved in a feeding mode at a concentrated food source such as the peripheral crown of a fruiting emergent. Leaping and diving as a means of tree transfer or descent seem relatively common in *Pan paniscus*. Like common chimpanzees (Reynolds and Reynolds, 1965), pygmy chimpanzees frequently leapt or jumped on thin saplings and rode them to the ground.

7. Conclusion

African apes share anatomical specializations that offer them the best of both the arboreal and terrestrial habitus (Kortlandt, 1975). Pygmy chimpanzees, common chimpanzees, and gorillas in the wild are skilled climbers as well as adept terrestrial quadrupeds. Arboreal adaptations are evidenced by grasping halluces and long, curved toes; powerful, long fingers; and relatively long upper limbs. A major terrestrial specialization of African apes is seen in the knuckle-walking adaptations of the wrist and fingers (Tuttle, 1967; Jenkins and Flengle, 1976; Susman, 1979). African apes have conserved their arboreal hominoid heritage and added to it the terrestrial inevitability of an animal with large body size. Interestingly, the compromise has had an apparent price in the form of an increased cost of (knuckle-walking) quadrupedalism. Taylor and Rowntree (1973) have demonstrated that the cost of locomotion in chimpanzees is 50% higher than would be predicted from the relationship between cost of quadrupedal running and body size in other quadrupedal primates. The long foot, long fingers, and muscular anatomy adapted to arboreality may be less efficient on the ground when the animals engage in terrestrial quadrupedalism.

Both *Pan paniscus* and *Pan troglodytes* divide the time they spend feeding and traveling between the ground and the trees. Most would agree that knuckle-walking quadrupedalism and terrestriality in the African apes is a secondary adaptation, one that was "superimposed" on the arboreal common ancestor of chimpanzees and gorillas [for a differing view see Kortlandt (1975)]. A survey of the literature on the natural history of primates indicates that the diverse locomotor repertoire of chimpanzees is probably unequalled among other primates. Chimpanzees regularly engage in quadrupedalism (ground and trees), quadrumanous climbing and opportunistic scrambling, bimanual suspension, leaping and diving, and bipedalism. The association of these behaviors with particular substrates and in particular contexts suggests that the "diversity" of chimpanzee locomotion and its morphological counterparts have evolved to allow access to the widest possible variety of arboreal and terrestrial resources. The range of habitats exploited by chimpanzees also attests to their "flexibility.'' Chimpanzees range from humid lowland rain forests in Uganda (Reynolds and Reynolds, 1965), to gallery forest and savanna woodland in Tanzania (Goodall, 1968; Nishida, 1972), to dry, hot grassland habitats in Senegal (McGrew et al., 1981). None of the other extant Hominoidea ranges over such a variety of habitats.

The range of habitats inhabited by chimpanzees and the broad range of locomotor modes they exhibit provide a model that is potentially useful for studying ecological determinants of behavior and social organization. Preliminary indications are that group size, diet, and other aspects of behavior differ at the pygmy chimpanzee sites of Wamba and in the Lomako Forest. Studies of common chimpanzees likewise indicate that forest chimpanzees (Reynolds and Reynolds, 1965) differ from those that inhabit disturbed, regenerating forest (Kortlandt, 1962) and savanna (McGrew et al., 1981).

The combination of arboreal and terrestrial behavior and adaptation in chimpanzees also provides a model of how an arboreal species (the presumptive common ancestor of the African apes) might solve the problems incurred in coming to the ground. The common ancestor of common chimpanzees, pygmy chimpanzees, and gorillas successfully solved the problem by evolving quadrupedal knuckle walking, which may well have enabled it to enhance its ability to exploit more patchy and widely dispersed food resources (see Rodman, 1984). The earliest hominid ancestor, by 3.5 million years or so, also began to adapt to life on the ground (Lovejoy, 1981), and like African apes today, it, too, retained for some time (perhaps the subsequent 2 million years) the capacity to climb trees (Stern and Susman, 1983). Why early hominids elected a bipedal rather than knuckle-walking (or some other) solution to the problem of terrestriality is perhaps the key question. There is wide divergence of opinion, but most present hypotheses incorporate the need for freeing the hands for some form of carrying (e.g., Hewes, 1961; Kortlandt, 1962; Lovejoy, 1981). In this regard it is interesting to note that many who have observed bipedal locomotion in free-ranging chimpanzees (common and pygmy) have noted that bipedality is strongly linked to food carrying (this study; Kortlandt, 1962; Goodall, 1968; Bauer, 1977; Susman et al., 1980) or carrying objects during display (Kortlandt, 1962, 1975; Goodall, 1968). While it is important to keep in mind that early hominids were not chimpanzees (of either variety), it is safe to assume (based on studies of paleoecology, most of which suggest Plio-Pleistocene environments in Africa similar to those of today) that they encountered selection pressures broadly similar to those imposed on living apes. As such, further studies of ape locomotion, its diversity, its context, and its relationship to morphology will provide the basis for understanding the behavioral pathways of ape and human evolution.

ACKNOWLEDGMENTS. A great many people have contributed to the success of the Lomako Forest Pygmy Chimpanzee Project. The Chicago Zoological Society, the L. S. B. Leakey Foundation, the Explorers Club of New York, and the State University of New York provided the pilot funds for our initial survey in 1979. In the initial year of our long-term project (1980) we were privileged to receive support from the National Geographic Society and the National Science Foundation. The NSF has generously supported our project since then (grants BNS 79-24162, BNS 80-841292, BNS 82-18236). We are grateful to these organizations for their support.

Since the very first meeting with my Zairian colleagues in 1978 our project has had the unfailing endorsement and help from the Institut de Recherche Scientifique (IRS), Kinshasa. We thank Dr. Nkanza Ndolumingu, Dr. Kankwenda M'Baya, and Dr. Iteke Bochoa for their colleagueship and support.

The staff of the U. S. Embassy has provided much logistical help and friendship. We are especially indebted to Fred Keller for his assistance in transporting our equipment and supplies in Kinshasa, and to the Consular Section for help with our passports and visas. In Kinshasa we also received much assistance from Hai Keller, Page Goodall, and Janice Westcott. Up country we are grateful to Harry and Luann Goodall for their hospitality and logistical support in Mbandaka. The Fathers of Mission Baliko, Boende, and the Peace Corps, Boende, have provided friendship, food, lodging, and Landrover repairs. We are especially indebted to Père Jos for his friendship, humor, and hospitality. Peace Corps volunteers Brenda Finucane, Cheryl Hayes, Paula Loscoco, and Aaron Zee have been friends to all.

The people of Bokoli, our friends and fellow chimpanzee-watchers, deserve a great deal of the credit for the success of the project. Without our guides Lofinda Bongende, Ikwa Nyamaolo, and Bofaso Bosenja and our camp supervisor Lokuli Isekofaso we could not have achieved what we have. The Christaans, Franz and Leontyne, have been special friends of the project and we are grateful for their hospitality and help. Noel Badrian, Richard Malenky, and Nancy Handler have been exceptional fieldworkers and friends. Noel and Richard deserve credit for the photographs in this paper.

Cultures Zairoises in Kinshasa, at Watsi, and at Bokoli have provided our contact with the outside world. We thank Mr. R. De Ron, Mr. L. Hejja, and Citizen M. Engulu for their help.

I have driven to distraction numerous people in the offices of the research foundation at Stony Brook as they have administered this project over the last 3 years. I thank Mary Ellen Rosenblatt, Karen Warren, Robert Schneider, and Gene Schuyler for their help. Joan Kelly, who patiently and skillfully runs our offices, deserves special thanks for putting up with me and this project among her many responsibilities. Last, and most all, I thank Sandee for her understanding and for bearing the main burden of my annual sojourns in the remote, out-of-touch reaches of central Zaire.

References

Bauer, H. R., 1977, Chimpanzee bipedal locomotion in the Gombe National Park, East Africa, *Primates* **18**(4):913–921.

Bingham, H. C., 1932, *Gorillas in a Native Habitat,* Carnegie Institute Publication.

Carpenter, C. R., 1940, A field study in Siam of the behavior and social relations of the gibbon (*Hylobates lar*), *Comp. Psychol. Monog.* **16**:1–212.

Coolidge, H. J., 1933, *Pan paniscus:* Pigmy chimpanzee from south of the Congo River, *Am. J. Phys. Anthropol.* **XVIII**(1):1–57.

Elftman, H., 1944, The bipedal walking of the chimpanzee, *J. Mammal.* **1944**:67–71.

Fleagle, J. G., 1976, Locomotion and posture of the Malayan siamang and implications for hominoid evolution, *Folia. Primatol.* **26**:245–269.

Fleagle, J. G., Stern, Jr., J. T., Jungers, W. L., Susman, R. L., Vangor, A. K., and Wells, J. P., 1981, Climbing: A biochemical link with brachiation and with bipedalism, *Symp. Zool. Soc. Lond.* **48**:359–375.

Fossey, D., 1972, Vocalizations of the Mountain Gorilla (*Gorilla gorilla beringei*), *Anim. Behav.* **20**:36–53.

Goodall, J., 1968, The behavior of free-living chimpanzees in the Gombe Stream Reserve, *Anim. Behav. Monogr.* **1**:161–311.

Gregory, W. K., 1928, Were the ancestors of man primitive brachiators? *Proc. Am. Philos. Soc.* **67**:129–150.

Hewes, G. H., 1961, Food transport and the origin of hominid bipedalism, *Am. Anthropol.* **63**:687–710.

Horn, A. D., 1976, A Preliminary Report on the Ecology and Behavior of the Bonobo Chimpanzee (*Pan paniscus,* Schwarz, 1929) and a Reconsideration of the Evolution of the Chimpanzee, Ph. D. Dissertation, Yale University, New Haven, Connecticut.

Jenkins, E. S., and Fleagle, J. G., 1976, Knuckle walking and the functional anatomy of the wrist in living apes, in: *Primate Functional Morphology and Evolution* (R. H. Tuttle, ed.), Mouton, The Hague, pp. 213–227.

Jenkins, F. A., 1972, Chimpanzee bipedalism: Cineradiographic analysis and implications for the evolution of gait, *Science* **178**:877–879.

Kano, T., 1979, A pilot study on the ecology of pygmy chimpanzees, *Pan Paniscus,* in: *The Great Apes* (D. A. Hamburg and E. McCown, eds.), Benjamin/Cummings, Palo Alto, California, pp. 123–135.

Kano, T., 1983, An ecological study of the pygmy chimpanzees (*Pan paniscus*) of Yalosidi, Republic of Zaire, *Int. J. Primatol.* **4**:1–31.

Kano, T., 1984, Observations of physical abnormalities among the wild bonobos (*Pan paniscus*) of Wamba, Zaire, *Am. J. Phys. Anthropol.* **63**:1–11.

Keith, A., 1923, Man's posture: Its evolution and disorders, *Br. Med. J.* **1**:451–454, 499–502, 545–548, 587–590, 624–626, 669–672.

Kortlandt, A., 1962, Chimpanzees in the wild, *Sci. Am.* **206**:128–138.

Kortlandt, A., 1972, *New Perspectives on Ape and Human Evolution*, Stichting Voor Psychobiologie, Amsterdam.

Kortlandt, A., 1975, Ecology and paleoecology of ape locomotion, in: *Symposium 5th Congress of the International Primatological Society* (S. Kondo, M. Kawai, A. Ehara, and S. Kawamura, eds.), Japan Science Press, Tokyo, pp. 361–364.

Kuroda, S., 1980, Social behavior of the pygmy chimpanzees, *Primates* 21(2):181–197.

Lovejoy, C. O., 1981, The origin of man, *Science* 211:341–350.

MacKinnon, J., 1971, The orang-utan in Sabah today, *Oryx* 11:142–191.

MacKinnon, J., 1974, The behavior and ecology of wild orang-utans (*Pongo pygmaeus*), *Anim. Behav.* 22:3–74.

McGrew, W. C., Baldwin, P. J., and Tutin, C. E. G., 1981, Chimpanzees in a hot, dry and open habitat: Mt. Assirik, Senegal, West Africa, *J. Hum. Evol.* 10:227–244.

Nishida, T., 1968, The social group of wild chimpanzees in the Mahali Mountains, *Primates* 9:167–224.

Nishida, T., 1972, Preliminary information on the pygmy chimpanzee (*Pan paniscus*) of the Congo Basin, *Primates* 13(4):415–425.

Nissen, H. W., 1931, A field study of the chimpanzee, *Comp. Psychol. Monogr.* 8(1).

Reynolds, V. F., 1965, *Budongo. An African Forest and Its Chimpanzees*, Natural History Press, New York.

Reynolds, V. F., and Reynolds, F., 1965, Chimpanzees of the Budongo Forest, in: *Primate Behavior. Field Studies of Monkeys and Apes* (I. Devore, ed.), Holt, Rinehart, and Winston, New York, pp. 368–424.

Rodman, P. S., 1973, The Synecology of Bornean Primates, Ph.D. Dissertation, Harvard University.

Rodman, P. S., 1984, Foraging and social systems of orangutans and chimpanzees, in: *Adaptations for Foraging in Nonhuman Primates and Apes* (P. S. Rodman and J. H. G. Cant, eds.), Columbia University Press, New York, pp. 134–160.

Schaller, G. B., 1963, *The Mountain Gorilla. Ecology and Behavior*, University of Chicago Press, Chicago.

Stern, J. T., and Susman, R. L., 1983, The locomotor anatomy of *Australopithecus afarensis*, *Am. J. Phys. Anthropol.* 60:279–317.

Sugardjito, J., 1982, Locomotor behavior of the Sumatran orang-utan (*Pong pygmaeus abelii*) at Ketambe, Gunung Leuser National Park, *Malay Nat. J.* 35:57–64.

Susman, R. L., 1979, Comparative and functional morphology of hominoid fingers, *Am. J. Phys. Anthropol.* 50:215–236.

Susman, R. L., Badrian, N. L., and Badrian, A. J., 1980, Locomotor behavior of *Pan paniscus* in Zaire, *Am. J. Phys. Anthropol.* 53:69–80.

Taylor, C. R., and Rowntree, V. J., 1973, Running on two or four legs: Which consumes more energy, *Science* 179:186–187.

Tuttle, R. H., 1967, Knuckle-walking and the evolution of hominoid hands, *Am. J. Phys. Anthropol.* 26(2):171–206.

Tuttle, R. H., 1974, Darwin's apes, dental apes, and the descent of man: Normal science in evolutionary anthropology, *Curr Anthropol.* 15:389–398.

Vangor, A., 1979, Electromyography of Gait in Non-human Primates and Its Significance for the Evolution of Bipedality, Ph.D. Dissertation, State University of New York at Stony Brook.

Washburn, S. L., 1968, *The Study of Human Evolution*, Oregon State System of Higher Education, Eugene, Oregon.

CHAPTER 16

Pan paniscus and Pan troglodytes

Contrasts in Preverbal Communicative Competence

E. S. SAVAGE-RUMBAUGH

1. Introduction

Previous ape language studies have been undertaken with common chimpanzees (*Pan troglodytes*), orangutans (*Pongo pygmaeus*), and gorillas (*Gorilla gorilla*) (Gardner and Gardner, 1969; Rumbaugh, 1977, Savage-Rumbaugh, 1979; Miles, 1982; Patterson and Linden, 1981). Additionally, a brief attempt was made with a pygmy chimpanzee housed in the Stuttgart Zoo by Jordan and Jordan (1977), who employed a Premackian problem-solving paradigm. However, no one has previously attempted to place pygmy chimpanzees in a full-time communicative environment that entails close and constant interaction with human beings for the purpose of attempting to teach them complex symbolic tasks. In fact, apart from the brief work mentioned above by Jordan and Jordan (1977), no serious attempts have been made to investigate the cognitive capacities of these apes, nor to contrast them with other apes. This paucity of information is not due to lack of interest, but simply to lack of availability of these animals for research purposes. They were not reorganized as a distinct species until 1929 (Schwarz, 1929) and it was not until 1956 (Tratz and Heck, 1954) that the wide range of behavioral differences between common chimpanzees and pygmy chimpanzees began to be recognized. Shortly after this it also became apparent that their numbers in the wild were

E. S. SAVAGE-RUMBAUGH ● Language Research Center, Yerkes Regional Primate Research Center, Emory University and Georgia State University, Atlanta, Georgia 30322.

quite small and thus it was imperative that they be placed on the rare and endangered list. This has virtually eliminated all possibility of research with these animals. Because of our paucity of knowledge regarding these animals and because of their close biological relationship to humans (Zihlman, et al., 1978), it is important that every possible effort be expended to protect the present wild populations and to properly care for the present captive populations. Behavioral studies can aid in both these goals to the extent that they help clarify the cognitive and social behaviors that set these great apes apart from others. Strong behavioral differences between Pan troglodytes and Pan paniscus should serve to emphasize the specific need for continued and effective protection of Pan paniscus. Behavioral studies must, however, restrict themselves to areas of investigation that have a positive social and emotional impact upon the apes that are studied.

Individuals who have had first hand interactive experience with both Pan troglodytes and Pan paniscus (Yerkes and Learned, 1925; Tratz and Heck, 1954) have been left with the distinct impression that pygmy chimpanzees are considerably more intelligent and more sociable than Pan troglodytes. To date, however, no clear scientific evidence has been advanced to support the views regarding intellectual superiority. With the social behavior of Pan paniscus, though, a number of important and distinctive behaviors, including ventroventral copulation and prolonged and frequent interindividual gaze, have been documented (Savage and Wilkerson, 1978; Kano, 1980). There is also some evidence to suggest that Pan paniscus, in contrast to other great apes, are more bipedal (at least arboreally), engage in more arm swinging, diving, and leaping, have less fear of water, are less aggressive, readily share food, and share the responsibilities of infant rearing (Kano, 1980; Susman et al., 1980; Kuroda, 1980; Patterson, 1973). Many of the differences between Pan paniscus and Pan troglodytes, though far from substantiated at present, are very important because they suggest that, at least on the behavioral plane, pygmy chimpanzees may share more characteristics with humans than do common chimpanzees.

2. History of Subjects

There are two Pan paniscus subjects in the present symbol-training program being carried out at the Language Research Center in Atlanta, Georgia (a facility jointly sponsored by Georgia State University and the Yerkes Regional Primate Research Center of Emory University). One subject, Matata, is a 12-year-old, wild-born adult female who was brought to the United States in 1975 under an agreement with the government of

Zaire. At the time of capture she was 5–7 years of age. Previous studies of the sociosexual behavior of this female and the young male (Bosondjo) and the old female (Lokelema) also captured with Matata have been reported elsewhere (Savage and Bakeman, 1978; Savage and Wilkerson, 1978; Savage et al., 1977). The second subject is Kanzi, a captive-born male 1½ years of age. Kanzi is the son of Lorel (a captive-born female owned by the San Diego Zoo). Kanzi was stolen from his mother (without contest) 30 min after birth by Matata and has been reared by Matata as her son since that time. These two chimpanzees remain together 24 hours per day.

At the time Kanzi was born, Matata and Bosondjo (two wild-caught Yerkes animals) were housed at the Yerkes Field Station with two adult female pygmy chimpanzees, Laura and Lorel, on breeding loan from the San Diego Zoo. This group had been together for 1 year. (The female originally caught with Matata and Bosondjo died of natural causes related to old age.) During this period Matata had conceived and borne one infant of her own, Akili. This infant was 10 months old at the time Matata took Kanzi from his natural mother. Matata cared very adroitly for both infants, until Kanzi was 6 months of age. One hour of video tape was made of Kanzi and the rest of this *Pan paniscus* group every 2 weeks from birth on. When Kanzi was 6 months old and Akili was 9 months of age, Akili was sent to the San Diego Zoo. Kanzi and Matata were assigned to the language project and remained together.

Other subjects mentioned for purposes of comparison in the present study are Sherman and Austin two male *Pan troglodytes*, 9 and 10 years of age, respectively. These individuals have been in language training since the age of 2 and 3 years, respectively, and during this time have been exposed to an enriched cognitive-social environment (Savage-Rumbaugh and Rumbaugh, 1978, 1980; Savage-Rumbaugh et al., 1980). Also mentioned is a social group of five *Pan troglodytes* mothers and their offspring. This social group of chimpanzees was observed daily for 2 years (Savage et al., 1973; Savage and Malick, 1977; Savage, 1975) and provides the background for many points of contrast of maternal behavior between these two species.

3. Research Environment

Matata and Kanzi are housed in a large six-room indoor–outdoor enclosure. Even though Matata was wild-born, both she and Kanzi continually seek out and appear to enjoy and depend upon human companionship. Consequently, human teachers and caretakers are with them throughout the day. Mild distress is evidenced by Matata, even though she is an adult, at the departure of human teachers with whom she has formed close relationships. She seeks to maintain close proximity to these teachers as they work with her, often simply sitting with an arm or leg draped across them. When they stay until evening she pulls them into the large nest that she builds and goes to sleep next to them. Nesting with others, of all sex classes, has been reported in wild *Pan paniscus* and is peculiar to this species (Kuroda, 1980).

Similar sorts of attachment behavior toward humans are rarely observed in adult apes of other species (Yerkes and Yerkes, 1929). In instances where attachment to humans does occur in other ape species it is invariably preceded by a long period of rearing in which a human becomes a parental surrogate for the ape at a very early age (Patterson and Linden, 1981; Temerlin, 1975). Since Matata came to the project fully adult, with her own offspring to care for, her attachment to the humans around her is clearly not a result of human upbringing. Likewise, observations of her social behavior toward other pygmy chimpanzees housed with her at the field station prior to her inclusion in the Language Research Project revealed that she was well integrated into that group. She was the highest ranking female, the male's favorite partner, and a competent mother of two young infants. She was, in general, the focus of group attention and evidenced frequent affiliative behaviors toward all the other individuals in the group.

As a result of these affiliative behaviors, which are extremely similar in form and content to those of humans, it has been possible to introduce new individuals who have no previous experience working with apes to Matata and Kanzi. The ready extension of affiliative behaviors toward new individuals stands in marked contrast to the behaviors displayed by Sherman and Austin, two *Pan troglodytes* males, who have been reared by humans as subjects in the Language Research Project since they were 2 and 3 years of age. Although Sherman and Austin, as adolescents, have maintained remarkably cooperative relationships with teachers who have worked with them since they were juveniles, it is no longer safe to introduce new teachers to them. Their initial response to strangers is often quite aggressive. These tendencies to display considerable aggressive behavior toward strangers is not unique to Sherman and Austin. It is typical of adult *Pan troglodytes,* both male and female, and in both captive and wild environments (De Waal and Hoekstra, 1980; Goodall *et al.,* 1979). It would not be reasonable to work with a wild-caught female *Pan troglodytes* and her offspring in a way that entailed the close and continual interspecies (*Homo sapiens/Pan paniscus*) contact that characterizes the present working and living environment of Matata and Kanzi.

Kanzi, like Matata, is extremely affiliative and socially responsive to human interaction and contact. He enjoys being carried around by his teachers and he initiates frequent carrying. Matata not only permits this, but on occasion even encourages Kanzi to go to others by detaching his hands and shoving him in the direction of another individual. This is not a sign of lack of interest in Kanzi, nor a sign of atypical maternal behavior in Matata. While Matata was housed with other pygmies at the field station she also allowed them to carry, play with, and discipline both Akila and Kanzi. She also encouraged these infants to go to other pygmy chimpanzees and retrieved them only when they appeared quite distressed. By 1½ years of age, Akili spent most of his time playing with other individuals in the group, particularly the adult male. He returned to Matata when he was hungry, sleepy, hurt (bumped his head, tripped, etcetera) or when she was moving from one area to another. Kanzi is following a similar pattern of strong attraction to other individuals, though in his case these other individuals are not pygmy chimpanzees, they are humans. This pattern of allowing others (particularly the adult male) to participate extensively in the task of infant caretaking is not unique to Matata. It has also been observed in the San Diego group (Patterson, 1973). While it has not yet been reported for wild populations, this may be due to the tension among the wild pygmy chimpanzees created by the presence of observers, since in captive groups the infant returns quickly to the mother if there is the slightest suggestion that she is aroused.

Interaction with others appears to be initiated primarily (though not entirely) by the strong attraction of young infants toward other individuals. On numerous occasions Kanzi has leaped directly from Matata's ventral surface onto the body of an approaching person. This behavior was quite frequent around 9 months of age and when it was directed toward a person that Kanzi had not seen in several weeks, he often screamed loudly, both before and after he clung to this person. Only hugging and cradling would calm him at these times. In such cases Matata did not give any signs of encouraging Kanzi to cling to someone else. She herself was busy trying to calm him by cradling and hugging. She showed no aggressive behavior toward others as Kanzi greeted them by leaping onto them while screaming vociferously. Most human recipients of this behavior were quite frightened initially, as they thought that Kanzi's leaping onto them and screaming might provoke an attack by Matata. Matata, however, tended only to react as though Kanzi were unduly excited and needed to be calmed down.

On most occasions, when Kanzi decided to transfer himself from Matata to a person, he simply let go of her and climbed into the person's arms with no vocalization at all. On some occasions the transfer would be initiated by a gestural request from the person, but typically Kanzi was the initiator. When Kanzi was being carried by others, he at times refused to go back to Matata if she gesturally requested by holding her hand out toward him. On some occasions, Matata would even whimper and pull on Kanzi's fingers and toes and he would still ignore her. However, at the slightest indication that she was concerned about a noise or that she might leave his sight, he would leap onto her back, even though just a second ago he had ignored her completely.

When Kanzi is carried by Matata he rides either in the ventral position or, increasingly as he gets heavier, in the dorsal position. However, when he is carried by a human partner, he rides on their hip while they move bipedally as they would with a human child. This difference in carrying orientation provides opportunities for dramatically different types of communication to occur while being carried by humans. While on the teacher's hip, both Kanzi and the teacher can each respond to the direction of gaze of the other. Likewise, gestural directions (such as pointing or waving the arm in a particular direction) can easily be noted by both individuals. Additionally, because Kanzi's body becomes upright, oriented in the vertical, and because his weight is supported by the teacher, his hands are left free to gesture. In contrast, when riding on Matata ventrally his hands must be used for grasping, and when riding dorsally, even if he can let go to gesture, such a gesture could not easily be seen by his mother since he rides behind her field of vision. The human teachers, unlike his mother, have shown from the beginning considerable interest in and sensitivity to Kanzi's focus of attention. Consequently, if Kanzi seemed to regard a particular object with interest the teacher would carry Kanzi toward that object. Kanzi quickly learned to lean in the direction he wished to be carried and then to simply hold one arm out in that direction to indicate to the teacher where he wanted to go (Fig. 1). He also learned that jumping movements with his legs would get the teacher to move if she were standing still. No deliberate or conscious attempt was made to teach Kanzi to communicate his desires regarding travel. Rather, it seemed to occur as a natural outgrowth of the human bipedal posture, bipedal carrying behavior, and sensitivity to the direction of Kanzi's regard. Kanzi quickly learned that he could get the human caretakers to carry him whenever he wished, but that his mother often paid no mind to such signals and if he rode with her, he usually went wherever she wished. Consequently, he seemed to prefer being carried by the human caretakers in all calm situations. As an infant, Kanzi is interested in exploring everything. He asks his teachers to carry him to places in which his mother shows no interest and he consequently gets no chance to explore if he remains with her (the storage shelves, the bathroom sink, the observation window, etc.). While being carried about by his teachers Kanzi quickly learned, through observation, how to turn lights off and on and how to turn water on—two things his mother has yet to learn. Although Matata seems to be quite content to have others carry and entertain Kanzi for long periods, she is always keenly aware of his location at every moment. She never allows doors to be shut that might separate them as he is moved about the six-room enclosure (any one room of which can be closed off from the others by one or more doors). If it becomes necessary to discipline Kanzi (for example, to physically restrain him from stealing food), she always rushes to him immediately, though once there she generally observes and does not attempt to take him back since he does not appear to be in any danger.

Matata is tolerant of a wide variety of interactions between Kanzi and the teachers because of her capacity to understand the goals and intentions of the human caretakers with regard to Kanzi. Two examples of many illustrate the extent to which Matata can understand complex social exchanges quite readily. In the first instance a teacher was using a long knife to cut bits of food for Matata as they worked at a sorting task. Each time Matata was correct a piece of food was cut off for her. Kanzi played beside Matata as she worked, often jumping into the bowls in which she was sorting objects, jumping on her head from the ropes above, jumping on the teachers, and generally being a rowdy and distracting infant. At times he would even grab the sorting objects out of Matata's hands or out of the bowl and throw them on the floor. Matata and the teachers alike patiently tolerated these behaviors, as they were simply a reflection of Kanzi's age and he could not understand why his behaviors caused difficulties for others. At one point, as Kanzi was jumping into the bowls he suddenly jumped straight toward the cutting board just as the teacher was bringing the knife downward to chop off a slice of apple. Kanzi's movements were so rapid that the experimenter was unable to stop the downward motion of the knife before it thumped Kanzi soundly on the shoulder. It did not cut him (it was a dull knife) but the blow was surely a painful one. Kanzi became furious, screaming and lashing out at the experimenter in attempts to bite. The experimenter, fearful (not of Kanzi), but that an attack from Matata would result, looked at Matata with an expression of dismay and pulled the knife back. Matata, having closely observed the entire set of events, simply pulled Kanzi to her and tried to quiet him even though Kanzi kept threatening the teacher. When the teacher reached out toward Matata, Matata hugged her and again tried to quiet Kanzi. Kanzi tried to bite the teacher even as his mother was hugging her. He remained angry for about 20 minutes.

It is quite surprising that Matata did not bite the experimenter in response to Kanzi's screams. The *Pan troglodytes* females observed by Savage (1975) would have responded to such an event (intended or not) with instant aggression. Matata is not afraid to respond in this way—she has attacked the teachers when Kanzi has screamed on other occasions in which it was impossible for her to see the events that preceded Kanzi's vociferous vocalizations and to judge, independently of the distress that Kanzi expresses, what happened.

However, a second example serves to show that even in some situations where she does not see the flow of events leading to Kanzi's distress, if she can be shown the problem, then at the instant she understands the situation she will cease attacks toward the teacher and attempt to help: Matata was working at the keyboard and Kanzi was nearby, learning to put items in and out of containers. His attention span was much shorter than hers and consequently each successful attempt on Kanzi's part was interrupted by 5 minutes or so of vigorous play. During one such play bout, a teacher (who had her long hair tightly tied in a bun on her head) was holding Kanzi over her head and tickling him. Kanzi began to playfully slap the teacher on her head and to grab and tug at her hair. Suddenly, his hand became entangled in her tightly bound hair and he could not pull it out. He panicked, screamed, and began to bite the experimenter. The experimenter, who did not realize that Kanzi's hand was caught (she thought Kanzi was grabbing her hair) began to scream at Kanzi and tried to shove him away. Amidst the confusion, with the experimenter and Kanzi both screaming angrily at each other, Matata rushed to Kanzi's aid and also began screaming at the experimenter and prepared to leap on her. At this point, a second experimenter who saw the problem and realized that Kanzi's hand was caught, gesturally pointed this out to Matata and verbally told the experimenter. Immediately, Matata restrained herself from attacking, and the experimenter, who was being bitten, stopped screaming and shoving at Kanzi, while Kanzi's hand was freed by the second experimenter who had seen the problem. Kanzi also seemed to understand at this point that someone was working to free his hand and he stopped biting, though he continued to scream. As soon as his hand was freed, he leaped to Matata's ventrum and then angrily and repeatedly threatened the experimenter in whose hair his hand had been caught. Matata, however, did not join him in this threat and when the experimenter reached out to hug Matata she responded, although Kanzi scratched the experimenter and screamed at her even while his mother was hugging her.

In this case, it appeared initially that Matata was about to attack, but when she saw that the teacher began holding still and was making no attempt to hold on to Kanzi, but was instead leaning over to facilitate the extraction of Kanzi's hand from her hair (which was very difficult), Matata responded to what she saw and understood, not to Kanzi's vocalizations.

Kanzi, on the other hand, being much younger and presumably unable to understand what had happened to him, continued to behave as though the experimenter had simply "grabbed"him with her hair unexpectedly. He remained angry with the experimenter for about 30 minutes, and continued to exhibit fear of close contact play with her a week after this incident. Matata behaved as if nothing had happened within minutes after the episode. This sort of comprehension on Matata's part is strikingly different from what one typically sees in *Pan troglodytes*. In over 2 years of daily observation of a captive group of common chimpanzees consisting of four mothers and their offspring (Savage, 1975), not one occasion occurred where a mother clearly looked for and interpreted the cause of her infant's screaming in a way that differed from the infant's own interpretation. Screaming seemed to be responded to rather automatically. If an infant's scream was directed toward another chimpanzee, the mother invariably threatened and/or attacked the chimpanzee toward whom the infant screamed, unless she herself was very low ranking, in which case she just tried to gather up the infant and hurry away, a response Matata has never shown. Matata's behavior, in fact, leaves little doubt that if Kanzi were severely threatened, she would act to protect him.

4. Development of Nonverbal Communicative Competence in Kanzi

Symbol training was begun with Matata in August 1981. Kanzi observed all training received by his mother, but he received no specific training. At 1 year of age, he still nursed frequently and was not responsive to any of the teacher's attempts to get him to attend to the keyboard. His mother allowed him to do virtually anything he pleased (including eating the food she asked for) and at this age he was quite unresponsive to any requests that he sit still, that he attend, etc. He interacted with the teachers as long as he was interested; otherwise he did what he wanted.

Between 6 months and 1½ years of age, Kanzi learned how to control his teachers' behavior in numerous ways. The remainder of the chapter will focus on Kanzi's spontaneous nonverbal communicative development during this 1-year period. Since Kanzi was not really old enough or interested enough during this period to participate in any formal training tasks, the communicative skills that he developed were ones that emerged spontaneously from his own desires to communicate.

The constant presence of human recipients of his communications perhaps fostered the emergence of some types of behavior that would not occur in a chimpanzee. The presence of humans in the environment was not as important to Kanzi as it was to Sherman and Austin or other language-trained apes, for whom humans were essentially parental surrogates. Consequently, when Sherman and Austin were encouraged to imitate the behaviors of their human companions they did so far more readily than Kanzi. Kanzi imitated only spontaneously, never upon request, and he imitated only those things that interested him. Where striking differences are observed between Kanzi and common chimpanzees, they are likely to result from both the fact that Kanzi is a *Pan paniscus* and from the fact that he is being raised in an environment that is designed to stimulate, to the fullest degree, the development of intentional communicative skills.

The point at which communicative intentionally evidences itself in human infants has been the subject of much recent study (Bates, 1979; Bruner, 1974–1975). This ability generally appears at about 9 months of age and is marked by a rather dramatic change in the infant's communicative behavior. Bates (1979) defines communicative intentionality as follows:

> Signaling behavior in which the sender is aware a priori of the effect that a signal will have on his listener, and he persists in that behavior until the effect is obtained or failure is clearly indicated. The behavioral evidence that permits us to infer the presence of communicative intentions include (a) alterations in eye contact between the goal and the intended listener, (b) augmentations, additions, and substitutions of signals until the goal has been achieved and/or exaggerated patterns that are appropriate *only* for achieving a communicative goal (Bates, 1979, p. 36).

Studies of the emergence of communicative intentionality in infants have repeatedly emphasized the role of the mother who responds to even the infant's earliest communications as though they were intentional (Gray, 1978). For example, when an infant reaches toward an object that is just out of his grasp, he will not (prior to the emergence of intentionality) signal to his mother in any way that he wants a particular object or that he wants her to get the object for him. He will only repeatedly strain to reach the object. The human mother, however, responds to such behavior by behaving *as though* the infant had signaled. She will reach for the object and hand it to the infant, often overlapping this behavior with a verbal marking such as, "Oh, you want the pull-toy. Here you go." A number of researchers who have studied the behavior of both mother and infant prior to the emergence of communicative intentionality attribute a large role to this interpretive behavior of the mother (Gray, 1978; Bullowa, 1979). According to this view, when the mother behaves as if nonintentional behavior was a signal, the child comes to learn the effect that such behavior has on the mother. Thus, while he does not, at first, intend that his gesture serve as a signal to his mother (he intends only to grasp the object), he learns that his mother monitors the gesture and that the gesture obtains the object for him through the action of his mother. This is easier for him than when he achieves the same result through his own direct action. At this point he stops monitoring the direct consequences of his own action (i.e., is his hand in line with the object?) and instead monitors the consequences of his action upon the behavior of his mother.

Pan troglodytes mothers, unlike human mothers, do not respond to their infants' actions as though they were intentional communications. In the *Pan troglodytes* group of mothers and infants observed by Savage (1975) no instance of a mother ever giving an object (other than food) to an infant in response to a gesture was ever observed. On one occasion a mother did remove an object that an infant was reaching for and on many occasions mothers allowed infants to *take* objects in their possession. Matata, by contrast *does* give objects to Kanzi, though this behavior is relatively infrequent (in that it may be seen only once or twice a week). What she does do reliably and frequently, however, is to respond to Kanzi's gestures for aid in moving from place to place while he is not clinging to her. Thus if Kanzi is climbing about on ropes and tables near Matata and he wants to move higher, lower, or to another table, Matata will aid in this movement. As with human infants, Kanzi did not initially signal his desires to his mother. Instead he would repeatedly try to grasp a rope over his head and fall back. Matata, who always quite closely watched Kanzi's acrobatics at this stage (4–11 months) (again in contrast to *Pan troglodytes* mothers, who frequently take their eyes off of their infants for long periods of time) would nearly always raise a foot or arm toward Kanzi and shove him toward the object he had been trying to reach (Figs. 1 and 2). Kanzi, like human infants, began to signal his desired intent to go to a particular location and to look back and forth between his locomotor goal and his mother. Prior to the emergence of such checking behavior, Kanzi had simply continued to try and get there on his own and if his mother did not see him, he did not vocalize or gesture to gain her attention.

Intentional directed communication regarding locomotor goals appeared rather suddenly during a 2-week period at 10 months. Gestural signals and visual checking came together and occasionally vocal signals accompanied gestures. Across a 2-week period at the end of August 1981, when Kanzi was 10 months of age, he seemed to begin to realize that his own actions on objects and space had consequences that could be anticipated in advance and consequently altered. He also seemed to realize that his actions affected others and to anticipate these effects and consequently alter his actions toward others. He began to "ask" his mother to pick him up, and to "ask" her to help him reach another shelf and to "ask" her to put him down or let him go, etcetera. He simultaneously began to make similar requests of his teachers and to attend to their requests. Although we have no way of determining whether Matata's previous responsiveness to Kanzi's locomotor attempts fostered the development of such communicative intentionality, it is clear that Kanzi did *not* go through the *object* giving and taking stages that are described for human infants (Gray, 1978) and have been linked to the emergence of communicative intentionality in human infants. Kanzi's human companions likewise did little to foster the *emergence* of this behavior, since Kanzi did not begin to signal his desired direction of travel to his teachers *as* he was being carried until approximately 12 months of age, or well after the time when communicative intentionality had emerged in other settings. Additionally, similar signaling of desired help in getting from one shelf or rope to another was observed and documented in Akili (Kanzi's older half-brother, who was later sent to the San Diego Zoo) at the same age while Matata was still housed with the other pygmy chimpanzees at the field station and was rearing both Kanzi and Akili. These observations suggest that, in the pygmy chimpanzee, with an infant who (in contrast to a human infant) is highly mobile and motivated to practice climbing, leaping, diving, somersaulting, spinning, and brachiating, that aid in locomoting, as opposed to aid in acting upon objects, may serve as the focus of initial mother–infant communicative interactions. Matata's attentiveness to the locomotor activities and desires of her infants is significantly greater than that observed in *Pan troglodytes* mothers, and while more *Pan paniscus* mothers must be observed, we see, at least in this case, a form of maternal responsiveness that, although clearly less sensitive than that formed in human mothers, is nevertheless of such a type and quality as to foster the appearance of intentional communicative skills in infants. The very rapid appearance of this phenomenon in Kanzi, however, suggests that strong developmental factors were also operating.

As noted above, during this period of emergence of intentional communication it was not just communicative awareness that blossomed in Kanzi. It was a more general awareness of causality in all forms. Kanzi's behavior changed across all domains, with the emergent realization that all his actions had effects or consequences that could be predicted in advance of actually executing the action. This predictability allowed reflection and choice of alternatives prior to action in many different situations. Although it is difficult to specify precise behavior that unequivocally demonstrates such an awareness, excerpts of notes made by teachers working with Kanzi during this 2-week period reflect some of the behavior that altered during this period.

The biggest change is in how Kanzi responds to others. When I touch him on the head or shoulder to get his attention he does not turn around with the sporadic jumpy movements that he did. Previously, his such sporadic responses seemed to be a reflexive reaction to being touched, that is, Kanzi jerked around quickly to see what was producing that particular sensation on his back. Now he turns with deliberation and looks directly at me waiting to see what I want. He turns as though he knows I have touched him for the purpose of gaining attention, not just in response to a sensation. That is, he attributes purpose to my behaviors. He seems suddenly to be exhibiting more purpose in the majority of his own behaviors. Only a short time ago he was either clinging to Matata or rushing rapidly toward something that caught his attention. He did not seem to plan to go toward something before he moved, he just saw something of interest and rushed toward it, often stumbling over objects in his path, or if he was moving in 3 dimensional space, just working out his route as he went. If he was moving toward a person he did not seem to be aware that landing on their head or grabbing their hair for support had any effect on the person. Now, his movements toward things appear more deliberate, planned. He constantly visually checks my face as he moves towards me and his actions toward me have a deliberateness that was absent a short time ago, although it is hard to say how one knows that such deliberateness is present. His behavior now suggests that he is repeatedly attempting to determine the "meaning" of what is done to him instead of just how things feel. For example, a short time ago when he would dangle above my head playfully, if I began to pull on his feet and play bite, it was necessary each time that I begin very slowly and take into account how each of my actions felt toward him, making certain that the tactile experience that I was providing was always received as pleasant. Only by very slowly escalating the contact could I make a long play session, otherwise Kanzi moved away. It was also necessary to repeatedly intersperse each action with a playful expression and to attempt, by moving my head and body to gain eye contact frequently. Now, Kanzi initiates the eye contact and if I respond with a playful expression we can immediately begin playing. As he plays, he now modulates his tugging and pulling on my hair to his biting. He pulls on my hair in a more playful way, instead of just grabbing at it, and he lets go in response to my facial expression, instead of my having to pull his hand away. Before, when playing, he would flail and kick at me in an undirected manner, with little regard for where his kicks landed. Now he kicks in a directed manner and is gentle as he monitors the effect of his kick on me. He likes to kick me all about my face. In response to his pleasure at this, I held his feet and made them slap my face. He laughed loudly and would reinitiate the game by holding his feet toward my hands and looking at my face.

Once intentional communication emerged in Kanzi it began to manifest itself in ways that were strikingly different from the forms of nonverbal intentional communication seen in Sherman and Austin. The most salient differences between Austin and Sherman (*Pan troglodytes*) and Kanzi (*Pan paniscus*) are described below.

1. Kanzi frequently combines gestures with vocalizations. (For ex-

ample, if he wants to be carried to the refrigerator he will gesture in that direction and accompany the gesture with long, drawnout "aaahhee" vocalization.) Such gestural–vocal combinations are infrequent in Austin and Sherman and in *Pan troglodytes* in general. Sherman and Austin may whimper as they reach toward an object and they may make a breathy "hhha" sound and point to a food they desire in these instances. When they are whimpering, the noise itself is a result of frustration. The whimpering noise cannot be combined at will with a gesture. The combination occurs only when the context is such that both a gesture and the whimper are evoked by differing environmental factors. The breathy "haa" sound is a voluntary vocalization that both Austin and Sherman have been taught. They can produce it at will and do use it with gestures, though in most cases either a gesture or a sound is used, not both. Kanzi, by contrast, accompanies many gestures quite spontaneously with vocalizations. He does not need to be frustrated and the types of vocalizations he employs are quite variable. He vocalizes quite frequently and spontaneously, just as do young human children before they begin to talk. Although none of his vocalizations could be characterized as babbling, he is able to make a vocal sound whenever he wishes. It is not necessary that the context be such as to evoke a vocalization. In addition, he vocalizes when teachers laugh, when they talk in an excited manner, and when they talk somewhat loudly. He seems to want to make noise as they do and it is not easy to conduct long conversations that do not include Kanzi. Sherman and Austin do not tend to vocalize under such circumstances in response to human vocalizations, though if their teachers yell in alarm or "pant hoot" in excitement, Sherman and Austin do vocalize. However, the types of vocalizations that Kanzi uses to accompany his gestures are simply not in the repertoire of Sherman and Austin or other *Pan troglodytes*.

In general, it is accurate to say that the gestural–vocal combinations produced by Kanzi are distinctly different from those seen in *Pan troglodytes,* in that the vocalizations appear to be voluntary and used intentionally to draw attention to Kanzi and to what he wants. (Matata has exhibited some similar behavior, but in her case such combinations are relatively infrequent.) Gestural–vocal combinations have been observed in the following contexts:

(a) To have the teacher transport him from one area to another.
(b) To request assistance in reaching high places.
(c) To ask one teacher to transfer him to another.
(d) To get the teacher to approach him.
(e) To get the teacher to assist him in manipulating some object that he is having difficulty with (e.g., opening a bottle).

Figure 1. Kanzi spontaneously uses gestures to show his teacher where he wants to go.

(f) To get the teacher to transfer objects from the teacher to him.

(g) To get one teacher to make Matata or the other teacher give him an object or food that he wants and they are withholding.

2. When Kanzi is frustrated in any sort of activity he fusses and whines in a manner remarkably similar to that of human infants. *Pan troglodytes* infants typically whimper or scream when they are frustrated, but whining and fussing noises are not part of their repertoire (Savage, 1975). The range of situations in which Kanzi exhibits this behavior is also not limited to deprivation of food or contact as is the case with *Pan troglodytes*. Kanzi fusses when he cannot get one object inside another, when he cannot get two things apart, when we will not let him have his ball or other favored toy, when he wants to go somewhere by himself and we insist on carrying him, when he wants to be carried and we will not carry him, when we will not let him go to a particular location for play, etcetera.

3. Shortly after Kanzi began to demonstrate the forms of intentionality described above, he also spontaneously began to engage in a primitive sort of pointing. He would approach unusual objects and repeatedly touch them with his extended index finger. Such pointing was occasionally ac-

Figure 2. Kanzi gestures for the teacher to go to the door, indicating the desire for the teacher to transport him there.

companied by vocal signals and visual checking. When the pointing behavior was responded to by visual orientation on the part of another, it
was often repeated. Both Matata and the human teachers often glanced
at the objects to which Kanzi pointed. When a teacher pointed at the
same object in response, Kanzi's interest would often intensify and he
would point again and again at the object and then wait for the teacher
to point. Matata usually did not respond by also pointing at the object.
When she did do more than just glance at the object, it was invariably
the case that the object at which Kanzi pointed was alive (typically some
type of bug) and Matata would then also touch the object and inspect it
closely. Depending on what it was, she would ignore it, knock it away,
eat it, play with it, etcetera.

   Pan troglodytes typically do not point, though they may direct one
another's gaze by glance or bodily orientation, and at times use a limb to
indicate general direction (Woodruff and Premack, 1979). However, these
behaviors are not as specific as pointing. Sherman and Austin both learned
to point with the index finger at 4–5 years of age following training in
symbolic tasks that required a high degree of specificity. In their case,
however, pointing occurs as a communicative device to indicate which

one of a set of things they want, or which thing another is to choose. Kanzi does not yet use pointing for these purposes; however, he has received no symbolic training that would enourage pointing, and yet it is appearing at a far younger age than was seen in Sherman and Austin. After pointing at objects, Kanzi sometimes pushed the experimenter's hand toward the object to request that the experimenter act upon the object. This was most frequently seen with new objects for which the experimenter had not previously demonstrated any action and thus the use or function of that object was unknown to Kanzi.

4. Kanzi leads his teachers by the hand to get them to go to locations that he wants. He also leads them around in circles in a game of chase. If he is leading them to a particular location, he may, on arrival there, pull on their hand to tell them to sit down. He has not acquired this pattern from the teachers, since they did not lead him, they carried him. However, Matata does at times lead Kanzi by the hand. When Kanzi leads his teachers by the hand the size discrepancy between the two makes it difficult for the teacher, who must lean over and take very small steps. Although Sherman and Austin may occasionally take their teacher by the hand and tug in a certain direction, they then always let go. They never lead the teacher step by step. Yet both of them, as a result of their size (they are too heavy to carry), have considerable experience in being led by the hand by their teachers. Matata, like Kanzi, frequently takes the teacher by the hand and leads her to where she wants the teacher to be.

5. Conclusion

In summary, although after 1 year we are just beginning to develop a good social setting for continued long-term, cognitive-linguistic studies of a *Pan paniscus* mother and her infant, we are already observing behavior that is quite different from that seen in *Pan troglodytes* in a similar setting. In particular, these differences lie in the areas of gestural–vocal nonverbal communication and the complexity of interindividual social interactions. We are only beginning to understand and describe these phenomena at this point, but clearly the pygmy chimpanzee offers many intriguing points of behavioral comparison with other ape species.

It is also quite clear that in most, if not all, of the behavioral differences we have described between Matata and Kanzi on the one hand and Sherman and Austin on the other, the behavior of the pygmy chimpanzees is considerably more reminiscent of our own species. Each individual who has worked with both species in our lab is repeatedly surprised by their communicative behavior and their comprehension of complex social contexts that are vastly different from anything seen among *Pan troglodytes*.

ACKNOWLEDGMENTS. The research on which this chapter is based was supported by grants from the National Institute of Child Health and Human Development (HD-06016) and from the Division of Research Resources National Institutes of Health (RR-00165).

References

Bates, E., 1979, *The Emergence of Symbols, Cognition and Communication in Infancy*, Academic Press, New York.

Bruner, J. S., 1974–1975, From communication to language—A psychological perspective, *Cognition* 3(3):255–287.

Bullowa, M., 1979, Introduction: Prelinguistic communication. A field for scientific research, in: *Before Speech. The Beginning of Interpersonal Communication* (M. Bullowa, ed.), Cambridge University Press, Cambridge pp. 1–62.

De Waal, F. B., and Hoekstra, J. A., 1980, Contexts and predictability of aggression in champanzees, *Anim. Behav.* 28:929–937.

Gardner, R. A., and Gardner, B. T., 1969, Teaching sign-language to chimpanzees, *Science* 165:664.

Goodall, J., Bandura, A., Bergman, E., Busse, C., Matama, H., Mpongo, E., Pierce, A., and Riss, D., 1979, Inter-community interactions in the chimpanzee population of the Gombe Stream National Park, in: *The Great Apes* (D. A. Hamburg and E. R. McCown, eds.), Benjamin/Cummings, Menlo Park, California, pp. 13–53.

Gray, H., 1978, Learning to take an object from the mother, in *Action, Gesture, and Symbol: The Emergence of Language* (Andrew Lock, ed.), Academic Press, London, pp. 159–182.

Jordan, C., and Jordan, H., 1977, Versuche zur Symbol-Ereignis-Yerknupfung bei einer Zwergschimpansen (*Pan paniscus*, Scwarz, 1929), *Primates* 18(3):515–529.

Kano, T., 1980, Social behavior of wild pygmy chimpanzees (*Pan paniscus*) of Wamba: A preliminary report, *J. Hum. Evol.* 9:243–260.

Kuroda, S., 1980, Social behavior of the pygmy chimpanzee, *Primates* 21(2):181–197.

Miles, L., 1982, Sign language studies with an orangutan, Paper presented at IXth Congress of the International Primatological Society, Atlanta, Georgia.

Patterson, T., 1973, The Behavior of a Group of Captive Pygmy Chimpanzees (*Pan paniscus*), Masters Thesis, University of Georgia.

Patterson, F., and Linden, E., 1981, *The Education of Koko*, Holt, Rinehart, and Winston, New York.

Rumbaugh, D. M., 1977, *Language Learning by a Chimpanzee: The LANA Project*, Academic Press, New York.

Savage, E. S., 1975, Mother–Infant Behavior in Group-Living Captive Chimpanzees, Ph.D. Dissertation, University of Oklahoma.

Savage, E. S., and Bakeman, R., 1978, Sexual morphology and behavior in *Pan paniscus*, in: *Proceedings of the Sixth International Congress of Primatology, Cambridge, England*, Academic Press, New York, pp. 613–616.

Savage, E. S., and Malick, E., 1977, Play and socio-sexual behavior in a captive chimpanzee (*Pan troglodytes*) group, *Behavior* **60**:179–194.

Savage, E. S., and Wilkerson, B. J., 1978, Socio-sexual behavior in *Pan paniscus* and *Pan troglodytes:* A comparative study, *J. Hum. Evol.* **1**:327–344.

Savage, E. S., Temerlin, J. W., and Lemmon, W. B., 1973, Group formation among captive mother–infant chimpanzees (*Pan troglodytes*), *Folia Primatol.* **20**:453–473.

Savage, E. S., Wilkerson, B. J., and Bakeman, R., 1977, Spontaneous gestural communication among conspecifics in the pygmy chimpanzee (*Pan paniscus*), in: *Progress in Ape Research* (G. H. Bourne, ed.), Academic Press, New York, pp. 97–116.

Savage-Rumbaugh, E. S., 1979, Symbolic communication—Its origins and early development in the chimpanzee, *New Directions Child Devel.* **3**:1–15.

Savage-Rumbaugh, E. S., and Rumbaugh, D. M., 1978, Symbolization, language and chimpanzees: A theoretical re-evaluation based on initial acquisition process in four young *Pan troglodytes*, *Brain Language* **6**:265–300.

Savage-Rumbaugh, E. S., and Rumbaugh, D. M., 1980, Language Analogue Project, Phase II: Theory and tactics, in: *Children's Language*, Vol. II (K. Nelson, ed.), Gardner Press, New York, pp. 267–307.

Savage-Rumbaugh, E. S., Scanlon, J., and Rumbaugh, D. M., 1980, Communicative intentionality in the chimpanzee, *Behav. Brain Sci.* **3**:620–623.

Schwarz, E., 1929, Das Vorkommer der Schimpansen auf den linken Kongo-Ufer, *Rev. Zool. Bot. Afr.* **XVI**(4):425–426.

Susman, R. L., Badrian, N. L., and Badrian, A. J., 1980, Locomotor behavior of *Pan paniscus* in Zaire, *Am. J. Phys. Anthropol.* **53**:69–80.

Temerlin, M. K., 1975, *Lucy: Growing up human, Science and Behavior Books*, Palo Alto, California.

Tratz, E., and Heck, H., 1954, Der Afrikanische Anthropoide "Bonobo," eine neue Menschenauffengattung, *Saugetierkd. Mitt.* **2**:97–101.

Woodruff, G., and Premack, D., 1979, Intentional communication in the chimpanzee: The development of deception, *Cognition* **7**:333–362.

Yerkes, R. M., and Learned, B. W., 1925, *Chimpanzee Intelligence and Its Vocal Expressions*, Williams and Wilkins, Baltimore.

Yerkes, R. M., and Yerkes, A. W., 1929, *The Great Apes*, Yale University Press, New Haven, Connecticut.

Zihlman, A., Cronin, J. E., Cramer, D. L., and Sarich, V. M., 1978, Pygmy chimpanzee as a possible prototype for the common ancestor of humans, chimpanzees, and gorillas, *Nature* **275**:744–746.

CHAPTER 17

Will the Pygmy Chimpanzee Be Threatened with Extinction as Are the Elephant and the White Rhinoceros in Zaire?

KABONGO KA MUBALAMATA

Wild animals such as the elephant (*Loxodonta africana*), the white rhinoceros (*Ceratotherium simum*), the zebra (*Equus burcelli*), and the cheetah (*Acinonyx jubatus*) being killed and illegally exploited for their meat and secondary products (horns, tusks, skins, furs). The same is true for primates, especially the pygmy chimpanzee (*Pan paniscus*), which is being heavily exploited in a number of different ways, most notably as a source of food, for the pet trade, and for biomedical research.

The pygmy chimpanzee (*Pan paniscus;* called "Eliya" in Lontomba in the Bikoro area, "Mokumbusu" or "Sokomuntu" in Lingala and Swaili, and "ejá" in Limongo) is a primate species found in the dense equatorial forest of Zaire. It is confined to the left bank of the Zaire River, where it is far less widespread than the common chimpanzee (*Pan troglodytes*).

A considerable number of field studies and scientific publications, dating to the 1930s, indicate the wide scientific interest in *Pan paniscus*. For example, we can refer to the valuable service that primates in general and chimpanzees in particular have rendered medical research programs. According to Held and Whitney (1978), they played a major role in the work leading to the perfection of an antipoliomyelitis vaccine and they have yielded important information on other diseases, such as malaria, yellow fever, measles, intestinal viruses, tuberculosis, mental disorders,

---

KABONGO KA MUBALAMATA ● Institut de Recherche Scientifique, Centre de Recherche de Lwiro, Bukavu, Zaire.

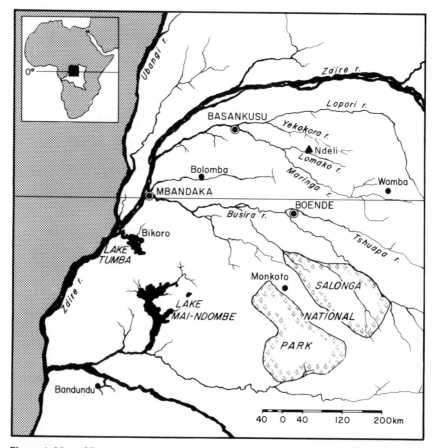

Figure 1. Map of Equateur, showing the Bikoro field station. The site of the Lomako Forest Pygmy Chimpanzee Project (directed by R. Susman) and the Wamba Pygmy Chimpanzee Project (directed by T. Kano) are also shown.

and viral oncogenesis. Morris and Morris (1966) have pointed out the use since the 1960s of chimpanzees in the space programs of the US and USSR.

To these experiments we can add those carried out on five pygmy chimpanzees that officially left Zaire in 1975 for the Yerkes Primate center in Atlanta. According to a scientific report of Dr. Frederick King, Director of Yerkes, to his counterpart, Dr. Kankwenda M'Baya, General Delegate to the Institute of Scientific Research in Kinshasa, the pygmy chimpanzees on loan to Yerkes and their offspring are involved in two research programs: one on reproductive biology and another on behavior. Some interesting results have been achieved in these respective fields. In biology, for example, menopause in a healthy older female has been observed, an observation not yet reported for common chimpanzees. Urinary hormone analyses conducted on the Yerkes group is documenting a number of physiological differences between *Pan paniscus* and *Pan troglodytes.*

The behavioral studies involve studies of pygmy chimpanzee intelligence and social organization. Sue Savage-Rumbaugh has noted a much greater gestural communication repertoire, a different pattern of sexual behavior, and different mother–infant interactions in *Pan paniscus* compared to *Pan troglodytes.* Pygmy chimpanzees at Yerkes seem to be faster learners than their common chimpanzees counterparts. It is interesting to note that techniques used to teach computer language to the Yerkes chimpanzees are now being used to develop communication skills of retarded children.

In addition to the above considerations, scientists, including primatologists, anthropologists, zoologists, and paleontologists, have an ongoing interest in *Pan paniscus* as a key element in reconstructing the story of human evolution. Some consider it to be the closest living relative of humans (Zihlman, 1979), while others accord it a prominent although not exclusive role in the search for the common ancestor of humans and the African apes (Susman, 1980).

In spite of all of the postive interest in *Pan pansicus* by the scientific community, the pygmy chimpanzee is a victim of widespread human depredation resulting from modifications of its natural habitat (including industrial exploitation of the forests in the Lomako area, and extension of coffee and cacao plantations at Wamba and at Djolu in the lower Tshuapa) and uncontrolled hunting in the Lake Tumba area at Botwali, Nkoso, and elsewhere throughout lower Equateur (Fig. 1). Pygmy chimpanzees are frequently hunted for export to foreign zoological gardens (especially in Europe and Asia) and to scientific laboratories, primate centers in the East and West, and for the pet trade (mostly in Belgium). These harmful practices are well supported by the foreigners in Zaire and condoned by many Zairian nationals.

A preliminary investigation in 1981 in the Bikoro Zone, especially in the locality of Botwali and around it (Lake Tumba) indicated that the hunting of *Pan paniscus* is based on three needs:

1. Food: chimpanzee flesh is eaten by the local people and can be purchased very cheaply.
2. Magico-religious practices: chimpanzee bone is often sought for local religious practices because traditional beliefs hold that chimpanzee bone provides a superhuman force.

3. Commercial exploitation: young chimpanzees captured for sale to foreigners bring relatively high prices (a live chimpanzee can sell for 500 zaires—a plantation worker in Equateur earns roughly 120 zaires per month.)

It is disturbing that of the above three causes of pygmy chimpanzee exploitation, the latter is growing the fastest among the Ntomba people of the Lake Tumba region. This activity is promoted by the "banunus," the name generally given to the illicit traders of Bandundu and to those of certain neighboring countries of Zaire. An illicit market has been established directly between the banunus and the Ntomba hunters, in which cash and trade items are used to pay for captured young pygmy chimpanzees. This illicit trading is done in three stages:

Poacher	→	Trader	→	Promoter
Ntomba hunter: using guns and poisoned arrows		Banunus: middlemen buy animals with local currency and goods		Foreigners and Zairian nationals: buy animals for resale abroad with currency and goods

In this illicit trade the first link of the chain occurs in the dense equatorial forest that is the natural habitat or refuge for the pygmy chimpanzee. In the capture of the young pygmy chimpanzee, the hunter kills the mother with a shotgun or poisoned arrows. Next the "banunu" middleman offers a relatively high price to the hunters in the form of money (a minimum of 300 zaires) and different gifts and receives cash from the promoter. The promoter is the last link of the chain: he could be African, Asiatic, American, or European. He is the one who determines the exact destination and the fate of the animal(s) outside of Zaire.

Considering the low frequency of this species, probably due to a low birth rate, its very limited geographic distribution (left bank of the Zaire River), and its preference for an unaltered, unpopulated habitat, it can be seen that as long as human depredation continues (destruction of the natural habitat and killing for food and the illicit traffic in this primate), the pygmy chimpanzee will be threatened with extinction in the lower region of Equateur and of the Tshuapa, just as the elephant and white rhinoceros have been all but eliminated in many of the forests and savannas of central and eastern Zaire.

In order to bring the massacre of pygmy chimpanzees to an end, it is our opinion that the importance of this species and the need for its protection must be publicized in the area of Botwali and the surrounding area. First, we are trying to inform and educate the local population that the protection of *Pan paniscus* really is important for all Zairians. We emphasize such facts as the endemic nature of *Pan paniscus* in the area, their relatively sparse population compared to ordinary monkeys, and their preference for undisturbed forests. By the same token, we have asked different village chiefs to prohibit all action contrary to the protection measures outlined (hunting, illicit trade, destruction of the forests, etcetera) in their respective areas. Other researchers working in Zaire under the auspices of the Institute of Scientific Research in the Lomako Forest and at Wamba are pursuing a similar program.

In addition, we suggest urgent action on a combined scale by the authorities in Zaire on the one hand and by the international organizations dedicated to the protection of wildlife on the other. The former will reinforce the protection measures designated for the pygmy chimpanzees by keeping its natural habitat intact (creation of special natural reserves) and by banishing illegal hunting in the central basin of Zaire. The World Wildlife Fund, the International Union for the Conservation of Nature and Natural Resources, and other organizations must act to limit commercial trade in animals to zoos, primatology centers, and scientific laboratories. Additional international efforts must be made against individuals and third parties seeking purchases in the black-market trade.

It is only through quick and efficient intervention that the pygmy chimpanzee will be able to survive. Otherwise, we are running a high risk in the near future of adding this great ape to the list of extinct animals.

References

Held, J. R., and Whitney, R. A., 1978, Epidemic diseases of primate colonies, in: *Recent Advances in Primatology, Volume 4. Medicine* (D. J. Chivers and E. H. R. Ford, eds.) Academic Press, London, pp. 25–41.

Morris, R., and Morris, D., 1966, *Men and Apes,* McGraw-Hill, New York.

Susman, R. L., 1980, Acrobatic pygmy chimpanzees, *Nat. Hist.* **89:**32–39.

Zihlman, A. L., 1979, Pygmy chimpanzee morphology and the interpretation of early hominids, *Nature* **75:**165–168.

Author Index

Subject Index